Plant Genebank Utilization for Trait Discovery in Millets

Kuldeep Tripathi • Jyoti Kumari
Sushil Pandey • Badal Singh
Sunil Archak • Gyanendra Pratap SIngh
Editors

Plant Genebank Utilization for Trait Discovery in Millets

Volume IV

Springer

Editors
Kuldeep Tripathi
Division of Germplasm Evaluation
ICAR-National Bureau of Plant Genetic Resources
New Delhi, Delhi, India

Jyoti Kumari
Division of Germplasm Evaluation
ICAR-National Bureau of Plant Genetic Resources
New Delhi, Delhi, India

Sushil Pandey
Division of Germplasm Conservation
ICAR-National Bureau of Plant Genetic Resources
New Delhi, Delhi, India

Badal Singh
Division of Germplasm Evaluation
ICAR-National Bureau of Plant Genetic Resources
New Delhi, Delhi, India

Sunil Archak
Germplasm Exchange and Policy Unit
ICAR-National Bureau of Plant Genetic Resources
New Delhi, Delhi, India

Gyanendra Pratap SIngh
ICAR-National Bureau of Plant Genetic Resources
New Delhi, Delhi, India

ISBN 978-981-96-4042-3 ISBN 978-981-96-4043-0 (eBook)
https://doi.org/10.1007/978-981-96-4043-0

This Springer imprint is published by the registered company Springer Nature Singapore Pte Ltd.
The registered company address is: 152 Beach Road, #21-01/04 Gateway East, Singapore 189721, Singapore

Foreword

Millets, often called as "crops of the future," play a vital role in the sustenance of human agriculture. Millets, known as Nutri Cereals, have potential to increase farmers' income and ensure food and nutritional security. Plant gene banks play a crucial role in preserving the genetic diversity of crops, including millets, which are recognized for their resilience and nutritional value. The utilization of plant gene banks for trait discovery in millets has gained importance due to the growing need to address challenges like climate change, food security, and agricultural sustainability. The strategic use of plant gene banks for trait discovery in millets can significantly contribute to enhancing millet productivity, resilience, and nutritional quality. By leveraging genetic diversity, advanced breeding techniques, and international cooperation, gene banks play a pivotal role in ensuring that millets continue to serve as a vital crop for food security, particularly in regions vulnerable to climate variability.

The book "Plant Genebank Utilization for Trait Discovery in Millets" aims to highlight the status of millet genetic and genomic resources and their potential to develop varieties resilient to climatic extremes in order to enhance food security and nutrition. The individual chapters bring forth the valuable role of genetic resources of individual millets in sustainable agriculture and offer valuable insights for future research and innovation in millet research. I congratulate the editors as well as the authors for the commendable exercise carried out in compiling and collating the excellent information in a handy format for the benefit of the researchers and students.

Secretary (DARE) & Director General (ICAR) Himanshu Pathak
Assistant Director General (International Relations)
Indian Council of Agricultural Research
Department of Agricultural Research
and Education, Krishi Bhavan
New Delhi, India
e-mail: dg.icar@nic.in

Preface

Millets are a group of small-seeded hardy grains cultivated for thousands of years in many parts of the world. They are a great source of nutrition, high in fiber and rich in vitamins, minerals, and proteins. Despite their ancient cultivation and resilience in adverse climates, millets cultivation is in general overshadowed by more prominent cereals like wheat, rice, and maize. This book endeavors to provide the detailed account of millet genetic resources, crucial for enhancing food security and promoting sustainable agriculture in an era of climate uncertainty.

Millets include a variety of small-seeded grasses cultivated across many parts of the world, notably in Asia and Africa. This book aims to describe the genetic variability in ten major as well as minor millet species including pearl millet, sorghum, finger millet, foxtail millet, little millet, barnyard millet, proso millet, kodo millet, brown top millet, teff, and fenio. Understanding millet genetic diversity is key to uncovering the traits that enable developing new varieties to succeed in challenging environments, thus offering promising solutions for sustainable farming. During recent years, consumers' and cultivators' interest in millets has increased, owing to their nutritional advantages and adaptability to less fertile environments respectively. With the world facing issues like climate change, water scarcity, and a growing population, millets stand out as a sustainable agricultural option.

This reference work aims to bring together the vast array of knowledge on the genetic diversity, breeding methodologies, and potential applications of various millet crops to bolster food security and nutrition. The contents are designed as a ready reckoner for researchers, students, policymakers, and agricultural practitioners.

This reference book is an outcome of collaborative effort by millet crop experts. Authors have managed to provide a balanced structure of scientific basis and practical applications, making it a valuable resource for students and researchers in agriculture, botany, genetics, plant genetic resources, and related fields. We express our gratitude to the contributors who have contributed specialized knowledge and insights.

We believe that this reference book enhances the interest of stakeholders in millet genetic resources, underscoring the significance of their conservation and use in breeding programs. By incorporating the potential of millets, we can make significant strides toward achieving global food security and climate resilience. We acknowledge the steadfast support of our families and colleagues.

New Delhi, Delhi, India

Kuldeep Tripathi
Jyoti Kumari
Sushil Pandey
Badal Singh
Sunil Archak
Gyanendra Pratap SIngh

Contents

Editors and Contributors

About the Editors

Kuldeep Tripathi is working as a Scientist at the Division of Germplasm Evaluation, Indian Council of Agricultural Research (ICAR)-National Bureau of Plant Genetic Resources, New Delhi. He has made enormous contributions to Plant Genetic Resources management, including pre-breeding and genetic enhancement of pulses that work seems quite unparalleled. His classic research work at ICAR-NBPGR since 2015 on legumes has brought to his credit 04 Indian pulse core sets developed in lentil, pea, grasspea, and cowpea (from entire Indian national gene-bank collections) as the main developer; 26 legume genetic stocks registered. He has also contributed in development of varieties of mungbean (1), lentil (1), and pillipsera (1) for commercial cultivation. He published >80 research papers in peer-reviewed national and international journals. Besides this, he has also been teaching as Post Graduate School faculty of ICAR-IARI (Indian Agricultural Research Institute) for the last nine years and has contributed to teaching and guiding MSc and PhD students.

Jyoti Kumari, PhD, Principal Scientist, ICAR-NBPGR, New Delhi, obtained her master's and doctoral degrees from Indian Agricultural Research Institute, New Delhi, in the discipline of Genetics in 2002 and 2005, respectively. Her major scientific contribution includes mega characterization and evaluation of ~22,000 wheat germplasm and development of core (2226 acc.) and minicore (224 acc.) set from the entire wheat germplasm conserved in the Indian national genebank. She has identified several accessions with unique traits in wheat, barley, maize, lentil, and pearl millet. She also identified Quantitative Trait Locus (QTLs) for terminal heat tolerance, salt tolerance, seed size, and seedling root architecture traits using genome wide association strategy. She has contributed in augmentation of rice and wheat germplasm from the drought- and heat-prone areas of Madhya Pradesh, Rajasthan, and Gujarat. She has published more than 140 publications, including 102 research papers in national and international journals, 18 book chapters, 3 books, and 1 catalogue. She has developed four varieties in lentil and mungbean and registered 40 genetic stocks with Plant Germplasm Registration Committee. She is fellow of Indian Society of Genetics and Plant Breeding since 2012 and recipient of prestigious award, Dr. Harbhajan Singh Memorial Award from the Indian Society of

Genetics and Plant Breeding. As a faculty of Division of Genetics, she has guided nine MSc and one PhD student of Post Graduate School, ICAR-IARI, New Delhi.

Sushil Pandey, Principal Scientist, is working as millet curator in the Division of Germplasm Conservation, ICAR-National Bureau of Plant Genetic Resources. He has more than 23 years of experience in the field of Seed and Plant Genetic Resources. He started his career as Scientist in National Research Centre for Soybean, Indore, where he developed the "Guidelines for Production of Nucleus and Breeder Seed in Soybean – 2002." Thereafter he worked as Senior Scientist at ICAR-Indian Agricultural Research Institute, New Delhi, and worked extensively on seed production of cereals and vegetables and developed hybrid seed production technology in rice and tomato. He has been working as millet curator for 15 years in ICAR-National Bureau of Plant Genetic Resources, New Delhi, and has contributed significantly in upgrading the National Genebank facility. Dr. Pandey has been involved in the formulation of ICAR mega project, "Consortium Research Platform on Agrobiodiversity," and has been coordinating the implementation, monitoring, and evaluation of different sub-projects involving plants, animals, fish, microbes, insects, veterinary type culture collection, documentation of fungal biodiversity (fungal barcoding), and Island biodiversity during first phase of 2012–2017; second phase during 2017–2021; and the third phase during 2021–2026. The Consortium Research Platform has more than 50 centers located in different parts of the country. Dr. Pandey has also worked with Food and Agriculture Organization (FAO) and International Center for Agricultural Research in the Dry Areas (ICARDA) and provided several consultancies to their Seed and Plant Genetic Resources Services in different countries of Asia and Africa. He is a Fellow of Indian Society of Seed Technology and also faculty member in the discipline of Seed Science and Technology and is presently involved in the large-scale characterization program of millets for the development of core sets, triggering their enhanced utilization.

Badal Singh is working as a Scientist at the Division of Germplasm Evaluation, ICAR-National Bureau of Plant Genetic Resources, New Delhi, and has more than 7 years of experience in the area of Plant genetic resources management. He is graduation in BSc (Agriculture) from Institute of Agricultural Sciences, Banaras Hindu University (BHU), and post-graduation and PhD in Economic Botany and PGR from ICAR-IARI, New Delhi. He joined ICAR Agricultural Research Services at ICAR-NBPGR, RS, Shimla, and worked on temperate agri-horticultural genetic resource management. He has been working for past 5 years on genetic resource management of potential crops, which include millets and nutri-cereals of the Himalayan region. Presently he is engaged in genetic resource evaluation of millets. He is also engaged in teaching at Graduate School, ICAR-IARI. He published >16 research papers in peer-reviewed national and international journals.

Sunil Archak is a Principal Scientist and ICAR-National Fellow at the ICAR-NBPGR, New Delhi, in the areas of PGR Informatics and PGR Policy. He has 26 years of experience working with various aspects of Plant Genetic Resources including large-scale characterization of germplasm. He was instrumental in establishing National Genomic Resources Repository and a PGR Informatics Facility at NBPGR. He has developed many web-apps and mobile apps for enhancing access to PGR information including PGR Portal. Sunil Archak holds a dual faculty membership at IARI in the disciplines of *Plant Genetic Resources* as well as *Bioinformatics*. He has organized trainings on DNA fingerprinting, genomic data analysis, conservation of genomic resources, climate analogue tools, etc. He represents India in International Treaty on Plant Genetic Resources for Food and Agriculture (FAO-ITPGRFA) and has participated in the Treaty Governing Body meetings at Kigali (2017), Rome (2019), Delhi (2022), and Rome (2023). He is currently the Co-Chair of the Treaty Working Group on Multilateral System (MLS) enhancement. He also represented India in Commission on Genetic Resources for Food and Agriculture (FAO-CGRFA) and Convention on Biological Diversity (CBD) meetings.

Gyanendra Pratap SIngh, PhD, Director, ICAR-NBPGR and former Director, ICAR-IIWBR, Karnal, have more than three decades of agriculture research including 10 years of teaching and more than 8 years of research management and administration experience. Dr. Singh is instrumental in the development of 60 wheat and 3 barley varieties and 1 potato variety, benefitting multitude farmers, consumers, and industries. Dr. Singh largely focused in intrinsic and strategic research on abiotic stress that led to publication of 300 peer-reviewed research articles of high impact, 18 books, 65 book chapters, 57 technical bulletins, 55 popular articles, and 5 Policy/Strategy papers. His strenuous efforts on advancing the R&D led to several accolades at national and international level by honoring with 17 individual awards. He is an elected fellow of three prestigious academies, i.e., Indian National Science Academy, India (INSA), National Academy of Sciences, India (NASI) and National Academy of Agricultural Sciences, India (NAAS), and also fellow of three registered professional societies. His acumen on leading research programs and visionary perspective garnered recognitions like Gene Stewardship Award from the Borlaug Global Rust Initiative, USA.

Contributors

B. Amasiddha ICAR-Indian Institute of Millets Research, Hyderabad, Telangana, India

Lalit Arya ICAR-National Bureau of Plant Genetic Resources, New Delhi, Delhi, India

Division of Genomic Resources, ICAR-NBPGR, New Delhi, India

S. Avinash ICAR-Indian Institute of Millets Research, Hyderabad, Telangana, India

Akansha Bajpai ICAR-National Bureau of Plant Genetic Resources, New Delhi, India

C. Deepika ICAR-Indian Institute of Millets Research, Hyderabad, Telangana, India

Seva Nayak Dheeravathu ICAR- Indian Institute of Millets Research, Hyderabad, India

Balakrishna Domathoti ICAR- Indian Institute of Millets Research, Hyderabad, India

Sneka Chinna Durai Siddha Central Research Institute, Chennai, India

K. N. Ganapathy ICAR-Indian Institute of Millets Research, Hyderabad, Telangana, India

Sunil Shriram Gomashe ICAR-National Bureau of Plant Genetic Resources, Regional Station, Akola, Maharashtra, India

K. Hariprasanna ICAR-Indian Institute of Millets Research, Hyderabad, Telangana, India

Himabindu Kudapa International Crops Research Institute for the Semi-Arid Tropics (ICRISAT), Hyderabad, India

Maggiga Revanth Kumar Department of Genetics and Plant Breeding, Sam Higginbottom University of Agriculture Technology and Sciences, Prayagraj, Uttar Pradesh, India

Mithilesh Kumar Agricultural Research Station, Mandor, Agriculture University, Jodhpur, Rajasthan, India

Pavan Kumar Malav Division of Plant Exploration and Germplasm Collection, ICAR-NBPGR, New Delhi, India

Sweta Mishra Dr Rajendra Prasad Central Agricultural University, Samastipur, Bihar, India

C. Nandini Zonal Agricultural and Horticultural Research Station, Hiriyur, KSNUAHS, Shivamogga, Karnataka, India

R. Narasimhulu Regional Agricultural Research Station, ANGRAU, Nandyal, Andhra Pradesh, India

D. Naresh International Crops Research Institute for the Semi-Arid Tropics (ICRISAT), Hyderabad, India

Chithra Devi Pandey Division of Germplasm Conservation, ICAR-National Bureau of Plant Genetic Resources, New Delhi, Delhi, India

Susheel Pandey Division of Germplasm Conservation, ICAR-NBPGR, New Delhi, India

Sushil Pandey Division of Germplasm Conservation, ICAR-National Bureau of Plant Genetic Resources, Pusa Campus, New Delhi, Delhi, India

P. Parashuram Centre on Rabi Sorghum, ICAR-Indian Institute of Millets Research, Solapur, Maharashtra, India

P. Pranusha ICAR-National Bureau of Plant Genetic Resources, Regional Station, Hyderabad, Telangana, India

Thirugnanasambandam Perumal Prathima ICAR- Sugarcane Breeding Institute, Coimbatore, India

P. Rajendrakumar ICAR-Indian Institute of Millets Research, Hyderabad, Telangana, India

Kirti Rani ICAR-NBPGR, Regional Station, Jodhpur, Rajasthan, India

P. Sanjana Reddy ICAR-Indian Institute of Millets Research, Hyderabad, India

Ankita Roy International Crops Research Institute for the Semi-Arid Tropics (ICRISAT), Hyderabad, India

Rahul Sager ICAR- Indian Institute of Millets Research, Hyderabad, India

Vaidurya Pratap Sahi Department of Genetics and Plant Breeding, Sam Higginbottom University of Agriculture Technology and Sciences, Prayagraj, Uttar Pradesh, India

Sobhan Sajja International Crops Research Institute for the Semi-Arid Tropics (ICRISAT), Hyderabad, India

Sangappa ICAR-Indian Institute of Millets Research, Hyderabad, India

C. Tara Satyavathi ICAR-Indian Institute of Millets Research, Hyderabad, Telangana, India

Dinesh Prasad Semwal ICAR-National Bureau of Plant Genetic Resources, New Delhi, Delhi, India

Neelam Shekhawat ICAR-NBPGR, Regional Station, Jodhpur, Rajasthan, India

Amit Kumar Singh Division of Genomic Resources, ICAR-NBPGR, New Delhi, India

Badal Singh Division of Germplasm Evaluation, ICAR-National Bureau of Plant Genetic Resources, New Delhi, India

G. P. Singh ICAR-National Bureau of Plant Genetic Resources, New Delhi, India

Kuldeep Singh International Crops Research Institute for the Semi-Arid Tropics (ICRISAT), Hyderabad, India

Mamta Singh ICAR-National Bureau of Plant Genetic Resources, New Delhi, Delhi, India

Monika Singh ICAR-National Bureau of Plant Genetic Resources, New Delhi, Delhi, India

Aviansh Singode ICAR- Indian Institute of Millets Research, Hyderabad, India

Sooganna ICAR-Indian Institute of Millets Research, Hyderabad, Telangana, India

Nandan Tayade ICAR-National Bureau of Plant Genetic Resources, Regional Station, Akola, Maharashtra, India

A. V. Umakanth ICAR-Indian Institute of Millets Research, Hyderabad, India

K. Venkatesh ICAR-Indian Institute of Millets Research, Hyderabad, Telangana, India

Manjusha Verma ICAR-National Bureau of Plant Genetic Resources, New Delhi, Delhi, India

Mani Vetriventhan International Crops Research Institute for the Semi-Arid Tropics (ICRISAT), Hyderabad, India

Pearl Millet

1

P. Sanjana Reddy

Abstract

Pearl millet is a climate-resilient and widely grown crop among the millets. Owing to C_4 photosynthetic system and cross-pollinating nature, pearl millet hybrids hold promise for changing climatic conditions. Genetic resources play an important role in creation and maintenance of new variability to make the crop profitable and address yield constraints. This chapter gives an insight into the origin, current status of germplasm, and characterization and utilization of genetic resources in crop breeding programs. The availability of genetic resources in the form of core collection, mini-core collection, wild species, and landraces and their use are discussed. The advances made with the utilization of genomic tools for identification of genes/markers, application of markers for genomic selection (GS), and potential of gene editing technologies are also included. Further, the genetic resources utilized in pre-breeding for improvement of yield and addressing biotic and abiotic constraints are discussed in detail.

Keywords

Pearl millet · Plant genetic resources · Germplasm · Breeding · Utilization · Pre-breeding · Trait-specific sources

P. S. Reddy (✉)
ICAR-Indian Institute of Millets Research, Hyderabad, India
e-mail: p.sanjana@icar.gov.in

K. Tripathi et al. (eds.), *Plant Genebank Utilization for Trait Discovery in Millets*, https://doi.org/10.1007/978-981-96-4043-0_1

1.1 Introduction

Pearl millet, *Pennisetum glaucum* (L.) R. Br., is a major millet crop that inhabits the dry regions of the semiarid tropics and spreads into the arid zone, which is characterized by sandy soils, a short growth season, and more frequent drought stress, thus well-adapted to the hottest and driest regions in India and Africa. The main causes of low productivity (500–700 kg ha^{-1}) in arid regions include low soil fertility, rainfed farming with irregular and insufficient rainfall (annual precipitation as low as 300 mm), little external inputs, use of mostly unimproved cultivars with a low harvest index (<20%), and a high frequency of disease and insect pests. In such circumstances, pearl millet is the most dependable food grain serving for both food and nutritional security. However, under productive situations found in semiarid locations, grain yield as high as three tons per hectare may be achieved.

Pearl millet is a cross-pollinated crop owing to its protogynous nature of flowering with high tillering, short life cycle, and a large genome size (~2352 Mbp) (Bennett et al. 2000). It is the sixth important cereal crop after wheat (*Triticum aestivum* L.), maize (*Zea mays* L.), rice (*Oryza sativa* L.), barley (*Hordeum vulgare* L.), and sorghum (*Sorghum bicolor* (L.) Moench). It is being grown in an area of 30.93 m ha (FAOSTAT, 2021) in the world providing food to almost 90 million poor people (Shivhare and Lata 2017). In terms of global statistics on millets, representing majorly pearl millet, India with 9.76 m ha occupies 31.6% of total global millet area followed by Niger, Sudan, Mali, Nigeria, and Chad. India is also the largest producer with 13.21 m tons, which is 43.9% of global millet production (30.09 m tons). In India, pearl millet is fourth important cereal crop after rice, wheat, and maize playing a vital role especially in the arid regions where other cereal crops cannot be successfully grown (Sanjana Reddy et al. 2021). It is consumed mainly in the western Rajasthan and parts of Gujarat, where it contributes to more than 50% of the total cereal consumption. Given the increase in global climatic temperatures by 5 °C above pre-industrial level by 2100 (Tollefson, 2020), crops like pearl millet with efficient C_4 photosynthetic system, shorter life cycle, and tolerance to abiotic stresses (heat and salinity) are termed as climate-smart as they meet the food and fodder requirements even under harsh environmental conditions.

Pearl millet is susceptible to a variety of biotic stresses, majorly diseases caused by bacterial, viral, and fungal infections similar to other cereal crops. Among the diseases, downy mildew, rust, smut, blast, and ergot caused by fungal pathogens are found to have a detrimental effect on pearl millet yield and production. One of the most destructive diseases of pearl millet, downy mildew (DM) known as "green ear," is caused by *Sclerospora graminicola* (Sacc. Schroet.). This disease causes the huge yield losses in India and Africa. It was unimportant in India until 1970, when the downy mildew epidemic severely reduced the yield of HB3, a well-known Indian pearl millet hybrid (Singh 1995), from roughly 8.2 million metric tons in 1970–1971 to 4.6 million metric tons in 1971–1972 (Kumar et al. 2012). Blast (*Pyricularia grisea*) and rust (*Puccinia substriata* var. *indica*) are other economically important fungal diseases affecting pearl millet grain and fodder yields. Abiotic stresses have a detrimental effect on the growth and potential yield of all crops, and pearl millet is no exception. It is susceptible to a variety of abiotic stresses much like other crops, even though it is commonly thought to be well-adapted to drought, salt, high temperatures, lodging, and poor soil (Shivhare and Lata 2017).

With regard to nutritional quality, it is equivalent to maize and superior to sorghum in protein content and quality, protein efficiency ratio, and metabolizable energy levels and does not contain tannins. Though it is deficient in essential amino acids, it contains 35% more lysine than sorghum (Rooney and McDonough 1987). It contains 5–6% oil (Jambunathan and Subramanian 1988) and higher levels of iron and zinc. Apart from being a main staple, its use for poultry feed is on the increase. It is an important fodder source and the only source of feed in dry months (Hash et al. 2003). The presence of all required nutrients makes it suitable for large-scale utilization in the manufacture of various food products such as traditional meals, baby foods, snacks, and dietary foods in both grain and flour form. Pearl millet has superior mineral profile in the grain and fodder compared to other cereals. Beside its nourishment qualities, the bioavailability of various minerals such as Ca and P and trace elements such as Zn, Fe, Cu, and Mn is low because of the presence of some inherent anti-nutritional factors, for example, phytate and polyphenols in grain and oxalic acid in fodder and forage, which are undesirable. However, a number of processing methods, including dehulling, milling, malting, blanching, acid treatment, dry heat treatment, and parboiling, have been developed to improve the food value, shelf life, and availability of minerals, protein, and starch in pearl millet products. Because of its impact on nutrition, degradation or reduction of phytic acid is highly beneficial (Sanjana Reddy et al. 2023).

Utilization of pearl millet is increasing in brewing industry, poultry, and animal feed, as well as for fodder and roofing material. It is also increasingly used in traditional foods and snacks such as porridge, chapati, *khichri*, *laddoo*, *mathi*, etc.; baked products such as cake, *nan khatai*, biscuits, etc.; extruded products such as *sev* and sweet vermicelli; flakes and pops; and many more. The National Dairy Research Institute, Karnal, has recently developed and launched pearl millet *lassi*, made from pearl millet flour. Pearl millet is also being used in preparing various types of health foods and food ingredients as it contains a relatively higher proportion of insoluble dietary fiber, which causes slow release of sugar, thus making the food products suitable for diabetic patients. Despite all these advantages, it is considered to be an inferior food crop due to lack of nutritional awareness and processing and economic incentives to farmers. Its per capita consumption is declining, whereas production is increasing, and this leads to high volatility in its prices and farmers' incomes.

1.2 The Crop

1.2.1 Origin and Taxonomy

It is widely believed that pearl millet originated in Africa, which is also considered as the primary center of diversity. According to Harlan (1975), the Sahelian Zone, which stretches from western Sudan to Senegal, is the center of origin. Pearl millet has been under cultivation for thousands of years (Brunken 1977; Burton and Powell 1968; Rachie and Majmudar 1980). In his study on the origin of crops (Vavilov 1949, 1950) proposed Ethiopia as the center of domestication for pearl millet. Two theories contradict the Ethiopian origin: (a) *Pennisetum americanum* ssp. *monodii*, a wild progenitor that is found in east of Sudan, was not observed in the highlands of Ethiopia, and (b) there is minimal diversity in the pearl millet from Ethiopia. Hence, Ethiopia is seen as the result of post-domestication introduction. Between 5000 and 4000 BCE, the Mande people lived close to the Niger River, where they domesticated a number of Western African crops, including pearl millet (Murdock 1959). Western Africa, south of the Saharan desert and north of the woodland zone, is currently home to the most morphological diversity (Anand Kumar and Appa Rao 1987). From Western Africa where it was first domesticated, it went to Eastern Africa, and about 2000 years ago, it reached Southern Africa and India. India is considered as the secondary center of diversity for pearl millet.

According to the shape of the seeds, cultivated pearl millet can be divided into four major races: *typhoides*, *nigritarum*, *globosum*, and *leonis* (Brunken 1977). However, there are also intermediate forms for any of these basic types. The race *typhoides* is the most diverse and extensively spread among the four races. It is the only fundamental race discovered outside of Africa and is the dominant race grown in Southern Africa (Appa Rao et al. 1986a) and India (Kumari et al. 2016). The race *nigritarum* can be found in northern Nigeria and western Sudan. According to Bono (1973), the race *globosum* is the most prevalent in central Ghana, Togo, Nigeria, Niger, and Benin. Sierra Leone is the home for the *leonis* race. The grain's substantially elongated characteristic may be an adaptation to the areas of Sierra Leone with comparatively high rainfall.

1.2.2 Taxonomic Classification

Taxonomists in the past have used different scientific binomials for pearl millet. In early 1900s, *Pennisetum typhoideum*, *Penicillaria spicata*, *Panicum spicatum*, and *Pennisetum alopecuroides* were the names given to pearl millet (Chase 1921). By the middle of the nineteenth century, a number of workers were using the binomials *P. typhoideum* L. C. Rich. and *P. americanum*. The *P. typhoides* (Burton and Powell 1968) was the name given to pearl millet in the 1960s, while *Pennisetum glaucum* (L.) R. Br. was mentioned in the *Manual of the Grasses of the United States* (Hitchcock and Chase 1951). This binomial is presently used by researchers studying pearl millet.

Based on its clustered spikelets that are encircled by an involucre, taxonomists place the genus *Pennisetum* in the tribe Paniceae, close to the genus *Cenchrus*. *Gymnothrix*, *Brevivalvula*, *Penicillaria*, *Heterostachya*, and *Eu-Pennisetum* are the five sections that make up the genus. The section *Penicillaria* contains all *Pennisetum* species with penicillate anther tips and pinnate involucral bristles (Leeke. 1907). The genus *Pennisetum* contains two species that are reproductively isolated, according to Brunken (1977). These include *P. americanum* (L.) Leeke, a diploid ($2n = 14$)

that is indigenous to the semiarid tropics of Africa and India, and *Pennisetum purpureum* Schumach, a perennial tetraploid ($2n = 28$) species that is found throughout the wet tropics of the world. Three subspecies make up the species *americanum*. The cultivated pearl millet falls within the subspecies *americanum*. The name *P. americanum* has been changed to *P. glaucum* (L.) R. Br. The current taxonomical classification places the cultivated pearl millet under family Poaceae, subfamily Panicoideae, tribe Paniceae, subtribe Panicinae, section *Penicillaria*, genus *Pennisetum*, and species *glaucum*.

Another system of classification proposed by Harlan and de Wet (1971), based on gene pools, groups the total variation into primary, secondary, and tertiary gene pools. All cultivars of *P. glaucum*, its subspecies, wild progenitor *violaceum* (monodii), and weedy form *stenostachyum* are included in the primary gene pool because these taxa naturally cross with the cultivated species. The apomictic and octaploid species *P. squamulatum* Fresen. and the allotetraploid, rhizomatous perennial species *P. purpureum* Schum., often known as napier grass, are included in the secondary gene pool. Their hybrids are sterile, although they can be crossed with cultivated pearl millet with ease. The tertiary gene pool contains all of the species that are cross-incompatible with pearl millet (Sharma et al. 2020a).

1.3 Status of Germplasm Conservation

Maintaining genetic resources and meeting the demands of worldwide crop improvement initiatives are made possible by the collection, conservation, and maintenance of germplasm. Globally, several genebanks together conserve over 73,000 accessions of pearl millet germplasm including wild and weedy relatives. The ex situ collections maintained in major genebanks globally are given in Table 1.1. In 1959–1962, the Indian Agricultural Research Institute (IARI) and The Rockefeller Foundation launched the initial efforts to pool global pearl millet germplasm. Approximately, 10,764 accessions were collected from 65 organizations, including various national and international institutes, universities, national agricultural research systems, etc., in an intensive campaign led by International Crops Research Institute for the Semi-Arid Tropics (ICRISAT). This includes 2178 accessions from Institute Français de Recherche Scientifique pour le Developpement en Cooperation; 2022 accessions from The Rockefeller Foundation, New Delhi, India; and 974 accessions from the International Bureau for Plant Genetic Resources (IBPGR), Rome, Italy. The gaps in the collection and the priority areas for systematic collection of germplasm in different countries were identified and focused, which led to the collection of 10,830 pearl millet samples during 76 collection missions in 28 countries. By the end of 2006, the genebank at ICRISAT had 21,594 accessions from 51 countries, including 750 accessions belonging to 24 species of genus *Pennisetum*. This can be classified as samples from institutions (10,201 accessions), farmers' fields (6537 accessions), commercial markets (1681 accessions), farmer's stores (1357 accessions), threshing floors (479 accessions), and wild species (750 accessions). Biological status of accessions indicates the presence of

Table 1.1 Ex situ collections of pearl millet germplasm conserved in the genebanks globally

S. No.	Genebank/institute	No. of accessions
1.	ICRISAT Genebank	25,537
2.	National Genebank, ICAR-NBPGR, New Delhi	8398
3.	Embrapa Milho e Sorgo, Brazil	7225
4.	Laboratoire des Ressources Génétiques et Amélioration des Plantes Tropicales, ORSTOM, France	4405
5.	Plant Gene Resources of Canada, Saskatoon Research Centre, Agriculture and Agri-Food Canada, Canada	3821
6.	Institut Sénégalais de Recherches Agricoles (ISRA) in Bambey, Senegal	2400
7.	Serere Agriculture and Animal Production Research Institute, Uganda	2142
8.	The Agricultural Research Station of the USDA at Griffin, Georgia	2063
9.	Institut national de la recherche agronomique du Niger	2052
10.	National Plant Genetic Resources Centre, National Botanical Research Institute, Namibia	1421
11.	Plant Genetic Resources Program, Pakistan	1377

Source: http://genebank.icrisat.org/; http://www.nbpgr.ernet.in; Vetriventhan et al. (2024), Singh and Nara (2023)

landraces (18,447), breeding materials (2268), advanced cultivars (129), and wild relatives (750) (Upadhyaya et al. 2007a). The characterization of pearl millet germplasm landraces was taken up in batches of 500–1000 every year at ICRISAT farm, Patancheru (17.53 N lat, 78.27 E long, 545 m asl), in alfisols, during the rainy and summer seasons from 1974 through 2013, for 21 morpho-agronomic traits following pearl millet descriptors (IBPGR and ICRISAT, 1993). Apart from ICRISAT, the Institut de recherche pour le développement (IRD, France) also maintains 3968 accessions of pearl millet from 16 countries. The Services Scientifiques Centraux of ORSTOM in Bondy, France, preserves a collection of more than 2700 accessions that represent cultivated, weedy, and wild relatives. The Canadian Genetic Resources Programme, Saskatoon, Canada, maintains 3821 accessions covering a few species with emphasis on *Pennisetum glaucum* (3390 accessions). Accessions of other species include *P. violaceum* (221), *P. macrourum* (1), *P. purpureum* (12), *P. orientale* (1), *P. pedicellatum* (11), *P. polystachion* (8), *P. ramosum* (3), *P. unisetum* (1), and other species (14). The Agricultural Research Station of the USDA at Griffin, Georgia, maintains 1314 accessions of which only 1 is a wild relative, 290 are of breeding/research material, and 552 are for other purposes. There are nearly 2400 accessions at the Institut Sénégalais de Recherches Agricoles (ISRA) in Bambey, Senegal. Besides, 500 accessions are conserved in Canada, 32 in Italy, 291 in Malawi, 36 in Pakistan, 32 in Tunisia, and 24 in Zambia (Acheampong et al. 1984; Mengesha et al. 1989). The SADC Plant Genetic Resources Centre (SPGRC) near Lusaka, Zambia, began conserving germplasm from SADC countries.

1.4 Characterization and Evaluation

Upadhyaya et al. (2016) evaluated 15,979 landraces originating in 34 countries to study the geographical distribution of pearl millet traits and diversity in the collection. The landrace collections from Pakistan, Ghana, Togo, and India were early flowering (33–40 days), while the Sierra Leone and the Central African Republic had landraces that were very late flowering (121–159 days). The accessions with short plant height (80–100 cm) were found in India, Zambia, and Sudan; with tallness (401–490 cm) in Chad, Burkina Faso, Nigeria, and the Central African Republic; with high tillering (11–35) in India and Yemen; with high panicle exertion (11–29 cm) in Ghana, Chad, India, and Yemen; with long panicles (75–135 cm) in Nigeria and Niger; with thick panicles (41–58 mm) in Namibia, Togo, and Zimbabwe; and with large seeds (16–19 g per 1000) in Togo, Benin, Ghana, and Burkina Faso. Collections from Ghana for flowering (36–150 days); Burkina Faso for plant height (80–490); India and Yemen for total tillers (1–35) and productive tillers (1–19) per plant; Niger for panicle exertion (45–21.0 cm), panicle length (9–135 cm), and panicle thickness (12–55 mm); and Zimbabwe for 1000-seed weight (3.5–19.3 g) were important sources for trait diversity.

Of the 1283 current collections maintained in the USDA Germplasm Resources Information Network, over 75% are attributed to Zimbabwe, India, Nigeria, and Burkina Faso. About 9000 accessions of pearl millet are conserved under long-term storage in the genebank of the National Bureau of Plant Genetic Resources (NBPGR), India. Of these, only 168 accessions belong to other countries with indigenous lines making up the majority of these accessions (8827) (Kumari et al. 2016). By 2017, there were 56,580 accessions of pearl millet available globally, conserved in 70 genebanks across 46 countries (possibly including duplicates). According to Yadav et al. (2017), the majority of these germplasm accessions consist of landraces, followed by breeding/research material and wild relatives.

Though a large number of accessions are conserved in genebanks, several gaps in the collection of germplasm from the African and Asian continents have been identified by a critical evaluation of the collection utilizing passport and characterization data and a variety of geographical information system (GIS) methods like FloraMap, DIVA-GIS, ArcGIS, and others, for geographical and trait-diversity deficits and taxonomic gaps in the collection. Using 5768 landraces from India, Pakistan, and Yemen in Asia, 134 distinct districts of 14 provinces in India and 12 districts of Punjab province in Pakistan were identified as the major geographical gaps (Upadhyaya et al. 2010a). When assessed for agroecological pattern of diversity among 5197 pearl millet landraces collected from 20 agroecological regions in India, it became clear that Malwa and Bundelkhand regions are underrepresented in genebank and should be explored further for additional diversity (Upadhyaya et al. 2007b). Gap analysis of 3750 georeferenced pearl millet germplasm accessions from East and Southern Africa (ESA) countries identified 34 districts located in 18 provinces of 4 East African countries and 76 districts located in 34 provinces of 7 Southern African countries as geographical gaps (Upadhyaya et al. 2012). Similar studies using 6434 landraces from WCA identified 62 districts in 13 provinces of

Fig. 1.1 Panicle size, shape, and color variability in pearl millet germplasm

Nigeria, 50 districts in 16 provinces of Burkina Faso, 9 districts in 6 provinces each of Mali and Mauritania, 8 districts in 8 provinces of Chad, and 7 districts in 3 provinces of Ghana as the major geographical gaps (Upadhyaya et al. 2010b). In Southern Africa, 11 districts of 7 provinces in Sudan and Uganda and 58 districts of 20 provinces of 7 countries were identified as geographical gaps.

The available germplasm varies greatly in terms of phenotypic and phenological features. Panicle variability in terms of size, color, and shape is given in Fig. 1.1. African landraces have been a good source of larger panicles, large seed sizes, and disease resistance, while Indian landraces have generally contributed to earliness, high tillering, high harvest index, and local adaptability. It has been possible to identify certain sources for improved grain quality, disease resistance, and tolerance to abiotic stresses (such as heat and drought) through systematic study and screening of germplasm. These sources of germplasm are still essential to global agricultural improvement initiatives. Utilization of genetic resources can be enhanced as trait-specific gene pools and core and mini-core collections are available in pearl millet (Yadav et al. 2017).

1.4.1 Utilization of Germplasm

Due to the availability of a large number of germplasm accessions and the prevalence of desirable traits in the unadapted genetic background, only a small fraction of germplasm has been utilized in the breeding program until now. In recent past, development of core and mini-core collections prompted breeders to use the trait-specific germplasm to extend the genetic base of commercial cultivars, which is vital for mitigating the effects of climate change and lowering the likelihood of disease epidemics (Upadhyaya et al. 2011). Crop improvement programs around the world exercises extensive use of germplasm for the traits including sweet stalk, yellow endosperm, high Fe and Zn, new dwarfing genes, CMS lines, early-maturing lines, and sources of resistance to biotic and abiotic stress challenges. Numerous novel and beneficial characteristics have been widely employed in academic research, including thin leaves, glossy leaves, brown midrib leaves, and leaf color variations (Upadhyaya et al. 2007a). The identification of excellent sources of resistance in the germplasm and breeding lines, primarily from West Africa, has been made possible by the development of efficient field screening tools at ICRISAT. In order to accelerate the germplasm collection, ICRISAT and NBPGR have been working collaboratively in recent years. They have catalogued genetic resources and conducted multilocational evaluations in a few chosen places that reflect various agroclimatic conditions. About 8460 accessions of pearl millet have been evaluated from 1986 to 1989 at NBPGR Headquarters, New Delhi; NBPGR Regional Station, Jodhpur; and AICMIP, Pune, for a set of 14 descriptors, and these include promising fodder types (267 accessions), grain types (156 accessions), dual-purpose types (53 accessions), early-maturing types (54 accessions), and bold seed types (20 accessions). Upadhyaya et al. (2007a) evaluated 20,844 accessions from 51 countries for 23 traits, including number of flowering days, plant height, tiller number, 100-seed weight, panicle length and shape, seed shape and color, and reported significant variations among the accessions studied. Therefore, in order to expand its cultivated gene pool, germplasm collections—both cultivated and wild—and their characterization using both traditional and cutting-edge approaches must be carried out.

As pearl millet landraces exhibit more intrapopulation variability, selection by plant breeders has led to improved open-pollinated varieties (OPVs); and yet, productivity increases have been modest. Traditional farmers have preferred these improved landraces. The development of composites in pearl millet, involving improved breeding lines and germplasm, primarily from Western and Southern Africa, has made extensive use of pearl millet germplasm (Rai and Anand Kumar 1994). The main purpose of pearl millet composites was to breed OPVs. Pedigree breeding has been used in the programs devoted to breeding hybrid parents. Although the genetic base of hybrids and hybrid parents has significantly shrunk, hybrid breeding approach is still effective. Long-term advancement in hybrid development depends on expanding the genetic base of elite hybrid parents by using more new germplasm.

The most widely used *Iniadi* germplasm in pearl millet breeding efforts globally comes from the Togo-Ghana-Burkina Faso-Benin region of Western Africa (Andrews and Anand Kumar 1996). Large grain (15–18 g per 1000 grains) with a globular shape, dark gray color, and bright outer exposed surface, starchy endosperm, compact and conical panicles, relatively less photoperiod sensitivity, earliness (70–85 days to mature), rapid grain-filling (23–25 days compared to 25–32 days for others), good resistance to downy mildew, and good combining ability for grain yield are typical characteristics of this germplasm. Nearly all of the hybrids undergoing advanced testing in the United States contain a significant amount of *Iniadi* germplasm, and most hybrids in India have some degree of this source in their parental lines. This germplasm is being introgressed more and more, which lowers the likelihood of creating high-yielding hybrids in the future by disrupting heterotic patterns in elite breeding materials.

Photoperiod-sensitive germplasm accessions have not been extensively employed in breeding programs outside their natural habitat. This is due to two factors: (i) the big less-sensitive population, of which only a small portion has been exploited so far, has mostly satisfied the phenotypic needs, and (ii) there haven't been many good outcomes from the high attrition rate in segregating generations resulting from sensitive × insensitive crosses. Some of the promising germplasm lines for various traits are given in Table 1.2.

Not much attention was given to the use of wild germplasm in trait-specific breeding programs. The rust and leaf spot resistance, cytoplasmic male sterility, and greater stalk strength and male fertility restorer genes from *P. purpureum* and apomictic genes from *P. squamulatum*, as well as rust and leaf spot resistance from *P. glaucum* ssp. *monodii* accessions, have all demonstrated intriguing results of a modest attempt to use wild pearl millet germplasm at Tifton, Georgia (Hanna 1992). It has been demonstrated that introgression across populations is a more efficient means of transferring this diversity than through elite inbred lines and that wild sorghum germplasm can provide valuable variation to highly adapted grain hybrid parents (Bramel-Cox and Cox 1989; Menkir et al. 1994).

1.4.2 Core and Mini-core Collections

In 1998, the 16,000 accessions at ICRISAT, Hyderabad, were first sorted by geographic distribution, and then 25 different groups were produced by Ward's method of hierarchical clustering based on 11 quantitative traits. In order to create a core collection of 1600, about 10% of the accessions from each of these 25 different groups were then chosen at random. A total of 501 accessions was added to the core collection representing 4717 accessions collected and characterized during the following 9 years. Thus, there were 2094 accessions in the updated core. Using information on 10 qualitative and 8 quantitative characteristics of 2094 core collection accessions from 46 countries, a mini-core collection of pearl millet consisting of 238 accessions was created (Upadhyaya et al. 2011). Mini-core collections have been effective in locating sources of salinity (Serraj et al. 2004) and drought tolerance (Kashiwagi et al. 2005), disease resistance (Sharma et al. 2010, 2012), and other economically significant features (Upadhyaya et al. 2014). As a result of their origins in various agroecologies, the agronomically desirable multiple-pathotype-resistant germplasm accessions found in this study are anticipated to be highly divergent and helpful in breeding programs aimed at creating pearl millet hybrids resistant to virulent, challenging-to-manage *S. graminicola* pathotypes. The worldwide pearl millet germplasm collection of ICRISAT was recently used to build the Pearl Millet Germplasm Association Panel (PMiGAP), which consists of 250 landraces, elite cultivars, and mapping population parents acquired from a wide geographical range in Africa and Asia (Yadav et al. 2011).

Table 1.2 Promising pearl millet accessions for agronomic and economically important traits

Character	Germplasm accession	References
Vigor (score)	IP 1024, IP 3467, IP 3986, IP 4021, IP 13523	Khairwal et al. (2007)
Days to 50% flowering (earliness)	IP 4006, IP 550, IP 3549, IP 4066, IP 9496, IP 9426, IP 13520, IP 9532, IP 18132, IP 17720, IP 4021 (Bhilodi)	Khairwal et al. (2007), Manga (2015), Sehgal et al. (2015), and Harinarayana et al. (1988)
Days to maturity	OPY 197, IP 3549, SNV 26	Khairwal et al. (2007)
Plant height (cm) (tall)	IP 1497, IP 947, IP 3541, IP 3749, IP 3616, IP 19190, IP 19405, IP 13150, and IP 5452	Khairwal et al. (2007) and Manga (2015)
Productive tillers per plant	IP 22269, RMFB-35, RMFB-34, IP 13645, IP 15273, and IP 17493	Khairwal et al. (2007), Manga (2015), and Appa Rao et al. (1998)
Total tillers per plant	IP 6148 and IP 15257	Manga (2015)
Sheath length (cm)	Dhodsar (Sunda Ram), Jhakrana, Sulkania, and SR 54	Manga (2015)
Blade length (cm)	Dhodsar (Sunda Ram), Jhakrana, Sulkania, and SR 54	Manga (2015)
Panicle length (cm)	IP 1763, IP 1745, IP 5447, IP 10273, IP 12848, IP 10456	Khairwal et al. (2007) and Manga (2015)
Panicle thickness (mm)	NSS 7776, NSS 7799, IP 4394, IP 7042, IP 7935, IP 5396, IP 18797, IP 15917, IP 13363, IP 19386, IP 19448, IP 16403, IP 12845	Khairwal et al. (2007), Manga (2015), and Sehgal et al. (2015)
Panicle density (score)	IP 14100, IP 8350, IP 14160	Manga (2015)
1000-grain weight (g)	IP 399, IP 11893, IP 19160, IP 10705, and IP 14160	Khairwal et al. (2007) and Manga (2015)
Grain yield per plant (g)	IP 128, IP 3416, IP 33, NSS 7911, IC-296718 HMS 1B	Khairwal et al. (2007)
Grain yield potential (score)	IP 17350, IP 3150, IP 17945, IP 7095, IP 22269, and IP 6098	Manga (2015), Kulkarni et al. (2006)
Green fodder yield per plant (g)	IP 3416, IP 17880, IP 104, IP 15304, IP 3616, IP 6125, IP 13885, IP 6146, IP 6897	Khairwal et al. (2007) and Manga (2015)
Overall plant aspect (score)	IP 16131, IP 22281, IP 15817, IP 17945, IP 7095	Khairwal et al. (2007)
Higher dry fodder yield	IP 22269 and IP 6098, IP 3616, IP 6105, IP 6112, and IP 6104	Kulkarni et al. (2006)
Dual-purpose lines	IP 3616, IP 6104, and IP 6112	Kulkarni et al. (2006)
Carotenoid content (mg/g fresh weight)	IP 164	Choudhary et al. (2021)
Total soluble sugars (mg/g fresh weight)	IP133, IP144, IP181, IP136, IP196, IP262	Choudhary et al. 2021
Proline content (mg/g fresh weight)	IP147	Choudhary et al. 2021
Protein content (mg/g fresh weight)	IP 130, IP 126, IP 3642, and IP 19877	Choudhary et al. (2021) and Yadav et al. (2017)

(continued)

Table 1.2 (continued)

Character	Germplasm accession	References
Yellow endosperm	IP 15536, IP 15533	Sehgal et al. (2015)
Blast	IP347, IP383, IP324, and IP351	Verma et al. (2021)
Popping traits (popping yield, puffing index, and size of popped grain)	IC 283734, IC 283745, IC 283763, IC 283842, IC 283908, 841-B, and PPMI 301	Chauhan et al. (2015)
Sweet stalk	IP 3471, IP 14439, IP 13817, IP 12128	Sehgal et al. (2015)
High Fe (mg kg^{-1})	IP 3329, IP 7536, IP 8972, IP 9301, IP 10394, IP 12682, IP 15817, IP 17620, IP 17690, IP 17707, IC 332727, IC 139900	Govindaraj et al. (2020) and Gunguniya et al. (2023)
High Zn (mg kg^{-1})	IP 3329, IP 3749, IP 4454, IP 10,471, IP 11,584, IP 11,784, IP 12,181, IP 13,900, IP 15,614, IP 17,217, IC 139903, IC 332715, IC 332703	Govindaraj et al. (2020) and Gunguniya et al. (2023)
High Mn (mg kg^{-1})	IP 7536, IP 8972, IP 9572, IP 10394, 13,384, 15,402, 15,614, 17,217, 17,620, 17,707, IC 332703	Govindaraj et al. (2020) and Gunguniya et al. (2023)
High Cu (mg kg^{-1})	IP 3329, IP 3626, IP 7208, IP 9351, IP 10394, IP 12507, IP 12682, IP 13384, IP 13900, IP 15614, IC 332715, IC 332716	Govindaraj et al. (2020) and Gunguniya et al. (2023)
High Mo (mg kg^{-1})	IP 3329, IP 13900, IP 15614, IP 3749, IP 10471, IP 11316, IP 11353, IP 11584, IP 11784, IP 17217	Govindaraj et al. (2020)
Low Ni (mg kg^{-1})	IP 11316, IP 12682, IP 13384, IP 7536, IP 11320, IP 9407, IP 11784, IP 14148, IP 9416, IP 10471	Govindaraj et al. (2020)
High Ca (mg kg^{-1})	IP 15402, IP 9416, IP 17707, IP 9351, IP 15614, IP 17620, IP 11584, IP 14148, IP 7536, IP 5316	Govindaraj et al. (2020)
High Mg (mg kg^{-1})	IP 15614, IP 9572, IP 7208, IP 17217, IP 7536, IP 3626, IP 3329, IP 4454, IP 12507, IP 10471	Govindaraj et al. (2020)
Low Na (mg kg^{-1})	IP 5316, IP 13384, IP 3626, IP 12181, IP 17690, IP 7838, IP 4454, IP 11353, IP 10471, IP 12682	Govindaraj et al. (2020)
High K (mg kg^{-1})	IP 9572, IP 10471, IP 12682, IP 11584, IP 3329, IP 12507, IP 14148, IP 15614, IP 11316, IP 11353,	Govindaraj et al. (2020)
High P (mg kg^{-1})	IP 9572, IP 15614, IP 3329, IP 7536, IP 12507, IP 14148, IP 7208, IP 10471, IP 12682, IP 17217, IC 370523, IC 332727, IC 139899	Govindaraj et al. (2020) and Gunguniya et al. (2023)
High S (mg kg^{-1})	IP 9572, IP 17217, IP 12507, IP 7536, IP 7838, IP 3329, IP 4454, IP 15614, IP 3626, IP 11784	Govindaraj et al. (2020)

1.4.3 Trait-Specific Germplasm

Evaluation of pearl millet genetic resources for important target traits is necessary to improve use of the genetic resources by plant breeders. Two hundred and twenty-one Indian pearl millet collections were examined and classified using 27 agro-morphological descriptors from the Indian National Genebank. Unique germplasm for diverse traits, including disease tolerance, early maturation, spike length, yield, dual-purpose type, popping characteristics, and antioxidant capabilities, was identified as a result of this study. An investigation of frequency distribution revealed that gray seeds, compact and cylindrical spikes, and early flowering (less than 40 days) were the most common characteristics. Multiple disease resistance was found in four pearl millet germplasm accessions: IC309064, IC393365, IC306465, and IC283866 (Kumari et al. 2016). Though a vast collection of germplasm is available across the world, only a narrow range of its germplasm could be exploited for improving its valuable agronomic traits, stress tolerance, and productivity (Passot et al. 2016), one of the reasons being the cross-pollinating nature of the crop. Some of the trait-specific germplasm is given in Table 1.3.

1.4.4 Utilization of Wild Species

Introgression of the genome of wild germplasm is used for creating diversity in cultivated germplasm. After several generations of backcrossing and selection, larger introgressions carrying favorable traits, as well as cryptic introgressions, are present throughout the genome. The success depends on both genotype and ploidy level. With clearly defined objectives and availability of good screening or selection techniques, a team approach is used for testing and evaluating large populations. With highly heritable characteristics and multiple cycles per year, there is a greater chance of success when using wild species to improve cultivated species. The transfer difficulty increases when employing wild species in the tertiary gene pool instead of wild subspecies in the primary gene pool.

The ICRISAT Genebank maintains 739 accessions of 23 wild species, while the USDA-ARS Southern Regional Plant Introduction Station, Griffin, Georgia, USA, holds over 90 accessions of 21 wild species (USDA-ARS, 2002), of which the majority of lines belong to *P. glaucum* ssp. *monodii* (syn. *Pennisetum violaceum*), followed by *Pennisetum pedicellatum* and *Pennisetum polystachion* (Rao et al. 2003). Given the abundance of genetic diversity found in pearl millet landraces, there has been a general lack of interest in employing wild species.

Pennisetum glaucum ssp. *monodii* contributed toward new source of cytoplasmic-nuclear male sterility (CMS). Also, populations resulting from crossing between cultivated pearl millet as male and the wild relative of GP1, *P. glaucum* ssp. *monodii* (=*P. violaceum*), as female have resulted in new CMS sources. The hybrid cultivar development in pearl millet [*P. glaucum* (L). R. Br.] began with the discovery of A1 CMS at Tifton, Georgia, USA (Burton 1958). This led to the release of the first grain hybrid in India in 1965 (Athwal 1965). The source of Av cytoplasm was determined to be the Senegalese germplasm accession (Marchais and Pernes 1985). At Tifton, Georgia, USA, Hanna (1989) discovered A4 CMS system in a wild, grassy *P. glaucum* (L.) R. Br. ssp. *monodii* (Maire) Brunken. The cytoplasm of *monodii* affects dry matter yields, offering a special chance to boost commercial forage hybrid productivity (Hanna 1997). In sub-Saharan West Africa, a few accessions of *P. glaucum* ssp. *monodii* (PS 202, 637, 639, and 727) are useful reservoirs of resistance against *Striga hermonthica*, a cereal parasitic plant. According to Wilson et al. (2004), PS 202 is also a source of resistance to downy mildew (DM), one of the most destructive fungal infections that affect pearl millet. The cytoplasm of *Pennisetum schweinfurthii* has also been transferred to pearl millet through pollination with pearl millet hybrids (Hanna and Dujardin 1985).

Napier grass, or *Pennisetum purpureum*, is a rhizomatous perennial that grows quickly, resists pests, and has high fodder yield. It crosses with cultivated pearl millet to produce sterile triploid hybrids that are vegetatively propagated. *P. purpureum* has also provided stiff stalk, earliness, longer panicle, greater leaf size, and restorer genes for the A1 CMS system (Dujardin and Hanna 1989). The species *Pennisetum schweinfurthii* is utilized for large seeds, *Pennisetum mezianum* for drought tolerance, *Pennisetum orientale* for forage and drought tolerance, and *Pennisetum pedicellatum* and *Pennisetum polystachion* for DM resistance (Rai et al. 1997). Utilizing wild *P. glaucum* ssp. *violaceum*, novel sources of blast resistance and flowering-stage heat tolerance have been introgressed in cultivated pearl millet backgrounds recently (Sharma et al. 2020b).

Apomictic but highly sterile hybrids were reported between pearl millet and *Pennisetum setaceum* (triploid) and *P. orientale* (tetraploid). The high sterility and low expression of apomixis in these species prevented their use as a source of genes for apomixis, despite the fact that interspecific hybrids are robust. The hexaploid obligate apomictic species *P. squamulatum* was crossed successfully to a tetraploid pearl millet, and several partially male-sterile, obligate apomictic interspecific derivatives were produced (Hanna 1992). *Pennisetum squamulatum*, a hexaploid species ($2n = 6x = 54$), was the third species to be tried in the germplasm transfer program. The hybrid between pearl millet × *P. squamulatum* was male and female sterile. When tetraploid pearl millet was used in a crossing and backcrossing program, hundreds of partially male-fertile, obligate apomictic, interspecific hybrids and back-crossed derivatives could be produced (Dujardin and Hanna 1983, 1985a). Research on *P. squamulatum* revealed that this species was an obligate apomict (Dujardin and Hanna 1984). As "bridges" to transfer genetic material, double-cross hybrids between *P. squamulatum*, napier grass, and pearl millet have also been created (Dujardin and Hanna 1984, 1985b). Altering complete genome sets has been the main goal of creating an apomictic pearl millet. However, it is important to consider transferring genes, chromosomal fragments, or even a single chromosome to pearl millet using gamma radiation, cell culture, and cytological and genetic methods.

1.4.5 Utilization of Landraces

Indian landraces are excellent sources of shorter plants, high tillering, and early maturity. African sources, however, are excellent sources of huge head volume and seed size, higher levels of disease resistance, and better seed quality. This is especially true for sources from the West African region. The genetic base of breeding programs has become more diverse due to the growing utilization of African germplasm at ICRISAT and in nearly all Indian national programs (Harinarayana and Rai 1989; Harinarayana et al. 1988). According to Upadhyaya et al. (2007a), donor parents such as 863B (IP 22303), P 1449-2 (IP 21168), ICMB 90111 (IP 22319), ICMP 451 (IP 22442), and IP 18293 have been discovered as sources of resistance to different pathotypes of DM disease in India.

Prior to the establishment of All India Coordinated Millets Improvement Project (AICMIP) in 1965 in India, a number of varieties (Co 2, Co 3, AKP 1, AKP 2, RSJ, RSK, and N-28-15-1 from Indian landraces and Co 1 and S 530 from African landraces) were developed through mass selection. Using mass selection in the germplasm assembled at the Indian Agricultural Research Institute in New Delhi, improved Ghana and Pusa Moti varieties of pearl millet were developed in the 1960s. The arrival of the male-sterile line Tift 23A from Tifton, Georgia, USA, in 1962 marked the beginning of the pearl millet hybrid era in India. Based on this line, five hybrids (HB 1, HB 2, HB 3, HB 4, and HB 5) were released between 1965 and 1969 and three hybrids (BJ 104, BK 560, and CJ 104) were released in 1977. Ten more hybrids were released between 1981 and 1984; however only three of them were noteworthy (MBH 118, GHB 27, and GHB 32). Four hybrids (ICMH 451, ICMH 501, MH 182, HHB 30) and three more (ICMH 423, MH 169, HHB 50) were released in 1986 and 1987, respectively, after 5141 A and J 104 broke down due to downy mildew. Only ICMH 451 has been widely cultivated out of all of these.

The *Iniadi* landrace is widely distributed in northern Togo and adjoining western Benin, southern Burkina Faso, and eastern Ghana regions (Andrews and Anand Kumar 1996). Besides early maturity (75–85 days), this landrace is among the germplasm least sensitive to longer days for flowering; has fewer tillers (one to two per plant); has compact, conical panicles with good exertion; and is typically lustrous, mostly dark gray with globular grains of large size (1000-seed weight, 13–20 g). The widespread use of this landrace in pedigree and population breeding in India, Southern and Eastern Africa, Western and Central Africa, and the United States has led to the development and release of several open-pollinated varieties (OPVs) and parental lines of hybrids (Andrews and Anand Kumar 1996). Development of ICTP 8203 by random mating five S2 progenies selected at ICRISAT from an *Iniadi* landrace from northern Togo is a milestone in demonstrating the utility of this landrace. This variety was released in India in 1988 (Rai et al. 1990). It was cultivated on 0.6–0.7 million ha at the peak of its adoption in Maharashtra State, India (Bantilan et al. 1998). It is still cultivated on 300,000 ha and retains its initial high level of DM resistance (<2% disease incidence in farmers' fields over two decades).

Among the progenies that contributed to the development of ICTP 8203, one S1 progeny (designated as Togo 13-4) during the inbreeding and selection process was found to be an excellent maintainer of an A_1-system male-sterile line, 81A. It also had a high level of DM resistance in the disease nursery (<5% disease incidence) at ICRISAT, Patancheru. An A/B pair of the male-sterile F_1 (81A × Togo 13-4) and the maintainer (B) line (Togo 13-4) was established in 1981 summer (dry) season. Selfing and selection in the maintainer progeny with concurrent backcrossing onto individual plants of the sterile backcross progeny was continued for six generations. It led to the identification of a family comprising six male-sterile BC_6 progenies and six maintainer S_8 lines, of which one pair was designated as 863A and 863B. This A/B pair was made available to research programs in the National Agricultural Research System and the private sector in India in 1987, who made extensive use of it in breeding hybrids. Line 863A or its sub-selections are the seed parents of three commercial hybrids developed and marketed by three private seed companies in India.

Extensive breeding efforts resulted in the development of elite material that was used in the development of diverse cultivars. However, periodically germplasm was evaluated for specifically targeted traits and identified lines were introgressed in the breeding programs. The 145 improved inbred lines involving 122 landraces from Western and Central Africa were diverse for flowering time, photoperiod sensitivity, and length of the panicle (Stich et al. 2010). The landraces from Cameroon, Togo, and Ghana were found to be good sources of earliness and bigger seeds. The landraces from Yemen known for early maturity, short stature, large seeds, and cold tolerance were exploited in breeding programs. From the study on 169 arid Indian landraces conducted by Yadav and Bidinger (2008), diversity was noticed for flowering time with most of them being early to flower. These landraces were suitable for dual purposes with more grain and fodder yield and used by several researchers for development of restorers for creating dual-purpose hybrids adapted for arid climatic conditions.

Climate change has resulted in extremities in the temperature and rainfall pattern. As the crop is grown in marginal regions, drought is major abiotic production constraint in South Asia and sub-Saharan Africa. The drought-prone production areas have been mapped to define the target population of environments or mega-environments (Gupta et al. 2013), which are highly variable in terms of their timing, intensity, and duration of drought (van Oosterom et al. 1995). With selection under drought stress, germplasm with drought-adaptive traits has been realized (Yadav et al. 2009, 2012a, b). Hybridization of adapted landraces with elite genetic material creates new gene combinations which are likely to possess adaptation to stress environments and high productivity (Patil et al. 2020; Presterl & Weltzien, 2003; Yadav & Rai, 2011).

1.5 Pre-breeding

Pre-breeding is an important stage in the plant breeding process that introduces genetic diversity and unique traits into elite germplasm. Pre-breeding aims to preserve the advantageous genes while effectively introducing new genetic variation into the elite cultivar without generating linkage drag. Therefore, it is a drawn-out procedure, especially if it involves wide crossing with wild relatives, progeny selection, discovery and translational research, and strategic crossing (Sukumaran et al. 2022).

Because pearl millet is grown for different purposes in different agroecologies, the essential and value-added qualities have been determined for each market segment in Asia and Africa in order to define product profiles. Important characteristics in pearl millet breeding programs include high yield in the desired maturity background, enhanced terminal drought and heat tolerance, resistance to downy mildew and blast, Fe and Zn fortification, low rancidity, lodging tolerance, and forage yield and quality. Pre-breeding can help to improve these traits by introducing genes and alleles for specific traits from wild species that are absent in the cultivated gene pool, as well as by providing new and diverse sources of genetic variability to immediately address the current production constraints. Pre-breeding programs, which are conducted in collaboration with pathologists, physiologists, and breeders, are essential to enhancing pearl millet's genetic gain. Several germplasm lines were identified as donors to genes governing biotic and abiotic stress tolerance in pearl millet (Table 1.3).

Table 1.3 Germplasm lines identified for biotic or abiotic stress tolerance in pearl millet

Trait	Donor germplasm	Reference
Downy mildew disease	863B (IP 22303); P 1449-2 (IP21168); ICMB 90111 (IP 22319); ICMP 451 (IP22442) and IP 18293; ICML 12; ICML 13; ICML 14; ICML 15 and ICML 16; IPW 151, 152, 153, 155; IP 18295; P 1449-2; YL-18; IP 18298; IP 8289; IP 22396	Upadhyaya et al. (2007a), Singh et al. (1990), Singh and Navi (2000) and Thakur et al. (2009)
	IP# 9, 29, 55, 87, 104, 172, 253, 262, 283, 336, 346, 352, 364, 365, 396, 486, 498, 505, 517, 545, 556, 558, 575, 682, 718; IP# 14537 resistant to 7 pathotypes; IP# 9645, 11943, 14542, 14599, 21438, 11930, 12374, 14522, 20715, 21187, 21201, 21244 resistant to 5–6 pathotypes	Sharma et al. (2007, 2015)
Blast	IP# 21525, 21531, 21536, 21540, 21594, 21610, 21640, 21706, 21711, 21716, 21719, 21720, 21721, 21724, 21987, 21988, and 22,160 (all *P. violaceum* wild species accessions); IP# 7846, 11036, and 21187 (cultivated pearl millet), ICMB 01333, 01777, 02111, 02444, 02777, 03999, 93222, 97222; ICMR 06222 and 07555; IP5964 and IP 8913; IP 11423; IP 11353; IP 3122; IP 7910 and IP 7941	Thakur et al. (2009), Sharma et al. (2013, 2020c) and Sankar et al. (2021)

(continued)

Table 1.3 (continued)

Trait	Donor germplasm	Reference
Rust resistance	IP# 21629, 21645, 21658, 21660, 21662, 21711, 21974, 21975, and 22038 (all *P. violaceum* wild species accessions); ICML 11; ICML 17; ICML 18; ICML 19; ICML 20; ICML 21; 700481-21-8; 700481-22-8; 70041-23-2; P 1564; P 1577; P 1581; P 1591; P 24; P 140; D 212 P1; P 2890; IP 548; P 615; 7042-1-4-4; IP 8695-4; IP 537-8; IP 2084-1; IP #16438 and 16762; P 310-17 and P 1449-3; IP18292; IP18293; 700651; ICML #12, 13, 14, 15, 16, and 22; ICMP #312, 423, and 85410; 7042S; 841A; IP #9, 55, 104, 253, 262, 336, 346, 498, 545, and 558	Singh et al. (1990), Singh et al. (1987b), Sharma et al. (2020c), Singh et al. (1997), Khairwal and Yadav (2005), Thakur et al. (2006) and Sharma et al. (2007)
Smut	ICML 5, ICML 6, ICML 7, ICML 8, ICML 9, ICML 10	Thakur et al. (1992b)
Striga resistance	PS 64, 132, 190, 202, 208, 212, 287, 427, 428, 459, 549, 555, 622, 637, 639, 727, 755	Wilson et al. (2000, 2004)
Ergot	ICML 1, ICML 2, ICML 3, and ICML 4	Thakur and King (1988)
Drought tolerance	IP# 8955, 9406, 3243, 3228, 3424, 3296, 3362, 3180, 3272, 3303, 3252, 3258, 11141, 3318, 3123, 3363, and 3244; IP# 133, 177, 164, 142, 120, 160, 136, 166, 192, 195, 106, 126, 121, 110, and 117; CZP 9802; 863B; PRLT 2/89-33ICMP 83720; ICMV 9413; and ICMV 94472	Yadav et al. (2003), Choudhary et al. (2021), Sehgal et al. (2015), Yadav (2004), and Dwivedi et al. (2010)
Salinity tolerance	IP# 22269, 6098, 6105, 3616, 6112, 6104, 3757, 6101, 3732, 6102; ICMB 02111, ICMB 94555, ICMB 95333, ICMB 00888, ICMB 01222; ICMP 451; IP 3732, IP 3757, IP8210; PRLT 2/89-33, 10876 and 10878 (Sudan), 18406 and 18570 (Namibia); ICMV 93753 and ICMV 94474 (India); 863-B; CZI 98-11, CZI 9621; HTP 94/54	Kulkarni et al. (2006), Krishnamurthy et al. (2007), Krishnamurthy et al. (2003), Sehgal et al. (2015), Ali et al. (2004), and Dwivedi et al. (2010)
Heat stress tolerance	IP# 19799, 19877, 19743, 21517, 3175, H77/833-2, H77/29-2, CVJ 2-5-3-1-3, 77/371XBSECT CP1, 96 AC-93, 1305, 77/371, Togo II, 99HS-18, G73107, 77/371	Yadav et al. (2012b), Sehgal et al. (2015), Upadhyaya et al. (2016b) and Yadav et al. (2014)

1.5.1 Pre-breeding for Tolerance to Abiotic Stresses

A strong emphasis on yield and yield components has led to a significant improvement in overall production and productivity; however, abiotic factors negatively affect its growth and potential yield (Shivhare and Lata 2017). A steady stream of novel genes and alleles giving resistance or tolerance to major abiotic stressors is required to gradually and sustainably improve breeding programs.

Pearl millet has attracted the greatest interest among the many millets when it comes to sources of resistance to drought, salt, and high temperatures (Dwivedi et al. 2012). ICRISAT has built phenotypic screens in an effort to identify germplasm that is resistant to various abiotic stresses (ICRISAT, 2008/2009; Krishnamurthy et al. 2007). The identification and use of pearl millet germplasm for resistance to abiotic stresses are covered in the following sections.

1.5.1.1 Drought

A great deal of research has been conducted to understand how pearl millet responds and adjusts to water shortage at different phases of development. Water stress during the germination stage or seedling emergence stage causes seedling death, which results in poor crop establishment, according to Farooq et al. (2009) and Lata et al. (2015). Severe moisture stress during the seedling stage is the primary cause of pearl millet's low yield in dry and semiarid conditions. It has been found that the impact of drought stress on pearl millet grain yield is negligible once the seedlings are established (Carberry et al. 1985; Soman and Peacock 1985; Soman et al. 1987).

It has been shown that the effects of drought stress on seedling germination rely on water availability and that the effects of low water stress are directly related to leaf formation and secondary root growth (Gregory, 1983). The effects of moisture stress were also investigated using various concentrations of polyethylene glycol PEG 6000. Several seedling measures, such as germination rate, root/shoot ratio, and root/shoot length, were found to be significantly impacted by these factors (Govindaraj et al. 2010). Drought stress at the vegetative phase shows little to almost nil reduction in crop growth and yield because of its asynchronous tillering and rapid growth rate, which allow it to recover fast (Bidinger et al. 1987). During the vegetative phase, additional drought stress delays the main shoot's flowering time. It can delay to reach flowering stage until the stress has passed due to its phenological plasticity (Henson and Mahalakshmi 1985). However, post-flowering or terminal drought stress has the greatest detrimental effects on pearl millet grain and fodder yield, as well as yield stability (Kholová and Vadez 2013; Winkel et al. 1997).

Pearl millet's ability to thrive in drought conditions is mostly due to its adaptability rather than its genetic yield potential (Bidinger et al. 1994). It is more challenging to improve for adaptation than for yield potential since adaptation is still relatively little understood from a physiological and genetic standpoint. Due to natural and artificial selection over thousands of years, pearl millet landraces that emerged in arid regions are more tolerant to drought stress (Yadav 2010, 2014; Yadav et al. 2000). Attempts were made to use these landraces in traditional pearl millet breeding methods. They have also been identified as an important source of germplasm for the development of inbred pollinator lines for hybrid breeding and drought-tolerant breeding lines (Yadav et al. 2009, 2012a).

Pearl millet's resistance to abiotic stress may benefit from a multitude of landraces that have been found in different regions of Ghana, northwest India, and West Africa. These landraces are planted preferentially in the desert regions of northwestern India (Gujarat, Rajasthan, and Haryana), where drought is a common occurrence, because they yield more grain and fodder than conventionally bred lines. For example, when compared to the controls Pusa 266 and ICTP 8203, CZP9802, the first open-pollinated pearl millet variety originating from Rajasthani landraces, has a high yield (14–33% higher grain yield and 18–36% higher fodder yield) (Dwivedi et al. 2012). Additionally, it has great drought adaptation. CZP9802 was found to be a good cultivar for arid zones of India and has the capacity to withstand terminal drought stress due to its remarkable traits of flowering in just 48 days and reaching maturity in roughly 75 days (Yadav et al. 2004). A relatively photoperiod-insensitive, early-maturing West African landrace known as *Iniadi*, with compact, conical panicles and bold grains, greatly aided in the development of several pearl millet cultivars around the world, including ICMV88904 (released as ICMV221, resistant to DM and terminal drought).

There is a vast amount of genetic variation available for drought-responsive and constitutive traits in the pearl millet germplasm worldwide. About 1532 accessions were characterized by Murty et al. (1967), who also noted differences in the days to flowering between exotic collections (53–85 days) and Indian collections (52–77 days). In a different study, days to blooming among 16,968 pearl millet accessions worldwide demonstrated a far greater diversity (33–140 days) (Harinarayana et al. 1988). The germplasm line IP 4021, also known as Bhilodi, collected from Gujarat, was the earliest to flower, with 33 days to 50% flowering in the rainy season and 34 days in the summer season. Moreover, out of 509 accessions examined, 2 promising drought-tolerant lines were found (Harinarayana et al. 1988). India's Rajasthan region produced landraces with a great deal of variation for different traits. For earliness, the landraces SR 15, SR 17, and SR 54 performed well. Bhilodi from Gujarat, Pittaganti from the Eastern Ghats of India, and Chadi from Rajasthan were three more landraces that demonstrated earliness (Kumar and Appa Rao 1987). Additionally, it was determined that the germplasm accessions IP 4066, IP 9496, and IP 9426 were promising lines for earliness. From the study on 105 landraces evaluated across varied environmental conditions, wide diversity for drought response was observed, and 15 landraces (IP 3243, IP 3228, IP 3424, IP 3296, IP 3362, IP 3180, IP 3272, IP 3303, IP 3252, IP 3258, IP 11141, IP 3318, IP 3123, IP 3363, and IP 3244) having a high degree of drought tolerance were identified for use in developing drought-tolerant cultivars (Yadav et al. 2003). Increased resistance to drought and high productivity were seen in crosses between elite genetic materials and adapted landraces (Patil et al. 2020; Presterl and Weltzien 2003; Yadav 2008; Yadav and Rai 2011). Combinations of desirable agronomic traits, drought adaptation, and yield potential determine the selection of landraces and elite breeding lines chosen for hybridization (Patil et al. 2020).

In order to identify promising lines for hybridization, study population structure, and find QTLs associated with environmental stress tolerance, pearl millet germplasm's degree and quantum of genomic diversity can be ascertained using DNA markers (Vadez et al. 2012). An elite inbred pollinator line (H 77/833-2) was enhanced in its drought tolerance using a donor PRLT 2/89-33 line, and an elite inbred seed parent maintainer genotype (ICMB841) was obtained utilizing the 863B line as a donor. Over the past few years, India has seen a notable increase in grain yields, productivity, and quality in pearl millet, largely attributed to marker-assisted backcrossing techniques. Development of a marker flanking QTLs governing yield and related traits under limited soil water conditions has been targeted by several researchers in pearl millet (Yadav 2010, 2014).

1.5.1.2 Salinity

Salinity affects about 20% of the world's farmed land and half of its irrigated areas (Shrivasata and Kumar 2015). In dry and semiarid regions, the effects of salt stress are very severe (Jha et al. 2021). Because pearl millet is thought to be quite tolerant to salinity, it can be used as a substitute in places afflicted by salinity. According to research reports, pearl millet exhibits a high degree of genotypic diversity in its ability to withstand salt stress compared to maize, wheat, and rice. Due to the high tolerance levels found in *P. glaucum* (Dua 1989) and other *Pennisetum* species (Muscolo et al. 2003), it is now possible to better understand the traits linked to tolerance and incorporate these tolerant genotypes into management plans that will maximize the productivity of soils affected by saline and salt water.

Plants are able to withstand the harmful effects of excess ions through a variety of physiological features (such as compartmentation, the synthesis and accumulation of suitable solutes in the cytoplasm, vigor to enable dilution of salt concentration by growth, and effective exclusion of Na+). It has been demonstrated that pearl millet displays a notable degree of genotypic heterogeneity for the overall plant response to salinity (Dua 1989). Reduced shoot N content, elevated K+ and Na + concentrations, and overall plant tolerance to salinity were generally correlated. The tolerance of pearl millet to salt has a lot of scope for improvement, and the concentration of Na + in the shoots might be considered a possible nondestructive selection criterion for screening during the vegetative stage (Krishnamurthy et al. 2007). Compared to the adult stage, the early vegetative growth stage has been shown to be more susceptible to salt stress (Cardamone et al. 2018). The two main processes influencing salt tolerance in pearl millet genotypes can be thought of as osmotic adjustment and effective free radical scavenging (Jha et al. 2021).

Large genotypic variation has been observed in pearl millet toward salinity stress tolerance. The landraces and wild relatives typically display genetic variation and are known to carry unique genes for features that are relevant to agronomy and environmental adaptation. A recent study found that at salinity levels of 8–12 dS/m, pearl millet grain production decreased by 12.90–22.43% (Yadav et al. 2020). In a study on 15 pearl millet accessions, there was an average decrease in shoot biomass productivity and grain yield by roughly three to four times (Krishnamurthy et al. 2011). In a different study, salinity caused significant reductions in biomass and grain yield for 11 lines of pearl millet (19.1% and 41.3%, respectively) (Toderich et al. 2018). Cultivars resistant to salt and sodicity in the soil can increase millet yield and cover more area with this crop. The germplasm line IP 22269 was found to be resistant to salt stress in field tests conducted in the northern region of Tunisia (Hajlaoui et al. 2021).

1.5.1.3 Heat Tolerance

For each degree Celsius that the average growing season temperature rises, crop yields are predicted to decrease by 17% (Lobell and Asner 2003). Global climate change is expected to cause temperatures to rise by a maximum of 4 °C by the end of the twenty-first century, putting plants at risk of heat stress (Driedonks et al. 2016). While pearl millet is a heat-tolerant crop, the germination rate, establishment, initial growth, and photosynthesis rate increases to a temperature of about 35 °C; above that, normal growth is affected (Arya et al. 2014). Yadav et al. (2011) conducted a thorough investigation on pearl millet's resistance to heat stress during the seedling stage. The morphological characteristics that determine a seedling's ability to withstand heat include the seedling height, leaves per seedling, emergence rate, seedling to seedling heat tolerance index, and seedling fresh and dry weights. It was discovered that hybrids outperformed the parents in terms of yield and heat tolerance (Joshi et al. 1997).

Yadav et al. (2004) reported significant genetic diversity in 105 landraces of pearl millet (*P. glaucum*) from northwest India. Farmers cultivate landraces over time, and as a result, they have gained the required adaptation genes and may provide a potential source of heat tolerance. In landrace-based topcross hybrids, heterosis manifested differently for different variables. Notably, considerable heterosis for biomass, grain yield, and fodder yield was noted in certain combinations of male-sterile seed parent x landrace-based pollinator (Yadav and Bidinger 2008). A significant genetic variation has been observed for tolerance to heat at the reproductive stage. Heat-tolerant lines have been identified, including ICMB 92777, ICMB 05666, ICMB 00333, ICMB 01888, ICMB 02333, and ICMB 03555 among the maintainer lines and ICMV 82132, MC 94, ICTP 8202, and MC-bulk among the populations (Yadav et al. 2012a). After 3–4 years of field screening (2009–2012), a germplasm accession IP 19877 with 61–69% seed set, compared to 71% seed set in a heat-tolerant commercial hybrid 9444 (used as a control), and five hybrid seed parents (ICMB 92777, ICMB 05666, ICMB 00333, ICMB 02333, and ICMB 03555) were identified (Gupta et al. 2015).

1.5.2 Pre-breeding for Biotic Stresses

The susceptibility of high-yielding lines to emerging pests and illnesses poses a severe threat to the sustainability of pearl millet cultivation.

1.5.2.1 Downy Mildew

The most common and significant disease causing huge economic losses in India, Burkina Faso, Chad, Mali, Nigeria, Niger, Mali, Senegal, Sudan, Tanzania, and Zambia is downy mildew (DM), which is caused by *Sclerospora graminicola* (Sacc.) J. Schröet. (Singh et al. 1993). Many breeding lines and germplasm accessions have been screened in greenhouses and fields, identifying a number of resistance sources that have been widely utilized to create DM-resistant hybrids (Singh et al. 1997). The pearl millet lines that exhibited the highest relative variation across locations and years were IP 18292 and IP 18293, which were resistant (<10% incidence) at 11 of the 17 locations and moderately resistant (>10%) at Bagauda, Cinzana, Durgapura, Kamboinse, Mandor, and Sadore. This suggests that the resistance of these lines is specific to certain DM pathotypes. On the other hand, lines

700651 and P 310-17 showed significantly less fluctuation over several locations and years, although having slightly lower average resistance values.

ICRISAT and the AICPMIP centers have screened a large number of germplasm accessions and breeding lines, leading to the identification of several resistant germplasm/lines (Singh 1995; Singh et al. 1997). Multilocational testing has validated the resistance stability of some of these lines, including P7 (ICML 12), SDN 503 (ICML 13), 700,251 (ICML 14), 700,516 (ICML 15), 700,651 (ICML 16), and 7042R (ICML 22) (Singh et al. 1994). High levels of resistance have been found in a number of other lines and germplasm accessions, including P 310-17, P1449-3, IP 18292, IP 18293, IP 18294, IP 18295, and IP 18298 (Singh et al. 1997; Thakur et al. 2004b). At the ICRISAT and AICPMIP centers, resistance breeding has made strategic use of a few of these sources. A pearl millet mini-core collection consisting of 238 accessions was screened against eight pathotypes (Sg 384, Sg 409, Sg 445, Sg 457, Sg 510, Sg 519, Sg 526, and Sg 542) of *S. graminicola* collected from various geographical locations in India in order to identify new and diverse sources of DM resistance. Significant differences for DM reaction were observed among pathotypes, mini-core accessions, and their interactions. Out of the 238 accessions, 68 had a DM incidence of less than 10% and were resistant to pathotype Sg 510, while 40 more were resistant to Sg 457. About 15, 27, 29, 30, and 34 mini-core accessions were resistant to pathotypes Sg 519, Sg 526, Sg 384, Sg 445, and Sg 542, respectively. In 62 accessions, resistance to 2 or more pathotypes was noted. Desirable agronomic features were also displayed by several of these accessions. In breeding programs, the diverse pathotype-resistant germplasm accessions with desirable agronomic traits may be helpful in developing pearl millet hybrids resistant to extremely virulent pathotypes of *S. graminicola*. The present work has found six accessions (IP 9645, IP 11943, IP 14542, IP 14599, IP 21438, and IP 14537) resistant to numerous pathotypes. These accessions could serve as donor parents for the purpose of pyramiding resistance alleles in the parental lines of pearl millet hybrids that have achieved commercial success. It has been discovered that a portion of the mini-core accessions resistant to DM are also resistant to pearl millet blast caused by *Pyricularia grisea* (teleomorph: *Magnaporthe grisea*), another significant disease. According to Sharma et al. (2013), mini-core accessions IP 11247 and IP 12650, which are resistant to two pathotypes of *S. graminicola*, are also resistant to pathotype Pg 118 of *M. grisea*. It has been observed that the DM-resistant accessions IP 3646, IP 9692, and IP 17396 are resistant to the *M. grisea* pathotype Pg 45. Some of the chosen DM-resistant accessions have also shown resistance to certain *M. grisea* pathotypes. According to Sharma et al. (2013), mini-core accession IP 11010 was shown to be resistant to two pathotypes of *M. grisea* (Pg 118 and Pg 119), while IP 21187 was resistant to four pathotypes (Pg 53, Pg 56, Pg 118, and Pg 119). Blast, which was formerly thought to be a minor pearl millet disease, is now posing a major danger to India's pearl millet crop. Since blast and DM are the two main diseases that affect pearl millet, mini-core accessions with resistance to numerous blast and DM pathotypes, such as IP 11010 and IP 21187, would be highly helpful in the pearl millet breeding program.

1.5.2.2 Blast

Pearl millet is susceptible to a highly damaging disease known as blast or leaf spot disease, which is caused by *P. grisea* Sacc. (syn. *M. grisea*) (Rai et al. 2012). The ideal growing and spreading environment for the disease is hot, humid weather, especially on rainy days. In order to create blast-resistant hybrid parent lines, stable resistance lines were identified via monitoring of virulence of pathogen populations and screening of genetic resources (Sharma et al. 2013; Yella Goud et al. 2016).

Based on reaction type, five *M. grisea* pathotypes (Pg118, Pg119, Pg56, Pg53, and Pg45) have been discovered. Accessions were assessed in a greenhouse condition against five *M. grisea* pathotypes to identify sources of resistance in the pearl millet mini-core collection to various pathotypes. Out of 238 accessions, 32 were found to be resistant to 1 or more pathotypes. Three accessions (IP 7846, IP 11036, and IP 21187) showed resistance to four of the five pathotypes. Of these accessions, 21 were derived from India. As a result, germplasm accessions from India seem to be prospective sources of blast resistance and might be tested against various M. grisea pathotypes to find further sources of blast resistance (Sharma et al. 2013). According to Hanna et al. (1987) and Wilson and Hanna (1992), efforts have been made in the United States to incorporate blast resistance in pearl millet into better cultivars and elite breeding lines. Wild species of *Pennisetum* could be good sources of resistance to disease genes. Additionally, pearl millet landraces offer a wealth of genetic diversity for resistance. The *P. glaucum* ssp. *monodii* accession from Senegal had Rr1 rust gene and was resistant to leaf blast as well (Hanna et al. 1987). Using field and greenhouse screening methods, Thakur et al. (2009) were able to locate the sources of hybrid parental lines that were resistant to blast disease. After being tested for blast resistance in a disease nursery, 211 elite hybrid parental lines—of which 126 were designated B-lines, 65 potential R-lines, and 20 designated R-lines—were found to be blast resistant. Of these, 45 lines (score $\leq$ 3.0 on a scale of 1–9) were then subjected to a greenhouse screening procedure. Under greenhouse screening, 25 of the 45 lines—8 designated B-lines, 14 potential R-lines, and 3 designated R-lines—were shown to be resistant to blast disease. In artificial conditions, 15 genotypes of pearl millet were screened against foliar blast. Two entries (PPMI 1084 and J 108) were resistant (score of 1.0–1.3), while three lines (PPMI 1087, PPMI 1089, and PPMI 660) showed high resistance (score of 0.0–0.4 on a scale of 0–9) (Prakash et al. 2016).

In both artificial and field screening conditions, a group of 93 genotypes—45 maintainer lines and 48 restorer lines—in Uniform Blast Nursery was screened against 3 isolates in Jaipur. The line ICMB 95444 was shown to be susceptible to blast with a mean score of 7.94 in the artificial screening across three pathotypes and 8.6 in the field screening at Jaipur. In the hotspot, one genotype had a lower blast score (2.35) than artificial screening (4.9). Ten genotypes were found to be resistant to Jaipur isolates in the Uniform Blast Nursery, but the resistant levels were slightly low in hotspots (Rao et al. 2021).

1.5.2.3 Rust

Rust (caused by *Puccinia substriata* var. *indica* Ramachar & Cumm) is often regarded as a disease that has little significance in grain crops. However, it has significance in fodder crops, as it lowers both quantity and quality. According to reports, rust infection of pearl millet fodder can reduce the amount of digestible dry matter by up to 51% (Monson et al. 1986).

The identification of stable resistance sources was made possible by greenhouse screening and field screening conducted during the late rainy season when disease pressure was high (Sharma et al. 2020b; Singh et al. 1990). Certain pearl millet germplasm accessions and breeding lines have been reported to be resistant to rust (Singh et al. 1997; Wilson 1993). But lines that were resistant in India become vulnerable in the United States, suggesting that the two countries have distinct physiological races (Tapsoba and Wilson 1996; Wilson 1991). In the disease nursery, about 214 advanced breeding lines, of which 126 were designated B-lines, 23 were designated R-lines, and 65 were potential R-lines, were assessed for rust resistance in the summer of 2008–2009 under natural epiphytotic circumstances. Three R-lines (ICMR 0699, ICMP451-P8, and ICMP 451-P6) and one B-line (ICMB 96222) were resistant, but the remaining four R-lines were susceptible (Sharma et al. 2009). Recurrent selection was used to improve the rust resistance of two exotic gene pools of pearl millet: Tiff #2, which included 740 accessions of cultivated pearl millet primarily from Africa (Hanna 1990), and Tiff #5, which included 114 accessions of wild *P. glaucum* ssp. monodii (Maire) Brunken and weedy *P. glaucum* ssp. *stenostachyum* (Klotzsch) Stapf and Hubb. from Niger, Senegal, and Mali (Hanna et al. 1993), were improved for rust resistance through recurrent selection (Tapsoba et al. 1997).

1.5.2.4 Ergot and Smut

Other important floral diseases include ergot (caused by *Claviceps fusiformis* Lov.) and smut (caused by *Moesziomyces penicillariae* Bref. Vanky), with losses in grain production commensurate with severity. Both infections enter the host through the stigma during flowering and are soilborne. Compared to open-pollinated cultivars, hybrid cultivars based on CMS lines are typically more prone to this disease. Ergot resistance was found to be related with short protogyny and quick pollination (Thakur and Williams 1980). A total of 2752 germplasm accessions from 19 nations and a few unidentified sources that were acquired from the global collection of ICRISAT's Genetic Resources Program were screened between 1977 and 1984. It was discovered that none of the accession had ergot resistance that was suitable. Nonetheless, 27 accessions with varied plant frequencies, 0%–10% ergot severity, and > 75% selfed seed set originated from India (10), Nigeria (11), Togo (3), and Uganda (3). After screening over 8350 lines from ICRISAT and AICPMIP breeding programs, it was shown that hybrids in particular were sensitive. Breeding material was screened for several years running without producing fruitful results. A total of 283 ergot-resistant lines and populations were identified and designated as ICRISAT Millet Ergot Resistant numbers.

1.5.2.5 Germplasm Utilization for Grain Micronutrients

According to studies, between 60% and 80% of people on Earth are believed to be deficient in iron, over 30% in zinc, and over 15% in selenium (Combs 2001; Kennedy et al. 2003). The issue of micronutrient deficiencies can be solved by biofortification breeding to increase the critical nutrient content of main food crops. The existence of variation in the main gene pool is a prerequisite for genetic improvement. A valuable germplasm resource for genetic improvement of these Fe and Zn in pearl millet has been identified in a set of 297 *Iniadi* germplasm accessions that originated from northern Togo (146) and neighboring eastern Ghana (83), southern Burkina Faso (63), and western Benin (5) regions of Western Africa. Of these, 27 accessions (20 from Togo and 7 from Ghana) had a Fe density of 95–121 mg/kg and a Zn density of 59–87 mg/kg. The accessions varied significantly for all micronutrients in a research comprising a core germplasm collection of diverse origin assessed in field trials at ICRISAT, India. For example, there was a more than twofold variance for Fe (34–90 mg/kg), Zn (30–74 mg/kg), and Ca (85–249 mg/kg). There were high heritability estimates (>0.81) for Fe, Zn, Ca, P, Mo, and Mg. The majority of the traits showed a smaller degree of genotype-by-environment interaction, suggesting substantial genetic control over grain nutrients. For each nutrient, the top 10 accessions were determined, and for multiple nutrients, 15 accessions from 5 different nations were discovered. For Fe and Zn, 39 accessions (including 15 with multiple nutrients) exceeded the Indian cultivars, and 17 of them exceeded the biofortification breeding target for Fe (72 mg/kg). The majority of these nutrients showed positive and significant association with 1000-grain weight, days to 50% blooming, and each other. This suggests that it is possible to improve these nutrients in desirable agronomic background. The primary genetic resources for creating biofortified and agronomically better cultivars would be the selected core collection accessions rich in particular and multiple nutrients (Govindaraj et al. 2020).

Previous studies have found that the protein content of germplasm can reach to an extent of 24.3% (Jambunathan and Subramanian 1988) and elite breeding lines can reach up to 19.8% (Singh et al. 1987a). However, because protein content has a negative link with grain yield, no significant efforts have been undertaken to boost it (Singh and Nainawatee 1999). The first approach in breeding programs is to select micronutrient-dense lines among existing breeding populations and varieties; however, deliberate selection from germplasm collections has a much greater potential for micronutrient enhancement than selection within available breeding lines or varieties (Gerloff and Gabelman 1983; Graham and Welch 1996).

1.6 Genomic Resources

As pearl millet is an allogamous crop, recessive alleles are revealed during inbreeding and are kept in the population in a heterozygous state. Many traits were found during the germplasm evaluation process. These comprise eight early-maturing genotypes (Appa Rao et al. 1988c), sweet stalk sources with high sugar content (Appa Rao et al. 1982), two new dwarfing genes (Appa Rao et al. 1986b), three glossy genes (Appa Rao et al. 1987), two midribless genes (Appa Rao et al. 1988b), and chlorophyll-deficient mutants (Appa Rao et al. 1984). Genes for maintaining sterility and restoring fertility can also be found in germplasm (Appa Rao et al. 1986a, b). It was discovered that high digestibility was linked to the brown midrib trait in brown midrib pearl millet genetic stock (Gupta et al. 1983).

Genus *Pennisetum* is polybasic with $x = 5, 7, 8$, and 9. In the $x = 5$ type, $2n = 10$ was observed in *P. ramosum* only. In the $x = 7$ type, $2n = 14$ was observed in *P. glaucum*, *P. violaceum*, *P. mollissimum*, and *P. schweinfurthii*, while $4x = 28$ was observed in *P. purpureum*. In the $x = 8$ type, $2x = 16$ and $4x = 32$ were observed in *P. mezianum*. In the $x = 9$ type, $2x = 18$ and $3x = 27$ were observed in *P. hohenackeri* and $3x = 27$ in *P. setaceum*. Tetraploid number $4x = 36$ was observed in *P. mecrorum*, *P. divisum*, *P. cenchroides*, *P. clandestinum*, *P. pedicellatum*, and *P. orientale*, while $5x = 45$ was observed in *P. villosum*. The species *P. squamulatum* and *P. polystachyon* showed $6x = 54$ chromosomes, while the highest chromosome number (68) was observed in the species *P. macrostachyum* only. Large grains with a waxy coating are valued in *P. schweinfurthii*; *P. setaceum* and *P. villosum* are valued as ornamentals; *P. hohenackeri* is valued for thatching and rope making; and *P. purpureum* and *P. clandestinum* are valued for pasture (ICRISAT 1989).

1.6.1 Genetic Diversity

Being cultivated in a variety of agroclimatic conditions and soil types worldwide, pearl millet accessions have a wide range of phenotypic and genotypic variations for various agronomic and stress tolerance traits (Shivhare and Latha 2017). Murty et al. (1967) and Upadhyay and Murty (1970) investigated The Rockefeller Foundation collection assembled at New Delhi, India, to analyze genetic diversity in pearl millet. They found that major variability occurs in Africa and that accessions from India showed low but distinct diversity. Variation pattern seems to be associated with agroclimatic conditions where early-maturing forms were found in low rainfall areas, especially in the fringes of deserts. Genetic diversity was observed for seed characters, physical and structural characteristics (McDonough and Rooney 1987), protein content and quality (Godawat and Kaul 1981), drought resistance (Blum and Sullivan 1986), disease resistance (Deshmukh et al. 1978; Basavaraju et al. 1981; Thakur et al. 1988, 1992a, b; Singh et al. 1993), insect resistance (Merkle et al. 1983), enzymes (Subba Rao et al. 1988; Lagudah and Hanna, 1990; Tostain 1992), mitochondrial DNA (Chowdhury and Smith 1988), and nucleotide polymorphism (Gaut and Clegg 1993).

Assessment of diversity in 21,594 pearl millet accessions from 50 countries for 23 morpho-agronomic traits revealed large variability for days to flowering (33–159 days), plant height (30–490 cm), tillers/plant (1–35), panicle length (5–135 cm), and 1000-seed weight (1.5–21.3 g). The phenotypic diversity index ranged from 0.427 (tiller number) to 0.772 (endosperm structure) with a mean of

0.528 (Upadhyaya et al. 2007a). A total of 335 accessions of *Pennisetum glaucum* subsp. *monodii*, originating from 13 different countries, were examined and their geographic gaps and diversity assessed. Upadhyaya et al. (2014) found that the collection exhibited a wide range in terms of total tillers per plant (16–609), productive tillers per plant (12–368), leaf length (21.3–58.8 cm), and leaf width (8.1–33.1 mm), all of which suggest a significant potential for fodder supply.

1.6.2 Application of Genomic Tools for Plant Genetic Resources Utilization

Genomic technologies include genome editing (GE) tools, high-throughput genotyping platforms, reference genome sequences, SNP arrays, and platforms for sequencing and resequencing.

The precision and speed of genomic tools and introgression will increase with the adoption of diagnostic markers, especially for traits with strong genotype-by-environment interactions. The Pearl Millet 1000 Genome Resequencing Project yielded more than 29 million genome-wide single nucleotide polymorphisms (SNPs) in the pearl millet plant (Varshney et al. 2017). As part of this endeavor, 31 wild *Pennisetum* accessions that were gathered throughout the Sahel, from Senegal to Sudan, were also sequenced. The resequencing provided important insights into the nature of genetic diversity in some wild *Pennisetum* species as well as the relative contraction and expansion of genes related to fitness and adaptation during the course of pearl millet's domestication history. Mariac et al. (2006) have demonstrated the effectiveness of microsatellite markers in genetic diversity study. They found that 46 wild accessions had more allelic diversity and a higher number of alleles compared to 421 cultivated lines from Niger. It is possible to select favorable genes and alleles for trait introgression and to functionally characterize the provided genes before the introgression. Marker systems based on next-generation sequencing, such as whole-genome resequencing SNPs and genotyping-by-sequencing SNPs, may be helpful in mapping traits. Additionally, these markers can help alleviate the negative linkage drag commonly observed in the wide cross derivatives by implementing genome-wide background selection. Using the already accessible marker library, pearl millet's genetic diversity—which was diminished throughout the domestication process—can be increased. Molecular markers can be used by genebanks to characterize and identify intra- and inter-accession variability among many regeneration cycles. Using advanced genomic resources and next-generation genotyping methods, the crop wild relatives of pearl millet can also be used to introduce variation at the gene, allele, or haplotype levels, hence increasing the allelic richness of the cultivated germplasm. These genomic technologies can be employed in accurate trait mapping, speed breeding, introgressions, and transdisciplinary ways to generate pearl millet cultivars resistant to the biotic and abiotic issues.

1.6.3 Genome Sequencing

Low-cost sequencing and resequencing technologies have facilitated assembling of plant genomes of many crops, including wild relatives. Resequencing of cultivars, landraces, and wild accessions is necessary to harness the genetic variation and identify the superior alleles for the desired traits because a single reference genome does not capture the diversity within a species. The availability of genome information has led to the development of numerous next-generation sequencing-based technologies for candidate gene discovery and allele mining. Through genotype-phenotype relationships, sequencing-based techniques offer the potential to discover novel variations for a large number of genes. For pearl millet, genetic and genomic sequencing data is now easily accessible. A useful resource for creating precision genetic maps is the pearl millet genome, which has a size of about 1.79 Gb and contains 38,579 genes, 88,256 SSRs, and 4,50,000 SNPs (Varshney et al. 2017). Re-sequenced, the Pearl Millet inbred Germplasm Association Panel (PMiGAP), created at ICRISAT in collaboration with Aberystwyth University, is a database containing about 29 million genome-wide SNPs. Drought resistance, grain Fe and Zn concentration, N utilization efficiency, endosperm starch components, grain yield, and other characteristics have all been mapped using PMiGAP (Srivastava et al. 2020).

1.6.4 Molecular Markers and Genetic Maps

Through marker-assisted selection, pre-breeding efficacy and efficiency are increased when specific traits have molecular markers associated with them. Diagnostic markers are molecular indicators that are connected to the genes of a desired trait and can be used indirectly to choose target traits (Singh et al. 2020). The Xicmp3092 marker showed a strong association with grain Fe content on LG 7, and markers Xpsmp2086, Xpsmp2213, and Xipes0224 showed association with grain Zn content on LG 4 and LG 6, respectively. These associations were developed using 250 SSR and 17 genetic markers associated with grain Fe and Zn content for 130 diversified lines tested across different environments (Srivastava et al. 2020).

1.6.5 Genome-Wide Association Studies

Genome-wide association studies (GWASs) offer a potent supplementary method for analyzing complex features and have the potential to overcome several of the limitations of traditional linkage mapping. Unlike QTL analysis, GWAS can discover genes associated with phenotypic traits at a greater resolution by utilizing historical recombinations in a variety of association panels. Because GWASs provide for the unrestricted access to a wide range of genetic resources for marker trait association without compromising marker availability, they have emerged as a potent tool for QTL mapping in plants. High-density SNP marker arrays have made it possible to conduct affordable GWASs utilizing natural populations. Genotyping by sequencing, or the next-generation genotyping approach, is becoming more popular for finding novel plant SNPs that can be utilized for GWASs because sequencing costs are constantly decreasing (Arruda et al. 2016). For example, a set of 250 full-sib progenies and 34 SSR markers were used for GWASs, and results revealed that the Xpsmp2248_162 marker allele at linkage group (LG)6 was found to have a high correlation with earlier flowering time and reduced plant height. On LG7, the marker allele Xpsmp2224_157 was highly correlated with plant height, while the Xpsmp2237_230 marker allele showed a strong positive correlation with grain yield. The Xicmp3058 193 marker allele demonstrated a high positive connection for fodder dry matter yield on LG6 (Srivastava et al. 2020).

1.6.6 Genomic Selection

More importantly, genomic selection (GS), a genomics-assisted breeding strategy, predicts each individual's genomic estimated breeding values (GEBVs) by combining large genotypic data, phenotypic data, and modelling with statistical tools (Crossa et al. 2017; Meuwissen et al. 2001). A representative sample of the breeding population, referred to as the "training population," is used to create a statistical model in genomic selection. The allelic effects of all marker loci are then determined using this model (i.e., genomic-assisted breeding values without phenotypic data), and these values can be used to preselect genotypes specific to traits (Heffner et al. 2011). Pre-breeding can be sped up by integrating genome-wide data with genomic selection, as demonstrated by Xu et al. (2012) and Spindel et al. (2016). This approach offers excellent specificity and predictability. By reducing the length of the selection cycle, GS can be used to quickly improve complex traits, including tolerance to biotic and abiotic stress and grain yield (Saint Pierre et al. 2016). In genomic selection, ascertainment bias is reduced by using markers created through genotyping by sequencing to capture genetic variability unique to the population or family (species) of interest. Complex trait selection, including tolerance to biotic and abiotic stress and grain yield (Saint Pierre et al. 2016), can be aided by GS. In genomic selection, ascertainment bias is reduced by using markers created through genotyping by sequencing to capture genetic variability unique to the population or family (species) of interest. When it comes to correcting for all genetic variation and choosing those with greater GEBV without phenotyping, GS is superior. High rates of outcrossing, heterozygosity, inbreeding depression, and residual heterozygosity in pearl millet create obstacles to inbred line development programs for the creation of association mapping panels and parental line/cultivar development that are utilized in GS training sets.

1.6.7 Genome Editing

The editing of genomes and their application in crop improvement initiatives are made possible due to recent developments in genomics. Genome editing techniques offer an alternative to pre-breeding, which entails genetic modification through recombination. Over the past 20 years, several genetic engineering technologies have been developed to replace traditional genetic engineering. These technologies include zinc finger nuclease, transcription activator-like effector nucleases, antisense, RNA interference, virus-induced gene silencing, oligonucleotide-directed mutagenesis, and clustered regularly interspaced short palindromic repeats/Cas9 (CRISPR/Cas9) (Zaman et al. 2019). As the products are not recognized as GMOs, they can accelerate pre-breeding programs through beneficial knockout mutations (Huang et al. 2016).

1.7 Future Prospects

A plant breeder's capacity to succeed is primarily determined by the amount of genetic variation at their disposal. Increase in population; dwindling resources impacted by climate change; and increase in demand for food, feed, and energy put immense pressure on agriculture and crops like pearl millet, but breeding has been lagging as compared to other cereals. Also, pearl millet has the potential to fill new niches around the world, especially in the semiarid zones of Central Asia and the Middle East, North and South America, and Australia, due to its superior tolerance to drought, high temperature, and salinity when compared to many other cereals, as well as its short duration and high nutritional value (National Research Council 1996).

Enhancing the yield potential of pearl millet in vulnerable dry areas is the main goal. While African sources, especially those from the West African region, offer excellent sources of larger head volume and seed size, higher degrees of resistance to diseases, and better seed quality, Indian landraces are an excellent source of early maturity, better tillering, and shorter height. The most valuable germplasm for adding desired variety has come from the landrace *Iniadi*, which is bold-seeded, productive, and early maturing. Plant breeders are also using less basic germplasm in their studies because of a lack of knowledge about features that are economically significant. Even if there is currently a well-established system in place for the distribution of genetic resources, there isn't enough or adequate input about the use of germplasm, and this needs to be addressed. Therefore, it would be beneficial to compile and disseminate additional catalogues that list the possible worth of the conserved germplasm. Databases have been built, but in order to make passport data easily accessible, it is important to link these databases and create a national database (Manga 2015).

In terms of growth habit, fodder yield, nutritional quality features, and resistance or tolerance to biotic and abiotic challenges, wild *Pennisetum* species offer a richness of genetic variety. However, it is challenging to maintain many wild and related species because of variations in their seed biology and dormancy and adaptability.

The potential for genetic development of pearl millet cultivars and hybrids for use as grain and fodder is hidden in unexplored genetic diversity. To boost the genetic gain of crop varieties and make pre-breeding a valuable scientific endeavor for breeders, focused and methodical efforts are required. In order to further utilize the adaption traits in hybrid breeding projects in Western Africa and for drought-prone ecologies in India, pre-breeding can be crucial in generating additional diversity. Enhancing the collaboration between public and private sectors is essential for a pre-breeding program to succeed.

Genomic techniques ought to be promoted in breeding projects for pearl millet. Examining the genetic diversity seen in wild species or germplasm accessions can provide valuable insights into the genetic and genomic characteristics of pearl millet. Additionally, genomic tools—such as genetic engineering, natural diversity panels, mapping populations, and GWAS—might be able to assist in analyzing the population's valuable variants. Targeted genes, genomic areas, or alleles can be substituted in the elite varieties by screening breeding lines or genetic stocks and using functional genomics and genomic selection. For the purpose of studying pearl millet diversification and domestication for a range of end applications, novel genomic variability will be made available by the sequencing of landraces, wild relatives, and better cultivars. A species' entire genomic structure is not represented by a single genome sequence. However, pan-genome analysis, which gathers genes at the phylogenetic level, offers the chance, with the aid of numerous individual sequencing of a species, to detect the genetic variability contained in the entire genome. Pearl millet accessions should undergo genomic selection for genotyping and phenotyping, with the model chosen based on population type and gene-environment interactions. Further, genome editing, a newly developed method in the genomic era, plays a vital role in the identification of genes and their function for crop development. Research on lowering anti-nutrients in pearl millet has a lot of potential to improve an already enriched cereal. Even though the genetic loci underlying many important agronomic traits have been identified with great success, other fields of study such as epigenomes and pan-genomes should be integrated to better understand genetic diversity and the superior alleles underlying complex agronomic traits (Singh and Nara 2023).

References

Acheampong E, Anisetty NM, Williams JT (1984) A world survey of sorghum and millets germplasm. IBPGR, Rome

Ali GM, Khan NM, Hazara R, McNeilly T (2004) Variability in the response of pearl millet (*Pennisetum americanum* (L.) Leeke) accessions to salinity. Acta Agronomica Hungarica 52:277–286

Anand Kumar K, Appa Rao S (1987) Diversity and utilization of pearl millet germplasm. In: Witcombe JR, Beckerman SR (eds) *Proceedings of the international pearl millet workshop, 7–11 April 1986, ICRISAT,* Patancheru, India. ICRISAT, Patancheru, pp 69–82

Andrews DJ, Anand Kumar K (1996) Use of the west African pearl millet landrace Iniadi in cultivar development. Plant Genetic Resources Newsletter 105:15–22

Appa Rao S, Mengesha MH, Subramanian V (1982) Collection and evaluation of sweet stalk pearl millet (*Pennisetum*). Econ Bot 36:286–290
Appa Rao S, Mengesha MH, Rajagopal Reddy C (1984) Characteristics and inheritance of viable chlorophyll mutants in *Pennisetum americanum* (L.) Leeke. Ind J Botany 7:1–5
Appa Rao S, Mengesha MH, Rajagopal Reddy C (1986a) Variation and adaptation of pearl millet germplasm in Tamil Nadu. Ind J Genetic Plant Breed 46:449–455
Appa Rao S, Mengesha MH, Rajagopal Reddy C (1987) Glossy genes in pearl millet. J Hered 78:333–335
Appa Rao S, Mengesha MH, Rajagopal Reddy C (1988a) Characteristics, inheritance and allelic relationships of midribless mutants in pearl millet. J Hered 79:18–20
Appa Rao S, Mengesha MH, Rajagopal Reddy C (1988b) New sources of early-maturing germplasm in pearl millet (*Pennisetum glaucum*). Indian J Agric Sci 58:743–746
Appa Rao S, Mengesha MH, Reddy KN (1998) Development and characterization of trait-specific genepools in pearl millet. Plant Genetic Resour Newsl 113:27–30
Appa Rao S, Mengesha MH, Rajagopal Reddy C (1986b) New sources of dwarfing genes in pearl millet (*Pennisetum americanum*). Theor Appl Genet 73:170–174
Arruda MP, Brown P, Krill A, Brown-Guedira G, Thurber C, Foresman B, Kolb F (2016) Genome-wide association mapping of fusarium head blight resistance in wheat using genotyping-by-sequencing. Plant Genome 9:1–14
Arya RK, Singh MK, Yadav AK, Kumar A, Kumar S (2014) Advances in pearl millet to mitigate adverse environmental conditions emerged due to global warming. Forage Res 40:57–70
Athwal DS (1965) Hybrid bajra-1 marks a new era. Indian Farming 15:6–7
Bantilan MCS, Subba Rao KV, Rai KN, Singh SD (1998) Research on high yielding pearl millet—background for an impact study in India. In: Assessing Joint Research Impacts: Proceedings of an International Workshop in Joint Impact Assessment of NARS/ICRISAT Technologies for the Semi-Arid Tropics. ICRISAT, Patancheru, India, 2–4 December 1996 (pp. 52–61). ICRISAT, Patancheru
Basavaraju R, Safeeulla KM, Murty BR (1981) Genetic variance and heritability for resistance to downy mildew in pearl millet. Ind J Genetic Plant Breed 41:137–143
Bennett MD, Bhandol P, Leitch IJ (2000) Nuclear DNA amounts in angiosperms and their modern uses—807 new estimates. Annal Botany (Lond) 86:859–909. https://doi.org/10.1111/nph.13264
Bidinger FR, Mahalakshmi V, Rao GDP (1987) Assessment of drought resistance in pearl millet (*Pennisetum americanum* (L.) Leeke). II. Estimation of genotype response to stress. Aust J Agric Res 38:49–59
Bidinger FR, Weltzien E, Mahalakshmi V, Singh SD, Rao KP (1994) Evaluation of landrace top cross hybrids of pearl millet for arid zone environments. Euphytica 76:215–226
Blum A, Sullivan CY (1986) The comparative drought resistance of landraces of sorghum and millet from dry and humid regions. Ann Bot 57:835–846
Bono M (1973) Contribution à la morpho-systématique des *Pennisetum* annuels cultivés pour leur grain en Afrique occidentale francophone. Agron Trop 2:229–355
Bramel-Cox PJ, Cox TS (1989) Use of wild germplasm in sorghum improvement. In: Proceedings of the 43rd Annual Corn and Sorghum Industry Research Conference, held at Chicago. Ill. American Seed Trade Association, Washington, D.C., pp 13–26
Brunken JN (1977) A systematic study of *Pennisetum* sect. *Pennisetum* (*Gramineae*). Am J Bot 64:161–176. https://doi.org/10.2307/2442104
Burton GW (1958) Cytoplasmic male-sterility in pearl millet (*Pennisetum glaucum* (L.) R. Br.). Agron J 50:230. https://doi.org/10.2134/agronj1958.00021962005000040018x
Burton GW, Powell JB (1968) Pearl millet breeding and cytogenetics. Advances in agronomy, vol 20. Academic Press, New York, pp 49–89
Carberry PS, Cambell LC, Bidinger FR (1985) The growth and development of pearl millet as affected by plant population. Field Crops Res 11:193–220.

Cardamone L, Cuatrín A, Grunberg K, Tomás MA (2018) Variability for salt tolerance in a collection of *Panicum coloratum* var. *makarikariense* during early growth stages. Tropical Grasslands-Forrajes Tropicales 6:134–147

Chase A (1921) The Linnaean millet concept of pearl millet. Am J Bot 8:41–49

Chauhan SS, Jha SK, Jha GK, Sharma DK, Satyavathi CT, Kumari J (2015) Germplasm screening of pearl millet (*Pennisetum glaucum*) for popping characteristics. Indian J Agric Sci 85(3):344–348. https://doi.org/10.56093/ijas.v85i3.47113

Chowdhury MKU, Smith RL (1988) "Mitochondrial DNA variation in pearl millet and related species." Theor Appl Genet 76:25–32.

Choudhary ML, Tripathi MK, Gupta N, Tiwari S, Tripathi N, Parihar P, Pandya RK (2021) Screening of pearl millet (*Pennisetum glaucum* (L.) R. Br.) germplasm lines against drought tolerance based on biochemical traits. *Current.* J Appl Sci Technol 40(23):1–12

Combs GF (2001) Selenium in global food systems. Br J Nutr 85:517–547

Crossa J, Pérez-Rodríguez P, Cuevas J, Montesinos-López O, Jarquín D, de los Campos G, Burgueño J, González-Camacho JM, Pérez-Elizalde S, Beyene Y, Dreisigacker S, Singh R, Zhang X, Gowda M, Roorkiwal M, Rutkoski J, Varshney RK (2017) Genomic selection in plant breeding: methods, models, and perspectives. Trends Plant Sci 22:961–975

Deshmukh HG, More BB, Utikar PG (1978) Screening germplasm of pearl millet against downy mildew, *Sclerospora graminicola* (Sacc.) Schroet. Ind J Agric Res 12:85–88

Driedonks N, Rieu I, Vriezen WH (2016) Breeding for plant heat tolerance at vegetative and reproductive stages. Plant Reprod 29:67–79. https://doi.org/10.1007/s00497-016-0275-9

Dua RP (1989) Salinity tolerance in pearl millet. Ind J Agric Res 23:9–14

Dujardin M, Hanna WW (1983) Apomictic and sexual pearl millet x *Pennisetum squamulatum* hybrids. J Hered 74:277–279

Dujardin M, Hanna WW (1984) Microsporogenesis, reproductive behavior, and fertility of five *Pennisetum* species. Theor Appl Genet 67:197–201

Dujardin M, Hanna WW (1985a) Cytology and reproduction of reciprocal backcrosses between pearl millet and sexual and apomictic hybrids of pearl millet × *Pennisetum squamulatum*. Crop Sci 25:59–62

Dujardin M, Hanna WW (1985b) Cytology and reproductive behavior of pearl millet napier grass hexaploids × *Pennisetum squamulatum* tri-specific hybrids. J Hered 76:382–384

Dujardin M, Hanna WW (1989) Crossability of pearl millet with wild *Pennisetum* species. Crop Sci 29:77–80

Dwivedi S, Upadhyaya H, Subudhi P, Gehring C, Bajic V, Ortiz R (2010) Enhancing abiotic stress tolerance in cereals through breeding and transgenic interventions. Plant Breed Rev 33:31–114

Dwivedi S, Upadhyaya H, Senthilvel S, Hash C, Fukunaga K, Diao X et al (2012) Millets: genetic and genomic resources. In: Janick J (ed) Plant breeding reviews. John Wiley and Sons, Inc., Hoboken, NJ, pp 247–374

FAOSTAT (2021) Faostat: Food and agriculture data (Rome: FAOSTAT). Available at: https://www.fao.org/faostat//#data/QCL

Farooq M, Wahid A, Kobayashi N, Fujita D, Basra SMA (2009) Plant drought stress; effects, mechanisms and management. Agron Sustain Dev 29, 185–212. https://doi.org/10.1051/agro:2008021

Gaut BS, Clegg MT (1993) Nucleotide polymorphism in the *adh 1* locus of pearl millet (*Pennisetum glaucum*) (Poaceae). Genetics 135:1091–1097

Gerloff GC, Gabelman WH (1983) Genetic basis of inorganic plant nutrition. In: Lauchli A, Bieleski RL (eds) Inorganic plant nutrition. Springer-Verlag, Berlin, Heidelberg, New York, Tokyo, pp 453–480

Godawat SL, Kaul BK (1981). Genetic variability in pearl millet. Indian J Agric Sci 51(3): 147–149.

Govindaraj M, Shanmugasundaram P, Sumathi P, Muthiah AR (2010) Simple, rapid and cost-effective screening method for drought-resistant breeding in pearl millet. Electr J Plant Breed 1(4):590–599

Govindaraj M, Rai KN, Kanatti A, Upadhyaya HD, Shivade H, Rao AS (2020) Exploring the genetic variability and diversity of pearl millet core collection germplasm for grain nutritional traits improvement. Sci Report 10:21177. https://doi.org/10.1038/s41598-020-77818-0. Erratum in: Scientific *Reports*, 2021 May 3; 11(1):9757. PMID: 33273504; PMCID: PMC7713302

Graham RD, Welch RM (1996) Breeding for staple-food crops with high micronutrient density. Agricultural Strategies for Micronutrients Working Paper 3. International Food Policy Research Institute, Washington, D.C., pp 1–72

Gregory PJ (1983). Response to temperature in a stand of pearl millet (Pennisetum typhoides S. & H.). III. Root development. J Exp Bot 34:744–756.

Gunguniya DF, Kumar S, Patel MP, Sakure AA, Patel R, Kumar D, Khandelwal V (2023) Morpho-biochemical characterization and molecular marker-based genetic diversity of pearl millet (*Pennisetum glaucum* (L.) R. Br). Peer J 11:e15403

Gupta SC, Monyo ES, Appa Rao S (1983) Registration of SDML 89107 brown midrib pearl millet germplasm. Crop Sci 33:882

Gupta SK, Rathore A, Yadav OP, Rai KN, Khairwal IS, Rajpurohit BS, Das RR (2013) Identifying mega-environments and essential test locations for pearl millet cultivar selection in India. Crop Sci 53:2444–2453.

Gupta SK, Rai KN, Singh P, Ameta VL, Gupta SK, Jayalekha AK, Mahala RS, Pareek S, Swami ML, Verma YS (2015) Seed set variability under high temperatures during the flowering period in pearl millet (*Pennisetum glaucum* L. (R.) Br.). Field Crop Res 171:41–53. ISSN 0378–4290

Hajlaoui H, Akrimi R, Hajlaoui F (2021) Screening field-grown pearl millet (*Pennisetum glaucum* L.) genotypes for salinity tolerance in the north of Tunisia. IOP Conference Series: Earth and Environmental Science 904:012046. https://doi.org/10.1088/1755-1315/904/1/012046

Hanna WW (1989) Characteristics and stability of a new cytoplasmic-nuclear male sterile source in pearl millet. Crop Sci 29:1457–1459. https://doi.org/10.2135/cropsci1989.0011183X002900060026x

Hanna WW (1990) Registration of Tift #2 S-1 pearl millet germplasm. Crop Sci 30:1376

Hanna WW (1992) Utilization of germplasm from wild species. In: Hanna WW (ed) Desertified grass lands: their biology and management. The Linnean Society of London, London, pp 251–257

Hanna WW (1997) Registration of Tift 8593 pearl millet genetic stock. Crop Sci 37:1412. https://doi.org/10.2135/cropsci1997.0011183X003700040100x

Hanna WW, Dujardin M (1985) Interspecific transfer of apomixis in *Pennisetum*. In: Presented at the XV International Grassland Congress, 24–31 Aug 1985. Japan, Kyoto

Hanna WW, Wells HD, Burton GW (1987) Registration of pearl millet inbred parental lines, Tift 85D2A1 and Tift 85D2B1. Crop Sci 27:1324–1325

Hanna WW, Wilson JP, Wells HD, Gupta SC (1993) Registration of Tift #5 S-1 pearl millet germplasm. Crop Sci 30:1417–1418

Harinarayana G, Rai KN (1989) Use of pearl millet germplasm and its impact on crop improvement in India. In: Collaboration on genetic resources: proceedings of a workshop on germplasm exploration and evaluation in India. ICRISAT, pp 89–92

Harinarayana G, Appa Rao S, Mengesha MH (1988) Prospects of utilizing genetic diversity in pearl millet. In: Paroda RS, Arora RK, Chandel KPS (eds) Plant genetic resources: Indian perspective. NBPGR, New Delhi, pp 170–182

Harlan JR (1975) Geographic patterns of variation in some cultivated plants. J Hered 66:184–191

Harlan JR, de Wet JMJ (1971) Toward a rational classification of cultivated plants. Taxon 20:509–517

Hash CT, Bhasker Raj AG, Lineup S, Sharma A, Beniwal CR, Folkertsma RT et al (2003) Opportunities for marker-assisted selection (MAS) to improve the feed quality of crop residues in pearl millet and sorghum. Field Crop Res 84:79–88

Heffner EL, Jannink JL, Iwata H, Souza E, Sorrells ME (2011) Genomic selection accuracy for grain quality traits in biparental wheat populations. Crop Sci 51:2597–2606

Henson IE, Mahalakshmi V (1985) Evidence for panicle control of stomatal behavior in water-stressed plants of pearl millet. Field Crop Res 11:281–290

Hitchcock AS, Chase A (1951) Manual of the grasses of the United States, 2nd edn. USDA Misc. Publ. 200. Government Printing Office, Washington, D.C.

Huang B, Gomez-Rodriguez J, Preite S, Garrett LJ, Harper UL, Schwartzberg PL (2016) CRISPR-Mediated Triple Knockout of SLAMF1, SLAMF5 and SLAMF6 Supports Positive Signaling Roles in NKT Cell Development. PLoS ONE 11(6):e0156072. https://doi.org/10.1371/journal.pone.0156072

ICRISAT (1989) Collaboration on genetic resources: summary proceedings of a joint ICRISAT/NBPGR (ICAR) workshop on germplasm exploration and evaluation in India, 14–15 Nov 1988. ICRISAT Center, Patancheru

Jambunathan R, Subramanian V (1988) Grain quality and utilization of sorghum and pearl millet. In: de Wet JMJ, Preston TA (eds) Biotechnology in tropical crop improvement. ICRISAT, Patancheru, pp 133–139

Jha S, Singh J, Chouhan C, Singh O, Srivastava RK (2021) Evaluation of multiple salinity tolerance indices for screening and comparative biochemical and molecular analysis of pearl millet [*Pennisetum glaucum* (L.) R. Br.] genotypes. Journal of Plant Growth Regulation:1–16. https://doi.org/10.1007/s00344-021-10424-0

Joshi AK, Pandya JN, Mathukia RK, Pethani KV, Dave HR (1997) Seed germination in pearl millet hybrids and parents under extreme temperature conditions. GAU Res J 23:77–83

Kashiwagi J, Krishnamurthy L, Upadhyaya HD, Krishna H, Chandra S, Vadez V, Serraj R (2005) Genetic variability of drought-avoidance root traits in the mini-core germplasm collection of chickpea (*Cicer arietinum* L.). Euphytica 146:213–222. https://doi.org/10.1007/s10681-005-9007-1

Kennedy G, Nantel G, Shetty P (2003) The scourge of "hidden hunger": global dimensions of micronutrient deficiencies. Food Nutr Agric 32:8–16

Khairwal IS, Yadav OP (2005) Pearl millet (*Pennisetum glaucum*) improvement in India—retrospect and prospects. Indian J Agric Sci 75:183–191

Khairwal IS, Yadav SK, Rai KN, Upadhyaya HD, Kachhawa D, Nirwan B et al (2007) Evaluation and identification of promising pearl millet germplasm for grain and fodder traits. J SAT Agric Res 5(1):1–6

Kholova J, Vadez V (2013) Water extraction under terminal drought explains the genotypic differences in yield, not the anti-oxidants changes in leaves of pearl millet (Pennisetum glaucum (L.) R. Br.). Functional Plant Biology, 40(1). pp. 44–53. ISSN 1445–4408

Krishnamurthy L, Serraj R, Rai KN, Hash CT, Dakheel AJ (2007) Identification of pearl millet [*Pennisetum glaucum* (L.) R. Br.] lines tolerant to soil salinity. Euphytica 158:179–188. https://doi.org/10.1007/s10681-007-9441-3

Krishnamurthy L, Zaman-Allah M, Purushothaman R, Irshad Ahmed M, Vadez V (2011) Plant biomass productivity under abiotic stresses in SAT agriculture. In Improving Crop Productivity in Stress Environments, pp 247–264. https://doi.org/10.5772/17279

Kulkarni VN, Rai KN, Dakheel AJ, Ibrahim M, Hebbara M, Vadez V (2006) Pearl millet germplasm adapted to saline conditions. SAT eJournal 2:1–4

Kumar AK, Appa Rao S (1987) Diversity and utilization of pearl millet germplasm. In: Witcombe JR, Beckerman SR (eds) Proceedings of the international pearl millet workshop. ICRISAT, Patancheru, pp 69–82

Kumar A, Manga VK, Gour HN, Purohit AK (2012) Pearl millet downy mildew: challenges and prospects, Rev Plant Pathol. 5:139–177

Kumari J, Bag MK, Pandey S, Jha SK, Chauhan SS, Jha GK et al (2016) Assessment of phenotypic diversity in pearl millet [*Pennisetum glaucum* (L.) R. Br.] germplasm of Indian origin and identification of trait-specific germplasm. Crop Pasture Sci 67:1223–1234. https://doi.org/10.1071/CP16300

Lagudah ES, Hanna WW (1990) Patterns of variation for seed proteins in the *Pennisetum* gene pool. J Hered 81:25–29

Lata C, Muthamilarasan M, Prasad M (2015) "Drought stress responses and signal transduction in plants," in The Elucidation of Abiotic Stress Signaling in Plants, ed. G. K. Pandey (New York, NY: Springer), 195–225

Leeke. (1907) Untersuchungen über Abstammung und Heimat der Negerhirse (*Pennisetum americanum* (L.) K. S. Schum.). Z Naturw 79:52–55

Lobell DB, Asner GP (2003) Climate and management contribution to recent trends in US agricultural yields. Science 299:1032

Manga VK (2015) Diversity in pearl millet [*Pennisetum glaucum* (L) R. BR.] and its management. Indian J Plant Sci 4:38–51

Marchais L, Pernes J (1985) Genetic divergence between wild and cultivated pearl millets (*Pennisetum typhoides*). I Male sterility *Zeitschrift für Pflanzenzüchtung* 95:103–112

Mariac C, Luong V, Kapran I, Mamadou A, Sagnard F, Deu M, Chantereau J, Gerard B, Ndjeunga J, Bezançon G, Pham JL, Vigouroux Y (2006) Diversity of wild and cultivated pearl millet accessions [*Pennisetum glaucum* (L.) R. Br.] in Niger assessed by microsatellite markers. Theor Appl Genet 114:49–58. https://doi.org/10.1007/s00122-006-0409-9

McDonough CM, Rooney LW (1987) Variation in the physical and structural characteristics of pearl millet varieties from the world collection. Cereal Foods World 32:665

Mengesha MH, Khanna PP, Chandel KPS, Kameswara Rao N (1989) Conservation of world germplasm collection of ICRISAT mandate crops. In: In Collaboration on Genetic Resources: Summary Proceedings of a Workshop on Germplasm Exploration and Evaluation in India. ICRISAT, Patancheru, pp 65–69

Menkir A, Bramel-Cox PJ, Witt MD (1994) Comparisons of methods for introgressing exotic germplasm into adapted sorghum. Theor Appl Genet 89:233–239

Merkle OG, Starks KJ, Casady AJ (1983) Registration of pearl millet germplasm lines with chinch bug resistance. Crop Sci 23:601

Meuwissen THE, Hayes BJ, Goddard ME (2001) Prediction of total genetic value using genome-wide dense marker maps. Genetics 157:1819–1829

Monson WG, Hanna WW, Gaines TP (1986) Effects of rust on yield and quality of pearl millet forage. Crop Sci 26:637–639

Mukesh Sankar S, Singh SP, Prakash G, Satyavathi CT, Soumya SL, Yadav Y et al (2021) Deciphering Genotype-By-Environment Interaction for Target Environmental Delineation and Identification of Stable Resistant Sources Against Foliar Blast Disease of Pearl Millet. Front Plant Sci 12:656158. https://doi.org/10.3389/fpls.2021.656158

Murdock GP (1959) Africa: its people and their cultural history. McGraw Hill, New York

Murty BR, Upadhyaya MK, Manchanda PL (1967) Classification and cataloguing of a world collection of genetic stocks of *Pennisetum*. Ind J Genetic 27:313–394

Muscolo A, Panuccio MR, Sidari M (2003) Effects of salinity on growth, carbohydrate metabolism and nutritive properties of kikuyu grass (*Pennisetum clandestinum* Hochst). Plant Sci 104:1103–1110

National Research Council (1996) Lost crops of Africa, Vol. I: Grains. National Academy Press, Washington, D.C.

Passot S, Gnacko F, Moukouanga D, Lucas M, Guyomarc'h S, Ortega BM et al (2016) Characterization of pearl millet root architecture and anatomy reveals three types of lateral roots. Front Plant Sci 7:829. https://doi.org/10.3389/fpls.2016.00829

Patil KS, Gupta SK, Marathi B, Danam S, Thatikunta R, Rathore A, et al. (2020). African and Asian origin pearl millet populations: genetic diversity pattern and its association with yield heterosis. Crop Sci 60, 3035–3048. https://doi.org/10.1002/csc2.20245

Prakash G, Srinivasa N, Mukesh Sankar S, Singh SP, Tara Satyavathi C (2016) Standardization of pearl millet blast (*Magnaporthe grisea*) phenotyping under artificial conditions. Ann Agric Res 37:200–205

Presterl T, and Weltzien E (2003) Exploiting heterosis in pearl millet population breeding in arid environments. Crop Sci 43, 767–776. https://doi.org/10.2135/cropsci2003.7670

Rachie KO, Majmudar JV (1980) Pearl millet. The Pennsylvania State University Press, University Park, London

Rai KN, Anand Kumar K (1994) Pearl millet improvement at ICRISAT—an update. Int Sorghum Millets Newsl 35:1–29

Rai KN, Kumar AK, Andrews DJ, Rao AS, Raj AGB, Witcombe JR (1990) Registration of 'ICTP 8203' pearl millet. Crop Sci 30:959

Rai KN, Appa Rao S, Reddy KN (1997) Pearl millet. In: Fuccillo D, Sears L, Stapleton P (eds) Biodiversity in trust, conservation and use of plant genetic resources in CGIAR centers. Cambridge University Press, pp 243–258

Rai KN, Yadav OP, Gupta SK, Mahala RS, Gupta SK (2012) Emerging research priorities in pearl millet. J SAT Agric Res 10:1–4. Available online at: http://ejournal.icrisat.org/index.htm

Rao NK, Reddy LJ, Bramel PJ (2003) Potential of wild species for genetic enhancement of some semi-arid food crops. Genet Resour Crop Evol 50:707–721

Rao KB, Motukuri SRK, Kumar KA, Babu CHVNP, Pathak V (2021) Pearl millet blast pathogen virulence study and identification of resistance donors on virulent isolate. J Pure Appl Microbiol 15:752–758. https://doi.org/10.22207/JPAM.15.2.27

Rooney LW, McDonough CM (1987) Food quality and consumer acceptance in pearl millet. In: Witcombe JR, Beckerman SR (eds) Proceedings of the International Pearl Millet Workshop. ICRISAT, Patancheru, pp 43–61

Saint Pierre C, Burgueño J, Crossa J, Fuentes Dávila G, Figueroa López P, Solís Moya E, Ireta Moreno J, Hernández Muela VM, Zamora Villa VM, Vikram P, Mathews K, Sansaloni C, Sehgal D, Jarquin D, Wenzl P, Singh S (2016) Genomic prediction models for grain yield of spring bread wheat in diverse agro-ecological zones. Sci Rep 6:27312

Sanjana Reddy P, Satyavathi CT, Khandelwal V, Patil HT, Gupta PC, Sharma LD et al (2021) Performance and stability of pearl millet varieties for grain yield and micronutrients in arid and semi-arid regions of India. Front Plant Sci 12:670201. https://doi.org/10.3389/fpls.2021.670201

Sanjana Reddy P, Parashuram P, Malathi VM (2023) Enhancing productivity and nutrition with biofortified pearl millet cultivars. Indian farming 73(1):30–32

Sehgal D, Skot L, Singh R, Srivastava RK, Das SP, Taunk J et al (2015) Exploring the potential of pearl millet germplasm association panel for association mapping of drought tolerance traits. PLoS One 10(5):e0122165. https://doi.org/10.1371/journal.pone.0122165

Serraj R, Krishnamurthy L, Upadhyaya HD (2004) Screening chickpea mini-core germplasm for tolerance to salinity. International Chickpea and Pigeonpea Newsletter 11:29–32

Sharma YK, Yadav SK, Khairwal IS, Bajri-Chomu D, Bajri-Osian D (2007) Evaluation of pearl millet germplasm lines against downy mildew incited by *Sclerospora graminicola* in western Rajasthan. SAT eJournal 3(1):1–2

Sharma R, Thakur RP, Rai KN, Gupta SK, Rao VP, Rao AS, Kumar S (2009) Identification of rust resistance in hybrid parents and advanced breeding lines of pearl millet. J SAT Agric Res 7:1–4

Sharma R, Rao VP, Upadhyaya HD, Reddy VG, Thakur RP (2010) Resistance to grain mold and downy mildew in a mini-core collection of sorghum germplasm. Plant Dis 94:439–444. https://doi.org/10.1094/PDIS-94-4-0439

Sharma R, Upadhyaya HD, Manjunatha SV, Rao VP, Thakur RP (2012) Resistance to foliar diseases in a mini-core collection of sorghum germplasm. Plant Dis 96:1629–1633. https://doi.org/10.1094/PDIS-10-11-0875-RE

Sharma R, Upadhyaya HD, Manjunatha SV, Rai KN, Gupta SK, Thakur RP (2013) Pathogenic variation in the pearl millet blast pathogen, *Magnaporthe grisea*, and identification of resistance to diverse pathotypes. Plant Dis 97:189–195. https://doi.org/10.1094/PDIS-05-12-0481-RE

Sharma R, Upadhyaya HD, Sharma S, Gate VL, Raj C (2015) New sources of resistance to multiple pathotypes of *Sclerospora graminicola* in the pearl millet mini-core germplasm collection. Crop Sci 55(4):1619–1628. https://doi.org/10.2135/cropsci2014.12.0822

Sharma R, Sharma S, Gate VL (2020a) Tapping *Pennisetum violaceum*, a wild relative of pearl millet (*Pennisetum glaucum*), for resistance to blast (caused by *Magnaporthe grisea*) and rust (caused by *Puccinia substriata var. Indica*). Plant Dis 104(5):1487–1491. https://doi.org/10.1094/PDIS-08-19-1602-RE

Sharma S, Sharma R, Govindaraj M, Mahala RS, Satyavathi CT, Srivastava RK, Gumma MK, Kilian B (2020b) Harnessing wild relatives of pearl millet for germplasm enhancement: challenges and opportunities. Crop Sci 61:177–200

Sharma S, Sharma R, Pujar M, Yadav D, Yadav YP, Rathore A et al (2020c) Utilization of wild *Pennisetum* species for improving biotic and abiotic stress tolerance in pearl millet (*Pennisetum glaucum* L.). Crop Sci 61:289–304. https://doi.org/10.1002/csc2.20408

Shivhare R, Lata C (2017) Exploration of genetic and genomic resources for abiotic and biotic stress tolerance in pearl millet. Front Plant Sci 7:2069. https://doi.org/10.3389/fpls.2016.02069

Shrivasata P, Kumar R (2015). Soil salinity: a serious environmental issue and plant growth promoting bacteria as one of the tools for its alleviation. Saudi J Biologi Sci 22:123–131. https://doi.org/10.1016/j.sjbs.2014.12.001

Singh SD (1995) Downy mildew of pearl millet. Plant Dis 79:545–550. https://doi.org/10.1094/PD-79-0545

Singh F, Nainawatee HS (1999) Grain quality traits. In: Khairwal IS, Rai KN, Andrews DJ, Harinarayana G (eds) Pearl millet breeding. Oxford and IBH, New Delhi, pp 157–183

Singh M, Nara U (2023) Genetic insights in pearl millet breeding in the genomic era: challenges and prospects. Plant Biotechnol Report 17:15–37. https://doi.org/10.1007/s11816-022-00767-9

Singh SD, King SB, Malla Reddy P (1990) Registration of five pearl millet germplasm sources with stable resistance to rust. Crop Science, 30:1165. https://doi.org/10.2135/cropsci1990.0011183X003000050061x

Singh SD, Wilson JP, Navi SS, Talukdar BS, Hess DE, Reddy KN (1997) Screening techniques and sources of resistance to downy mildew and rust in pearl millet. Information Bulletin no. 48. Patancheru 502 324, AndhraPradesh, India: International Crops Research Institute forthe Semi-Arid Tropics. 104 pp

Singh SD, Navi SS (2000) Genetic resistance to pearl millet downy mildew II. Resistance in wild relatives. Mycol Plant Pathol 30:167–171

Singh P, Singh U, Eggum BO, Anand Kumar K, Andrews DJ (1987a). Nutritional evaluation of high protein genotypes of pearl millet (Pennisetum americanum (L.) Leeke). J Sci Food Agric 38:41–48. https://doi.org/10.1002/jsfa.2740380108

Singh P, Singh U, Eggum BO, Anand Kumar K, Andrews DJ (1987b) Nutritional evaluation of high protein genotypes of pearl millet (*Pennisetum americanum* (L.) Leeke). J Sci Food Agric 38:41–48. https://doi.org/10.1002/jsfa.2740380108

Singh SD, King SB, Werder J (1993) Downy mildew disease of pearl millet. Information Bulletin No. 37. International Crops Research Institute for the Semi-Arid Tropics, Patancheru, A.P. 502 324, India

Singh SD, Alagarswamy G, Talukdar BS, Hash CT (1994) Registration of ICML 22 photoperiod insensitive, downy mildew resistant pearl millet germplasm. Crop Sci 34:1421

Singh K, Gupta K, Tyagi V, Rajkumar S (2020) Plant genetic resources in India: management and utilization. Vavilovskii Zhurnal Genet Selektsii 24:306–314. https://doi.org/10.18699/VJ20.622

Soman P, Peacock JM (1985) A laboratory technique to screen seedling emergence of sorghum and pearl millet at high soil temperature. Exp Agric 21:335–341

Soman P, Jayachandran R, Bidinger FR (1987) Uneven variation in plant to plant spacing in pearl millet. Agron J 79: 891–895

Spindel JE, Begum H, Akdemir D, Collard B, Redoña E, Jannink JL, McCouch S (2016) Genome-wide prediction models that incorporate de novo GWAS are a powerful new tool for tropical rice improvement. Heredity 116:395–408

Srivastava RK, Singh RB, Pujarula VL, Bollam S, Pusuluri M, Chellapilla TS, Yadav RS, Gupta R (2020) Genome-wide association studies and genomic selection in pearl millet: advances and prospects. Front Genet 10:1389. https://doi.org/10.3389/fgene.2019.01389

Subba Rao MV, Sujatha DM, Saideswara Rao Y, Manga V (1988) Chemotaxonomic characters in twelve species of the genus *Pennisetum* (Poaceae). Proceedings of the Indian Academy of Sciences. Plant Sci 98:11–112

Stich B, Haussmann BIG, Pasam R, Bhosale S, Hash CT, Melchinger AE, Parzies, HK (2010) Patterns of molecular and phenotypic diversity in pearl millet [Pennisetum glaucum (L.) R. Br.] from West and Central Africa and their relation to geographical and environmental parameters. BMC plant biology 10:216. https://doi.org/10.1186/1471-2229-10-216

Sukumaran S, Rebetzke G, Mackay I, Bentley AR, Reynolds MP (2022) Pre-breeding strategies. In: Reynolds MP, Braun HJ (eds) Wheat improvement. Springer, Cham, pp 317–331. https://doi.org/10.1007/978-3-030-90673-3_25

Tapsoba H, Wilson JP (1996) Pathogenic variation in *Puccinia substriata* var. *indica* in the southeastern United States and screening for resistance in pearl millet germplasm. Plant Dis 80:395–397

Tapsoba H, Wilson JP, Hanna WW (1997) Improvement of resistance to rust through recurrent selection in pearl millet. Crop Sci 37:365–369

Thakur RP, King SB (1988) Registration of four ergot-resistant germplasms of pearl millet. Crop Sci 28:382

Thakur RP, Williams RJ (1980) Pollination effects on pearl millet ergot. Phytopathology 70:80–84

Thakur RP, Singh SD, King SB (1988) Registration of four populations of pearl millet germplasm with multiple disease resistance. Crop Sci 28:381–382

Thakur RP, King SB, Rai KN, Rao VP (1992a) Identification and utilization of smut resistance in pearl millet. International Crops Research Center for the Semi-Arid Tropics, Patancheru

Thakur RP, Shetty KG, King SB (1992b) Selection for host-specific virulence in asexual populations of *Sclerospora graminicola*. Plant Pathol 41:626–632

Thakur RP, Sivaramakrishnan S, Kannan S, Rao VP, Hess DE, Magill CW (2004b) Genetic and pathogenic variability among isolates of *Sclerospora graminicola*, the downy mildew pathogen of pearl millet. In: Spencer-Phillips P, Jeger M (eds) Advances in downy mildew, vol 2. Kluwer Academic Publishers, London, pp 179–192

Thakur RP, Shetty HS, Khairwal IS (2006) Pearl Millet Downy Mildew Research in India: Progress and Perspectives. ISMN 47:125–130

Thakur RP, Sharma R, Rai KN, Gupta SK, Rao VP (2009) Screening techniques and resistant sources for foliar blast in pearl millet. J SAT Agric Res 7:1–5

Toderich K, Shuyskaya E, Rakhmankulova Z, Bukarev R, Khujanazarov T, Zhapaev R et al (2018) Threshold tolerance of new genotypes of Pennisetum glaucum (L) R Br to salinity and drought. Agronomy 8:e230

Tollefson J (2020) How hot will Earth get by 2100? Nature 580(7804):443–445, https://doi.org/10.1038/d41586-020-01125-x

Tostain S (1992) Enzyme diversity in pearl millet (*Pennisetum glaucum* L.) 3. Wild millet. Theoretical and Applied Genetics 83:733–742

Upadhyay MK, Murty BR (1970) Genetic divergence in relation to geographical distribution in pearl millet. Indian Journal of Genetics and Plant Breeding 30:704–715

Upadhyaya HD, Reddy KN, Gowda CLL (2007a) Pearl millet germplasm at ICRISAT genebank—status and impact. SAT eJ 3:1–5

Upadhyaya HD, Reddy KN, Gowda CLL, Ahmed MI, Singh S (2007b) Agroecological patterns of diversity in pearl millet (*Pennisetum glaucum* L.) germplasm from India. Indian Journal of Plant Genetic Resources 20:178–185

Upadhyaya HD, Reddy KN, Ahmed MI, Gowda CLL (2010a) Identification of gaps in pearl millet germplasm from Asia conserved at the ICRISAT genebank. Plant Genetic Resources 8(3):267–276. ISSN 1479-263X

Upadhyaya HD, Reddy KN, Ahmed MI, Gowda CLL, Haussmann BIG (2010b) Identification of geographical gaps in the pearl millet germplasm conserved at ICRISAT genebank from West and Central Africa. Plant Genetic Resources: Characterization and Utilization 1–7. https://doi.org/10.1017/S147926210999013X

Upadhyaya HD, Yadav D, Reddy KN, Gowda CLL, Singh S (2011) Development of pearl millet minicore collection for enhanced utilization of germplasm. Crop Sci 51:217–223. https://doi.org/10.2135/cropsci2010.06.0336

Upadhyaya HD, Reddy KN, Ahmed MI, Gowda CLL (2012) Identification of gaps in pearl millet germplasm from East and Southern Africa conserved at the ICRISAT genebank. Plant Genetic Resources: Characterization and Utilization 10(3):202–213

Upadhyaya HD, Dwivedi SL, Vadez V, Hamidou F, Singh S, Varshney RK, Liao B (2014) Multiple resistant and nutritionally dense germplasm identified from mini core collection in peanut. Crop Sci 54:2120–2130. https://doi.org/10.2135/crop-sci2014.01.0040

Upadhyaya HD, Reddy KN, Ahmed MI, Kumar V, Gumma MK, Ramachandran S (2016a) Geographical distribution of traits and diversity in the world collection of pearl millet (*Pennisetum glaucum* L.) landraces conserved at the ICRISAT genebank. Genet Resour Crop Evol:1–17. https://doi.org/10.1007/s10722-016-0442-8

Upadhyaya HD, Reddy K, Ramachandran S, Kumar V, Ahmed MI (2016b) Adaptation pattern and genetic potential of Indian pearl millet landraces conserved at the ICRISAT Genebank. Indian Journal of Plant Genetic Resources 29(2):97–113. https://doi.org/10.5958/0976-1926.2016.00015.2

Vadez V, Hash T, Kholova J (2012) II. 1.5 phenotyping pearl millet for adaptation to drought. Front Physiol 3:386. https://doi.org/10.3389/fphys.2012.00386

van Oosterom EJ, Mahalakshmi V, Bidinger FR, Rao KP (1995) Effect of water availability and temperature on the genotype by environment interaction of pearl millet in semi-arid tropical environments. Euphytica 89, 175–183. https://doi.org/10.1007/BF00034603

Varshney RK, Shi C, Thudi M, Mariac C, Wallace J, Qi P et al (2017) Pearl millet genome sequence provides a resource to improve agronomic traits in arid environments. Nat Biotechnol 35:969. https://doi.org/10.1007/s00122-011-1580-1

Vavilov NI (1949) The origin, variation, immunity and breeding of cultivated plants. Chron Bot 13:1–366

Vavilov NI (1950) The origin, variation, immunity and breeding of cultivated plants. Chron Bot 13:1–366

Verma R, Tripathi MK, Tiwari S, Tripathi N, Pandya RK, Parihar P (2021) Screening of pearl millet (*Pennisetum glaucum* L.) genotypes against blast disease on the basis of disease indexing and gene-specific SSR markers. Int J Curr Microbiol App Sci 10(02):2218–2227

Vetriventhan M, Kumar, Vinod, Dagunapur NR, Srinivas R, Jagadeesh K, Kanchanapally A et al (2024) Pearl Millet Germplasm Resources Conservation Status and Their Utilization in Crop Improvement. In: The Pearl Millet Genome. Compendium of Plant Genomes, 1 (1). Springer Nature, Switzerland AG, pp. 15–29. ISBN 978-3-031-56975-3

Wilson JP (1991) Detection of pathogenic races of *Puccinia substriata* var. *indica* in the U.S. Phytopathology 81:815

Wilson JP (1993) Two new dominant genes for resistance to pearl millet rust. Phytopathology 83:1395

Wilson JP, Hanna WW (1992) Disease resistance in wild *Pennisetum* species. Plant Dis 76:1171–1175

Wilson JP, Hess DE, Hanna WW (2000) Resistance to *Striga hermonthica* in wild accessions of the primary gene pool of *Pennisetum glaucum*. Phytopathology 90:1169–1172

Wilson JP, Hess DE, Hanna WW, Kumar KA, Gupta SC (2004) *Pennisetum glaucum* subsp. *monodii* accessions with *Striga* resistance in West Africa. Crop Prot 23:865–870

Winkel T, Renno JN, Payne W (1997) Effect of timing of water deficit on growth, phenology, and yield of *pearl millet* (*Pennisetum glaucum* L. R. Br.) in Sahelian conditions. J Exp Bot 48:1001–1011

Xu Y, Lu Y, Xie C, Gao S, Wan J, Prassana BM (2012) Whole genome strategies for marker-assisted plant breeding. Mol Breed 29:833–854

Yadav OP (2004) CZP 9802 – a new drought-tolerant cultivar of pearl millet. Indian Farm 54:15–17

Yadav OP (2008) Performance of landraces exotic elite populations and their crosses in *pearl millet* (*Pennisetum glaucum*) in drought and non-drought conditions. Plant Breed 127:208–210

Yadav OP (2010) Drought response of *pearl millet* landrace-based populations and their crosses with elite composites. Field Crop Res 118:51–57

Yadav OP, Rai KN (2011) Hybridization of Indian landraces and African elite composites of pearl millet results in biomass and stover yield improvement under arid zone conditions. Crop Sci 51:1980–1987

Yadav OP (2014) Developing drought-resilient crops for improving productivity of drought-prone ecologies. Indian J. Genet 74:548–552

Yadav OP, Bidinger FR (2008) Dual-purpose landraces of *pearl millet* (*Pennisetum glaucum*) as sources of high Stover and grain yield for arid zone environments. Plant Genetic Resources 6:73–78

Yadav OP, Weltzien-Rattunde E, Bidinger FR, Mahalakshmi V (2000) Heterosis in landrace-based topcross hybrids of *pearl millet* across arid environments. Euphytica 112:285–295

Yadav OP, Weltzien-Rattunde E, Bidinger FR (2003) Genetic variation in drought response of landraces of *pearl millet* (*Pennisetum glaucum* L. R. Br.). Indian J. Genet 63:37–40

Yadav OP, Bidinger FR, Singh DV (2009) Utility of pearl millet landraces in breeding dual-purpose hybrids for arid zone environments of India. Euphytica, 166:239–247

Yadav AK, Narwal MS, Arya RK (2011) Genetic dissection of temperature tolerance in *pearl millet* (*Pennisetum glaucum*). Indian J Agric Sci 81:203–213

Yadav AK, Narwal MS, Arya RK (2012a) Study of genetic architecture for maturity traits in relation to supra-optimal temperature tolerance in *pearl millet* (*Pennisetum glaucum* L. R.Br.). International Journal of Plant Breeding and Genetics 6:115–128

Yadav OP, Rai KN, Gupta SK (2012b) *Pearl millet*: genetic improvement in tolerance to abiotic stresses. In: Improving crop productivity in sustainable agriculture, pp 261–288

Yadav AK, Arya RK, Narwal MS (2014) Screening of *pearl millet* F1 hybrids for heat tolerance at early seedling stage. Adv Agric 2014:17

Yadav OP, Upadhyaya H, Reddy KN, Jukanti AK, Pandey S, Tyagi RK (2017) Genetic resources of pearl millet: status and utilization. Indian Journal of Plant Genetic Resources 30:31–47. https://doi.org/10.5958/0976-1926.2017.00004.3

Yadav RS, Hash CT, Bidinger FR, Devos KM, Howarth CJ (2004) Genomic regions associated with grain yield and aspects of post-flowering drought tolerance in pearl millet across stress environments and testers background. Euphytica 136:265–277

Yadav T, Kumar A, Yadav RK, Yadav G, Kumar R, Kushwaha M (2020) Salicylic acid and thiourea mitigate the salinity and drought stress on physiological traits governing yield in *pearl millet*-wheat. Saudi J Biol Sci 27:2010–2017

Yakubu H, Ngala AL, Dugje IY (2010) Screening of *millet* (*Pennisetum glaucum* L.) varieties for salt tolerance in semi-arid soil of northern Nigeria. World J Agri Sci 6:374–380

Yella Goud T, Sharma R, Gupta SK, Uma Devi G, Gate VL, Boratkar M (2016) Evaluation of designated hybrid seed parents of *pearl millet* for blast resistance. Ind J Plant Protect 44:83–87

Zaman QU, Li C, Cheng HT, Hu Q (2019) Genome editing opens a new era of genetic improvement in polyploid crops. The Crop Journal 7(2):141–150

Sorghum

2

Sunil Shriram Gomashe, Nandan Tayade,
and K. N. Ganapathy

Abstract

Sorghum [*Sorghum bicolor* (L.) Moench] is a major food and fodder crop in the semi-arid tropic and rainfed farming system. To counter the hunger and nutritional insecurity in the underdeveloped and developing nations, this crop has special importance in resilient agriculture. Sorghum crop is accomplished with tremendous plant genetic resources which include its traditional landraces, wild relatives and species, and different cultivars along with trait-specific registered genetic stocks. In the changing climate scenario and resource-limited farming systems, this genetic material of sorghum achieved a greater importance for the economic and conservative agriculture. In this context, this chapter provides the insights on origin, diversity, taxonomy, and biology of sorghum species. This also covers the existing genetic resources of sorghum with its conservation, evaluation, and characterization status. Potential utilization of the genetic resources in different ways for various determined aspects is also highlighted. Additionally, current trends in genomics research like molecular marker development, marker-assisted breeding, and next-generation sequencing have been deliberated as the future perspective in the sorghum improvement.

Keywords

Sorghum · Plant genetic resources · Genebank · Landraces · Germplasm · Utilization

S. S. Gomashe (✉) · N. Tayade
ICAR-National Bureau of Plant Genetic Resources, Regional Station,
Akola, Maharashtra, India
e-mail: sunil.gomashe@icar.gov.in

K. N. Ganapathy
ICAR-Indian Institute of Millet Research, Hyderabad, Telangana, India

K. Tripathi et al. (eds.), *Plant Genebank Utilization for Trait Discovery in Millets*, https://doi.org/10.1007/978-981-96-4043-0_2

2.1 Introduction

Sorghum [*Sorghum bicolor* (L.) Moench] is primarily a diploid ($2n = 2x = 20$) and self-pollinating species belonging to the family Poaceae with a genome size of 730 Mb. Sorghum, being a C4 plant having maximum efficiency of photosynthesis as well as high tolerance to abiotic stress, is adapted to a variable range of environments around the globe. For world's poorest and most food-insecure population, this crop serves as one of the key staple foods. Worldwide, rice, wheat, maize, and potatoes are the only other four food crops which get cultivated more than sorghum. Apart from human consumption, it is also used for feed and silage for animals, construction stuff, fencing, or brooms (Doggett 1988; House 1985; Rooney and Waniska 2000).

Sorghum has immense ability to cope with the detrimental condition which occurs in the hot and drought-prone agro-ecological zones specifically in semi-arid tropics (SAT) where cultivating other food crops is very difficult. It is mainly cultivated by farmers with limited resources and less input environment under conventional management conditions, therefore resulting in low productivity as compared to the United States and Mexico. Sorghum is grown both as a rainy and winter season crop in most of the parts of India. Biotic and abiotic constraints largely influence the yield and quality of the produce globally.

Regarding nutritional contents, sorghum grains are a rich source of starch and it is used in the pharmaceutical, food, textile, and paper industries. Grains of this crop are utilized as fodder for livestock in China, the United States, and South America, while in Asian and African continents, it is taken as human food (directly or after brewing). In contrast, the dried stalk is utilized for fodder, building materials, and fuel (Chandel and Paroda 2000). In western Africa and northern parts of India, sorghum is further cultivated primarily as fodder crop and is fed to animals as a silage and green chopped hay. Sweet sorghum is also used in formulating sorghum syrup and jaggery in India and as a biofuel-alcohol provider in Brazil. Green stems of sweet sorghum are sold for chewing in many places of Africa and India.

Sorghum therefore assumes prime importance in the economy and livelihood of people of many countries in Asia and Africa, where resource-limited farmers mostly inhabit. There may be a wide array of traditional, novel, and economically significant products produced from sorghum; the quality of products, such as ethanol, extrusion breakfast foods, etc., needs to be re-enhanced by utilizing the variability existing in the crop genetic resources.

2.2 The Crop

2.2.1 History and Origin

For a long period, sorghum as any plant seems relatively linked to humans and has evolved generously into a diverse type. Three fairly different categories can be recognized at present, the wild species, the cultivated forms, and the companion weeds of the cultivated forms.

The name sorghum appeared in literature after the New Latin word derived from Italian "sorgo" which literally means "Syrian grain." This crop has been grown in India since the prehistoric periods and there is a note of sorghum being cultivated in the region of Assyria as early as 700 BC. The origin and early domestication of sorghum is assumed to have started around 5000–8000 years ago in northeastern Africa or at the Egyptian-Sudanese border (Mann et al. 1983; Wendorf et al. 1992). There is hardly any doubt about the African origin of sorghum, as the greatest diversity of wild and cultivated sorghum types is present there (de Wet 1977; Doggett 1988; Kimber 2000), hence revealing Ethiopia as a center of origin of sorghum. The secondary center of origin belongs to the Indian subcontinent, as proof for early cereal production and cultivation detected at an archaeological site in western parts of Rojdi (Saurashtra) dating back to about 4500 years before 1950 A.D. (Vavilov 1992; Damania 2002).

2.2.2 Global Perspective of Sorghum

Globally sorghum is cultivated in almost 110 different countries, basically in Africa and Asia as well as in Europe, America, and Oceania (Popescu et al. 2018). Presently, the world's largest absolute sorghum produce (Table 2.1) is from the United States; then the Federation of Nigeria, India, and Mexico have a mediocre worldwide generation of 50 megatons annually (FAOSTAT 2019). For almost 500 million people in more than 30 different countries, exclusively in sub-Saharan Africa, sorghum is considered as the main staple food crop (Abreha et al. 2021). Presently there is a remarkable decrease in the area under sorghum cultivation and in grain yield in some major sorghum-producing countries. Nearabout 34% of the global cultivated

Table 2.1 World's major sorghum-producing countries during 2023–2024

Sr. no.	Country	Production*	Sr. no.	Country	Production
1	United States	8071	13	Mali	1500
2	Nigeria	6700	14	Cameroon	1200
3	Sudan	5000	15	Bolivia	1000
4	Mexico	4800	16	Tanzania	1000
5	India	4400	17	Chad	950
6	Ethiopia	4400	18	Egypt	750
7	Brazil	3600	19	South Sudan	750
8	China	3000	20	European Union	670
9	Argentina	2500	21	Ghana	350
10	Burkina Faso	1900	22	Senegal	330
11	Niger	1900	23	Togo	280
12	Australia	1800	24	Yemen	240

*Production in 1000 metric tons
Source: www.world agriculturalproduction.com/crops/sorghum.aspx

area is represented by India and Sudan which jointly account for 17.6% of the total world sorghum's production (FAO 2022). The decreasing universal yield of sorghum, with respect to other crops like rice and wheat, conceivably accelerated through capitalization of plants' inherent assets and substantial genetic material of undomesticated crop (Sasaki and Antonio 2009). The use of sorghum's genetic resources is crucial for developing improved sorghum cultivars.

2.2.3 Taxonomy of Sorghum

"Sorghum" which is categorized in the Poaceae family with Andropogoneae as tribe and Sorghinae subtribe (Clayton and Renvoize 1986) primarily is related to domesticated sorghum (*Sorghum bicolor* [L.] Moench subsp. *bicolor*); also it is predominantly cultivated for its grain (grain sorghum) and also for producing sweet or sugar syrup (sweet sorghum) or as an animal feed (forage sorghum). Sorghum has been classified into different basic and intermediate races based on its variation in floral biology and morphophysiological forms.

Taxonomically, in 1753 Carl Linnaeus first attempted to characterize sorghum under the name *Holcus*. The name sorghum was used as an alternative for *Holcus* by Adanson. Following this, Moench separated the genus *Sorghum* from the genus *Holcus* and then combined them, forming the *Sorghum bicolor*. The taxonomic classification settled by Moench matches the existing formal taxonomic approach. The genus *Sorghum* was classified into different races by Brotero (1804), Chiovenda (1912), Piper. (1915), and Stapf (1917). Snowden attempted the most specific classification of sorghum. Thirty-one cultivated and 17 wild-related species were described by Snowden who also stated 48 distinct types are collectively characterized by several noticeable traits. Cultivated sorghum is derived from its primordial form and classified into five major or basic (*bicolor*, *guinea*, *caudatum*, *kafir*, and *durra*) and ten intermediate or hybrid races, based on panicle architecture and spikelets; it constitutes all combination hybrids of the dominant races as described by Harlan and De Wet (1972). Based on mature spikelet alone, 15 races of domesticated sorghum can be distinguished easily, although panicle type is sometimes useful. The classification is based on five basic spikelet types (Table 2.2), races *bicolor*, *guinea*, *caudatum*, *kafir*, and *durra*, and is detailed in Harlan and De Wet (1972). This is backed by phylogenetic evaluation applying molecular markers (Tao et al. 1993; Deu et al. 1994; Cui et al. 1995; Deu et al. 2006; Perumal et al. 2007; Wang et al. 2013) although this is not always the case (Ritter et al. 2007). The following description of these races is predicated on the classification given by Harlan and De Wet (1972), De Wet (1978), and Dahlberg (2000).

On the basis of morphological characters like glumes and seed coloration along with endurance of pedicellate spikelets, the crop was categorized into two prime components as stated in Snowden's classification—*Eu-sorghum* and *Parasorghum* (Snowden 1955). Nevertheless, in sorghum five subgenera are presently recognized:

Table 2.2 Cultivated sorghum race characteristics

Sr. no.	Race	Distinct characteristics
1	*Bicolor* (B)	Open inflorescences (panicle) with pendulous branches, long, clasping glumes, elliptic grain
2	*Guinea* (G)	Large, open inflorescences with pendulous branches, long, separated glumes that expose grains, obliquely twisted grains
3	*Caudatum* (C)	Compact to open inflorescences, grains with one side flat, opposite side curved, shorter glumes that expose grains
4	*Kafir* (K)	Moderately compact, cylindrical inflorescences, elliptic spikelets, tightly clasping, long glumes
5	*Durra* (D)	Compact inflorescences, flat, ovate-shaped sessile spikelets, middle-creased lower glume, distinct texture on tip of lower glum

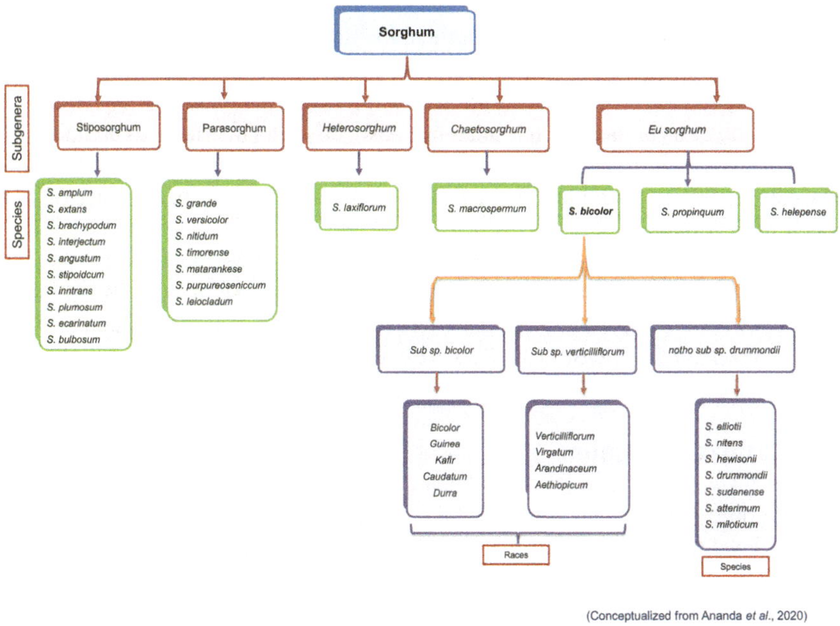

Fig. 2.1 Classification of *Sorghum* into different races

Stiposorghum, *Parasorghum*, *Heterosorghum*, *Chaetosorghum*, and *Eu-sorghum* (Harlan and De Wet 1972; De Wet 1978; Lazarides et al. 1991), differentiated on the basis of their morphological characters (Fig. 2.1). Regardless of its cultivation in East Africa, 17 sorghum species are found native to Australia, of which, 13 species were identified as endemic, accelerating the commitment to conserve crop's wild relatives. Except *Eu-sorghum*, native Australian species includes all sorghum subgenus.

2.2.3.1 Eu-sorghum

The genus *Sorghum* has *Eu-sorghum* as a vital component among two sections. Southern Asia and Africa are the major cultivating regions for this subgenus (Price 2005). According to existing authorized distribution, *Eu-sorghum* is treated as "true sorghum" and encompasses three species which are *S. halepense*, *S. bicolor*, and *S. propinquum*, along with a derived species called, *Sorghum* × *almum* Parodi. Majority of domesticated sorghum is included in *Sorghum bicolor*. Furthermore, the open type of inflorescence along with non-pendulous branching ending at the lateral part differentiated it from other species. *Sorghum bicolor* was divided into three distinct subspecies: subsp. *bicolor* (complete domesticated sorghums), subsp. *verticilliflorum* (wild ancestors of domesticated sorghums), and nothosubsp. *Drummondii* (Steud.). The subspecies *verticilliflorum* (formerly known as *arundinaceum*) comprises four different wild progenitor races which are *aethiopicum*, *arundinaceum*, *virgatum*, and *verticilliflorum*. Seven different weedy races have been recognized, on the basis of Snowden's (1955) classification, in nothosubsp. *Drummondii* (again identified as "Sudan grass"); it is generally taken as fodder to animals. Presently acknowledged races of subspecies *bicolor* are *bicolor*, *guinea*, *caudatum*, *kafir*, and *durra* on the basis of their panicle anatomy. Also, other ten hybrid races are included in this subspecies (Lazarides et al. 1991). Grounded on molecular confirmation, Mace et al. (2013) again identified *guinea-margaritiferum* as a specific race of subspecies *bicolor*.

A wild perennial, diploid rhizomatous species called *Sorghum propinquum* is widespread in Indian subcontinent and also in Southeast Asia which has a peculiar feature of shorter spikelets. Another tetraploid rhizomatous, perennial species named *Sorghum halepense*, called as "Johnsongrass," is prevailing in India and Southern Eurasia and is also identified as wild relative.

2.2.3.2 Chaetosorghum

The exclusive representative of subgenus *Chaetosorghum* is *S. macrospermum* E. D. Garber. This species is diploid ($2n = 40$) and annual with a sessile, tiny spikelet, with oval to elliptical-shaped caryopsis (Lazarides et al. 1991).

2.2.3.3 Heterosorghum

Heterosorghum member *Sorghum laxiflorum* F. M. Bailey, the only species in this subgenus , is native to New Guinea and Australia and has ancient origins (Price 2005).

2.2.3.4 Parasorghum

The genus *Parasorghum* comprises species which are *S. matarankense* E. D. Garber & Snyder, *S. grande* Lazarides, *S. leiocladum* (Hack.) C. E. Hubb., *Sorghum nitidum* (Vahl) Pers., *S. purpureosericeum* (Hochst. ex A.Rich.) Schweinf. & Asch, *Sorghum versicolor* Andersson, and *Sorghum timorense* (Kunth) Büse. All the seven *Parasorghum* species are indigenous and are native to Australia (Lazarides et al. 1991) except *Sorghum purpureosericeum* and *Sorghum versicolor*.

2.2.3.5 Stiposorghum

A total of ten species which all primarily originated in Australia are incorporated in the subgenus *Stiposorghum*: *S. amplum* Lazarides, *S. brachypodum* Lazarides, *S. angustum* S. T. Blake, *S. ecarinatum* Lazarides, *S. intrans* F. Muell. ex Benth., *S. bulbosum* Lazarides, *S. plumosum* (R. Br.) P. Beauv., *S. stipoideum* (Ewart & Jean White) C. A. Gardner & C. E. Hubb, *S. exstans* Lazarides, and *S. interjectum* Lazarides.

The wild lines of sorghums from all the domesticated races are firmly associated with race *bicolor*. This race is considered to be ancient within the five races. Crossing grain-type sorghum with Sudan grass (*S. bicolor* [L.] Moench ssp. *drummondii*) results in repeated production of *bicolor* sorghums (Harlan and De Wet 1972). The heterogeneous race *bicolor* includes various subraces, namely, sorgo, broomcorn, and Sudan grass. In the breeding of forage crop, *bicolor* race has been of great importance (Kamesawara Rao et al. 2004).

Contrary to race *bicolor*, cylindrical and compact spikelets are observed in race *kafir*. In eastern and southern savannah from Tanzania to South Africa, *kafir* sorghums are considered as major staple food. Hybrid races like *kafir-caudatum* in the United States, *guinea-kafir* in India, and *kafir-durra* which are included in race *kafir* are well considered to be vital genetic stocks and can be utilized as important breeding materials. Cytoplasmic male sterility (CMS) system can be adopted to obtain hybrid seed with *durra* cytoplasm and derivatives of race *kafir* as maintainers. As race *kafir* is photoperiod insensitive, it can acclimatize to a temperate environment as compared to other races (Grenier et al. 2001). Open-type panicles are common in regions with higher rainfall, whereas in drier areas, compact spikelets are common. It is commonly cultivated in Uganda, northeast Nigeria and Sudan. In terms of hybrid races, maximum current American hybrid grain sorghums are *kafir-caudatum*. For the large seed size and yellow endosperm, *durra-caudatum are* identified as a donor source. Compact panicles having ovate plane sessile spikelets were found to be the characteristic of race *durra*. *Durra* sorghums are commonly cultivated along the surroundings of West Africa, the southern Sahara, and some parts of India. Some of the most moisture stress-tolerant sorghums which can be cultivated in the moisture-limited regions are included in this race.

Grain repelling at maturity from a panicle with long and gaping glumes is the main descriptor of the race *guinea*. The race bears open and long panicles, and the branches are relatively drooping at maturity. Most commonly it is grown in African regions where the bold-seeded *guinea* grains are cultivated in the water-limited zone, while small-seeded are grown in the rainfed regions. The occurrence of mold damage under wet conditions seems minimal because of its hanging and open panicles with gaping glumes. In temperate regions, race *guinea* is more possibly to be sensitive to photoperiod (Grenier et al. 2001). Also, resistance to grain mold was recorded in the *guinea* race belonging to West Africa (Kamesawara Rao et al. 2004). In an investigation through SSR marker studies, it is clear that the *guinea* sorghums were disseminated in South Asia from West Africa (Folkertsma et al. 2005).

Probably from the earliest domesticated sorghums, especially *bicolor* sorghum, The existing cultivated races of sorghum are believed to have been derived probably from the earliest domesticated sorghums, especially bicolor sorghum. The region between Nigeria and Sudan, where the wild race *verticilliflorum* is abundantly present in the form of wild grass, is considered the first sorghum-cultivating region. Later, regions like tropical West Africa and Southeast Africa began its cultivation. The *guinea* race evolved due to selection efforts for tolerance to a wet tropical habitat. *Durra* sorghums are usually considered to have evolved farther in Africa. *Caudatum* sorghums are confined to the primary region of *bicolor* domestication.

2.2.4 Botanical Description and Floral Biology

Sorghum is an annual, tall plant that is monoecious, meaning it has both male and female reproductive structures. It produces few to many tillers from the stem nodes or base. Its roots are mainly concentrated near the soil surface but can grow deeper and spread laterally. The stem is upright and solid, with alternate leaves that are simple and tall. The leaf sheaths are usually waxy, with short white hairs near the base. The plant produces a panicle (flower cluster) at the top, emerging from the flag leaf sheath during the heading stage. The panicle can be long and loose or compact and open. Its branches are arranged in primary, secondary, and sometimes tertiary forms, bearing paired spikelets and groups of three spikelets at the edges of branches. Each spikelet can be sessile (attached directly) or pedicelled (on a small stalk). Sessile spikelets are usually long and have glumes (outer bracts) of equal length. The upper glume is shallow and has a keel (ridge), while the lower glume is hairy with vein-like ridges. Each spikelet contains two florets: the lower floret has only a lemma (a bract), and the upper floret is bisexual, with a cleft lemma, a twisted awn (bristle), a tiny palea (a small bract), three stamens, two lodicules (small scales), a superior ovary, and feathery stigmas with two long styles. Sessile spikelets are less persistent than pedicelled ones (Fig. 2.2). The grain, called a caryopsis, is partially enclosed by the glumes. It contains a round, blunt embryo and a seed coat and endosperm made up of the testa and pericarp. When the seeds germinate, they produce coleoptiles and roots.

2.2.5 Erosion of Genetic Diversity from the Traditional Areas

The practice of modern agriculture can negatively impact the genetic diversity of crop plants, as it adopts improved commercial crop varieties to maximize food production. Modern varieties get promoted for large-scale cultivation due to their high-yielding potential.

Shewayrga et al. (2008) pointed out that, in the past 30 years, several significant sorghum landraces have vanished either regionally or locally. In Ethiopia, primary factors accelerating genetic erosion include the replacement of landraces by alternative crops, the influx and extension of upgraded varieties, an increase in the production of economical crops, the inadequacy of a structure to resupply seeds of

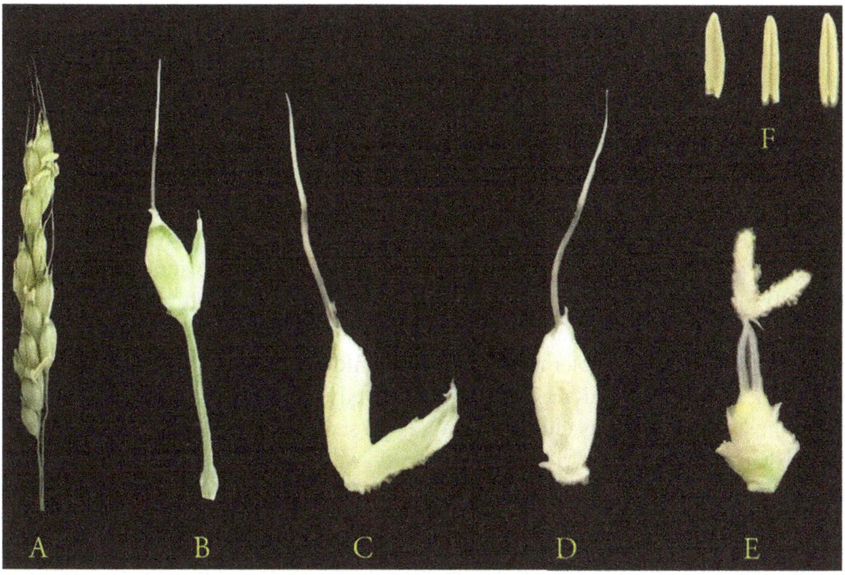

Fig. 2.2 Sorghum floral biology. (A, rachis; B, raceme; C, sessile spikelet, awn, and pedicellate spikelet; D, sessile spikelet; E, ovary and stigma; F, anthers)

landraces, marginal land holdings, continuously evolving land use and cropping patterns, and a lack of policy support.

For example, in India, an estimated 400,000 rice varieties were grown before colonialism; this number dropped to 30,000 in the mid-nineteenth century (Heal et al. 2004), and just 10 varieties of rice now occupy 75% of the land. Similar extinction of landraces due to modern rice varieties was also noted in Sri Lanka (Hargrove et al. 1988; Rhoades 1991), Indonesia, and Bangladesh (Hargrove et al. 1988). This loss of crop landraces is particularly severe in developed nations. Landraces have become almost extinct due to the widespread adoption of modern varieties in North America and northwestern Europe (Van de Wouw et al. 2010). The loss of landraces is negligible in places where there is no invasion of modern varieties. For example, in southern Africa, modern high-yielding cultivars were planted on only 14% of the total sorghum-growing area based on 1995 data (Maredia et al. 2000).

Similarly, in Niger, Deu et al. (2006) surveyed sorghum varietal changes in 71 villages using 28 SSR markers and found no major loss of sorghum landraces from 1976 to 2003. To estimate genetic erosion in sorghum landraces, Mekbib (2008) conducted interviews with farmers and compared sorghum collected in 1960 and 2000 in Ethiopia. It was observed that, at the farmer level, the four most crucial factors for landrace loss were decreased advantages from the landraces, abiotic stress (drought), reduction in land size, and adaptation of other crops. There was no rivalry but a complementation observed between improved varieties (IVs) and farmer varieties (FVs). Teshome et al. (2007) also surveyed sorghum varietal changes in

Ethiopia by interviewing farmers over an 8-year period (1992/1993 to 2000/2001). They found that, over this period, the total area planted to sorghum decreased drastically in all five farming communities studied and 51–72% of the farmers in these communities reduced the field size planted to sorghum due to increases in population, redistribution policies of land, seasonal shifts, and stagger cropping followed by interspecies crop rearrangement.

Among the developed countries, Australia also experienced sorghum landrace loss, where cultivars became extinct due to the widespread cultivation of hybrid sorghum, which was known for its resistance to sorghum midge and occupied more than 80% of the sorghum-cultivating area before 1995 (Jordan et al. 1998). The primary objective of Australian sorghum improvement work was the selection of sorghum lines showing tolerance to sorghum midge, and the relationship between resistance and genomic divergence was analyzed using RFLP assessments among 26 grain sorghum hybrids cultivated on a large scale in Australia. The results indicated that the narrowing of genetic diversity was purely due to the adoption of midge-resistant hybrids. This decrease in genetic diversity poses a serious threat regarding the genetic vulnerability of sorghum landraces and cultivars, as well as breeding for sorghum yield and improvement (Jordan et al. 1998).

2.3 Sorghum Germplasm Conservation Status

2.3.1 Importance and Need for Conservation

Plant genetic resources (PGR) can be defined as the "genetic material of crops or plants that is valuable as a resource for the current and future generations" (IPGRI 1993). The significance of PGR was recognized at an intergovernmental platform under the United Nations' Food and Agriculture Organization (FAO), and it was considered the "common heritage of mankind," which must be made accessible without any conditions (FAO 1983). The process of domestication, diversification, intensification, as well as advancement through deliberate or random selection by numerous generations of farmers is responsible for the evolution of genetic resources. These PGR provide essential and critical resources for the improvement of crops for both present and future generations (Rai 2002).

2.3.2 Comparative Status of Germplasm Resource Conservation

Sorghum germplasm collections vary in quantity and type across different nations and organizations of the world. Currently, the International Crops Research Institute for the Semi-Arid Tropics (ICRISAT) ranks first in storing the global collection of sorghum germplasm, with 42,869 representative accessions augmented from 92 nations (Table 2.3). The germplasm repository maintained at ICRISAT is estimated to represent about 80% of the genetic diversity available in sorghum (Eberhart et al. 1997). These accessions include landraces, cultivars, advanced breeding lines, and

Table 2.3 Sorghum germplasm conserved at different genebanks in the world

Country	Institute/organization	Total collection
Australia	Australian Tropical Crops and Forages Genetic Resources Centre (ATCFA)	4491
Brazil	Centre for Genetic Resources Preservation, Embrapa Milo et Sorgho (CNPMS), Embrapa Recursos Geneticos e Biotechnologia	10,812
China	Institute of Crop Science, Chinese Academy of Agricultural Sciences	18,263
Ethiopia	Institute of Biodiversity Conservation (IBC)	9772
France	Laboratoire des Ressources Génétiques et Amélioration des Plantes Tropicales, ORSTOM, Centre de Coopération Internationale en Recherche Agronomique pour le Développement (CIRAD)	7305
India	International Crops Research Institute for the Semi-Arid Tropics	39,553
India	ICAR-National Bureau of Plant Genetic Resources, Indian Institute of Millets Research	20,566, 23,086
Japan	National Agriculture and Food Research Organization (NARO)	5074
Kenya	National Gene Bank of Kenya, Crop Plant Genetic Resources Centred Muguga (KARI-NGBK)	5866
Mexico	Programa de Recursos Geneticos, Centro de Investigaciones Forestales y Agropecuarias (CIFAP-MEX), Estacion de Iguala, Instituto Nacionale de Investigaciones Agricolas (INIA-Iguala)	5500
Pakistan	Plant Genetic Resources Programme (PGRP)	1732
Russian Federation	NI Vavilov All-Russian Scientific Research Institute of Plant Industry	3963
The United States	Plant Genetic Resources Conservation Unit, Southern Regional, Introduction Station, University of Georgia, National Centre for Genetic Resources Preservation	43,710
Zambia	SADC Plant Genetic Resources Centre (SRGB)	3719

Source: Aruna et al. (2018)

wild or weedy relatives (Upadhyaya et al. 2014b). Among the five primary sorghum races, the ICRISAT genebank's collection is dominated by *durra* (21.21%), followed by *caudatum* (20.12%), *guinea* (12.89%), *bicolor* (4.59%), and *kafir* (3.49%) (Upadhyaya et al. 2014b). Following ICRISAT, the United States conserves 42,221 sorghum germplasm accessions in its National Plant Germplasm System (NPGS) (Dahlberg and Spinks 1995). In China, through extensive exploration and collection of landraces, 12,836 sorghum lines were gathered, with 10,414 of them conserved in the National Genetic Germplasm Resources Bank (Qingshan and Dahlberg 2001).

In the 1960s, the Indian Council of Agricultural Research (ICAR) and The Rockefeller Foundation's agricultural research program initiated the first major effort to collect global sorghum germplasm. A total of 22,701 foreign accessions were introduced into India, representing diverse traits from various countries. Researchers and collectors conducted multiple exploration tours across India, yielding significant collections. National Genebank of ICAR-NBPGR (National Bureau of Plant Genetic Resources) conserves 26,370 germplasm accessions of indigenous sorghum collected from across India. Additionally, sorghum germplasm lines with specific traits introduced from other countries are also preserved in the same genebank. Many of these lines possess desirable traits, such as resistance to insect pests, diseases, abiotic stresses, and male sterility. Genetic resources are fundamental for crop improvement and research. In situ conservation involves preserving germplasm in its natural habitat, while ex situ conservation involves storing germplasm outside its natural habitat, such as in genebanks (Upadhyaya et al. 2014b). The most effective method for conserving sorghum germplasm globally is through ex situ conservation. According to FAO (2009), global sorghum collections comprise 235,711 accessions.

2.3.3 Gene Pools

Sorghum (cultivated) ($2n = 2x = 20$), a diploid species, possesses 200 classified phenotypic, genotypic, and cytogenetic trait genes, making it genetically diverse (Rooney and Waniska 2000). It is sexually compatible with some of its wild and weedy relatives, with primary and secondary gene pools determined by the extent of cross-compatibility. The primary gene pool is found within the *Eu-sorghum* section and includes diploid species such as *S. bicolor subsp. verticilliflorum* and *S. propinquum*, and shattercane. Crosses within this gene pool yield entirely sterile progeny. In the primary gene pool, spontaneous outcrossing and high fertility levels can lead to recurrent introgression when distributions overlap under suitable conditions (Doggett and Majisu 1968; Baker 1972; Ejeta and Grenier 2005).

The secondary gene pool includes tetraploid species ($2n = 4x = 40$) of *Eu-sorghum*, such as Columbus grass (*Sorghum almum Parodi*) and Johnsongrass. Despite differences in ploidy levels, domesticated sorghum can outcross with members of the secondary gene pool, producing partially fertile tetraploids or infertile triploids (Arriola and Ellstrand 1997; Morrell et al. 2005). The tertiary gene pool comprises species from other sections of *Sorghum*. Under natural conditions, outcrossing between cultivated sorghum and tertiary gene pool members is highly improbable. Crosses formed through artificial means are typically abnormal, infertile, or lethal (Ejeta and Grenier 2005). However, controlled conditions using embryo rescue techniques have enabled successful crosses with certain Australian *Sorghum* species (Price 2005).

The genome of cultivated sorghum has been sequenced (Paterson et al. 2009). Its haploid genome size is approximately 730 megabase pairs (Mbp), larger than rice (510 Mbp) and *Arabidopsis* (155 Mbp). Molecular studies have identified unique traits and variations in endosperm composition and structure among sorghum relatives, which may help expand its gene pool. Managing gene flow between wild or weedy relatives and cultivated sorghum is vital. Traits like male sterility and maturity are particularly significant. Genes influencing maturity, as well as their interactions with temperature and day length, are crucial for floral initiation and reproductive success in cultivated sorghum. Additionally, genes affecting male sterility can influence outcrossing capacity. However, it is important to note that drought or high temperatures can alter these traits, potentially causing blooming or early flowering under atypical conditions. Self-pollination and seed set may also occur under male-sterile conditions.

2.4 Genetic Resources: Evaluation, Characterization, and Documentation

For the successful utilization and conservation of germplasm, it is requisite that it should be evaluated accurately, characterized, and documented against an applicable improvement system so that accessions having any pertinent traits can be efficiently regained and utilized in crop development programs (Gebrekidan 1982).

In India, the ICAR-Indian Institute of Millets Research (IIMR), Hyderabad, has been engaged in the collection and conservation of millet germplasm (>53,000) including pearl millet, sorghum, and other six small millets, in which sorghum crop tops the table with >33,000 germplasm accessions. The collaborative evaluation study conducted by ICAR-IIMR and ICAR-NBPGR aimed at characterizing the indigenous and exotic sorghum germplasm for qualitative and quantitative traits and defining trait-specific germplasm lines for its incorporation in breeding programs of sorghum.

A total of 2834 accessions were extensively characterized and evaluated for vital traits during 2018–2019. Earhead shape was observed as most diverse trait based on the calculated Shannon diversity index. Several trait-specific accessions for early flowering, high yield, etc. were identified. Character association analysis by Pearson highlighted that grain yield in sorghum was influenced in positive manner by the traits like earhead width, plant height, number of basal tillers, and 100-seed weight. The trait-specific germplasm accessions identified in the study could be used as trait-specific donors by breeders for sorghum crop improvement (Elangovan et al. 2022).

Similarly, 29,180 accessions of sorghum were screened at ICRISAT, for 23 significant agronomic attributes selected from the descriptor list maintained for sorghum (IBPGR-ICRISAT 1980) during *kharif* and *rabi* seasons. Cultivated sorghum races as well as sorghum wild forms possess high amount of diversity depicting high magnitude of variation with special distinctive features as such they are different species (Prasada Rao et al. 1995). A program called ICRISAT Data Management Retrieval System (IDMRS) was applied for extensive screening of the accessions and passport data was documented (Prasada Rao et al. 1995). Evaluated sorghum accessions and germplasm maintained at ICRISAT recorded to have resistance for biotic and abiotic factors. There were 21,661 sorghum accessions conserved to USDA-ARS S-9 location at Griffin which constitute nearabout 50 percent of entire US accessions of sorghum germplasm which were evaluated and characterized for 39 agro-morphic traits (Dahlberg and Spinks 1995).

ICAR-NBPGR (National Bureau of Plant Genetic Resources), under the Consortium Research Platform on Agro-biodiversity, conducted characterization and evaluation of entire sorghum germplasm as part of the Celebration of

Fig. 2.3 Sorghum mega-characterization program at the Agricultural Research Station, Washim (Maharashtra), India, during *rabi* 2022–2023

International Year of Millets 2023 (IYoM2023). Experimental trial was conducted during rabi 2022–2023 at the Agricultural Research Station, Washim, Dr. Panjabrao Deshmukh Krishi Vidyapeeth, Akola, Maharashtra. Experimental material consisted of a total of 24,950 sorghum germplasm accessions from indigenous and exotic collections from 51 different countries and 26 states and 2 union territories in India (Fig. 2.3). These materials include landraces (6262 accessions), germplasms (5429), breeding lines (1471), registered genetic stocks (104), parental lines (43), primitive cultivars (43), traditional cultivars (28), hybrids (10), and wild relatives (10). A wide variability for agro-morphological traits was recorded in sorghum germplasm and spike color and shape variability is depicted in Fig. 2.4.

2.4.1 Germplasm Evaluation and Maintenance

Sorghum, commonly cultivated in semiarid tropics, exhibits extensive genetic variation. To harness this genetic diversity, germplasm evaluation is essential in crop improvement programs. It is, therefore, crucial to assess genetic diversity based on qualitative and quantitative traits and isolate superior accessions with desirable characteristics for breeding programs. Plant growth, agronomic behavior, yield, and responses to various stresses are critical factors that determine the evaluation of germplasm. Generally, germplasm evaluation is carried out by germplasm holders in collaboration with plant breeders, pathologists, entomologists, and agronomists. The effectiveness of evaluation methods determines the inherent potential of the germplasm. Engaging the farming community in germplasm evaluation and considering their opinions and expectations is advantageous, as farmers possess significant traditional knowledge and directly get benefit from the outcomes of evaluation and breeding programs.

Although sorghum is a predominantly self-pollinated crop, it demonstrates varying degrees of outcrossing (Pedersen et al. 1998; Barnaud et al. 2008). Selfing is practiced for germplasm multiplication and maintenance. During selfing, panicles are enclosed for at least 21 days (until the dough stage) and then opened, but the selfing bags remain around the peduncles to identify selfed plants (Upadhyaya et al.

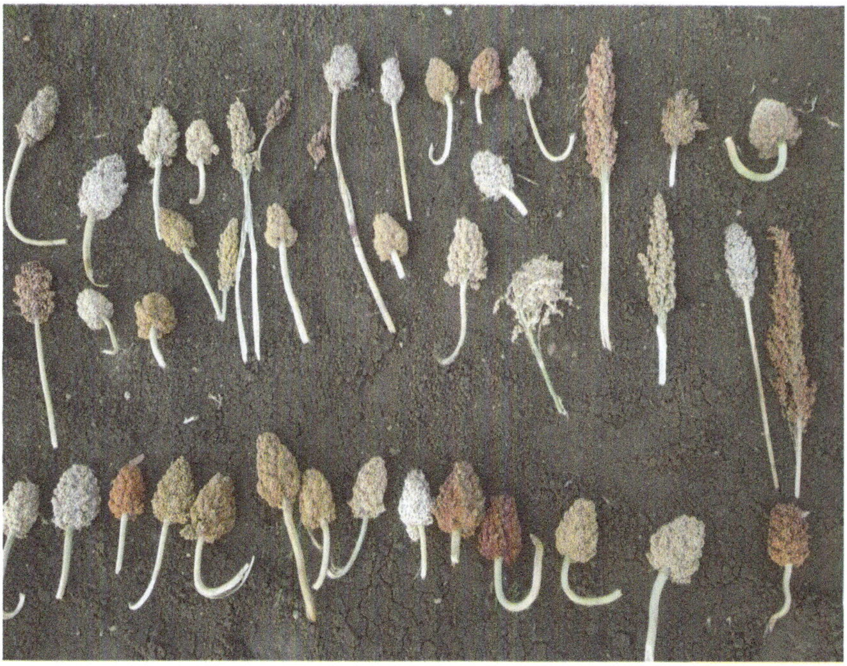

Fig. 2.4 Spike variability in sorghum germplasm

2014b). To maintain an accession, seeds from 50 selfed plants are bulked. For wild relatives, selfing is also practiced; however, some wild species fail to produce seeds. In such cases, these species are maintained in live form in a field genebank (Upadhyaya et al. 2014b).

2.4.2 Germplasm for Crop Improvement

A wide range of genetic diversity in sorghum is found in the Ethiopian-Sudan border region, recognized as the site of the first sorghum domestication. The Asian region also harbors significant sorghum variability due to its early introduction. These centers of diversity include local landraces and wild forms of domesticated sorghum, which serve as valuable sources of resistance or tolerance to insect pests, diseases, and various stresses. Traits that enhance grain yield, nutritional and fodder quality, and industrial applications can also be sourced from this germplasm. However, factors such as diminishing natural habitats, widespread adoption of improved cultivars, and commercial agricultural practices pose threats to this genetic diversity. To prevent the loss of primitive cultivars and wild species, efforts to explore, collect, and conserve sorghum germplasm have intensified globally. Consequently, germplasm collection and conservation have become integral to sorghum improvement programs at both national and international levels (Reddy et al.

2006). Plant breeders typically utilize only a fraction of the available sorghum germplasm, focusing on materials with distinct and useful traits of current significance (Gebrekidan 1982). Initially, pure line selection was the primary method for utilizing sorghum germplasm. Later, dwarf populations were isolated through selection, followed by the development of agronomically superior hybrids using cytoplasmic male sterility (Dahlberg et al. 1997). Advanced self-pollinated varieties and cultivars have been derived through direct or reciprocal hybridization of local germplasm with domesticated introductions (Prasada Rao et al. 1989). Important traits, such as long panicles, increased seed numbers per panicle, higher biomass, disease resistance, moisture stress tolerance, greater plant height, early maturity, and efficient dry matter partitioning, significantly impact yield improvement (Miller and Kebede 1984).

Most countries now have facilities to explore, conserve, characterize, and utilize plant genetic resources for various crops. Notable institutions managing sorghum germplasm include the National Bureau of Plant Genetic Resources (NBPGR) in India, the International Crops Research Institute for the Semi-Arid Tropics (ICRISAT), and the National Plant Germplasm System (NPGS) in the United States, along with countries like Ethiopia, Sudan, South Africa, and China (Reddy et al. 2006). ICRISAT holds the world's largest collection of sorghum germplasm, accounting for approximately 22% of the global total and representing about 80% of sorghum's genetic variation (Eberhart et al. 1997). This collection primarily comprises landraces (85.3%), followed by breeding material (13.2%), wild species accessions (1.2%), and modern cultivars (0.3%). The germplasm includes five primary races—*durra*, *kafir*, *caudatum*, *bicolor*, and *guinea*—along with ten intermediate races. However, *durra* (23.5%), *caudatum* (20.6%), and *guinea* (14.8%) dominate the collection. Ethiopia, Sudan, India, Zimbabwe, and Uganda also house diverse sorghum races and hybrids (Reddy et al. 2002).

The exchange and utilization of plant genomic resources are ongoing processes, benefiting agriculture worldwide. Emergencies like climate change, disease outbreaks, pest infestations, and the rising demand for food and nutritional security have necessitated the continuous search for novel genetic resources (De Jonge 2009). In India, the Indian Institute of Millets Research (IIMR) focuses on germplasm exploration, collection, evaluation, and characterization. Notable efforts include collecting germplasm from drought-prone regions of Maharashtra, such as Sangli and Solapur, to identify materials with moisture stress tolerance. Germplasm from these areas exhibits traits such as moderate to tall stature, stay-green characteristics, compact earheads, and lustrous white grains, making them suitable for drought conditions.

Genetic diversity within and among species allows plant breeders to select desirable variants for genetic stock development, hybridization programs, or direct variety improvement (Bhandari et al. 2017). Landraces harbor more genetic diversity than modern cultivated varieties, which are optimized for specific traits such as high yield, stress tolerance, and nutritional enhancement. Breeders use diverse parental materials to develop new varieties (Duvick 1986). Wild species and landraces, rich in untapped genes and alleles, are essential for enhancing modern sorghum varieties by introducing resistance to various stresses (Wilkes 1991). Sorghum germplasm utilization in breeding programs typically involves (i) generating pre-breeding materials for conventional breeding, (ii) developing genetic stocks for resistance to biotic and abiotic stresses and quality improvement, (iii) identifying and characterizing germplasm for hybrid development and male sterility induction, and (iv) incorporating genes of interest into popular cultivars to create modern varieties (Salgotra et al. 2012).

Utilizing the collected germplasm directly as varieties is the most efficient way to use the germplasm. About 34 germplasm lines from ICRISAT have been directly proposed as varieties in 17 different countries, with some of these released in more than 1 country (Upadhyaya et al. 2014b). The collected accession, IS 33844, is an excellent *Maldandi* type with large and lustrous grains and high yield potential. An example of germplasm utilization in sorghum has been the selection practiced in the accession IS 33844 which led to "Parbhani Moti" a released variety for the post-rainy season in Maharashtra, India (Reddy et al. 2006). At NBPGR, New Delhi, a total of 95 different trait-specific sorghum germplasm lines have been registered through germplasm registration system, by different developers and institutes (Table 2.4). These lines have single or multiple novel unique features; therefore further improvement in sorghum crops for various abiotic and biotic features can be accelerated by involving these registered germplasm lines.

2.4.3 Constraints in Utilization of Germplasm and Means to Overcome

Breeders generally avoid use of new collected germplasm and prefer to use their own working collections, largely due to nonavailability of reliable information on traits of economic importance (Upadhyaya et al. 2014a), fear of carrying linkage drag, and possibly due to breakdown of coadapted gene complexes (Ortiz et al. 1998). Other factors affecting the minimal use of genetic resources include (i) shortage of well-defined and time-bound commercial multi-environmental testing of germplasm and (ii) lack of robust, cost-effective tools to facilitate the efficient utilization of non-indigenous material in crop breeding (Dwivedi et al. 2009).

Absolutely, it is now the demand of time to accelerate the utilization of available genetically diverse germplasm with desirable traits to fulfill the current challenges in agricultural system. Curtailing the large collection in terms of core (~10% of total collection) or mini core (~1% of entire collection or ~10% of core collection) collections, defining divergence of the total collection of a particular species in the genebank, will be beneficial as a key factor to review population arrangement and variation, identify novel sources of diversity, and discover agronomically profitable and genetically variable germplasm for incorporation in crop enhancement programs. Reduced sets like mini core (Frankel 1984) and core (Upadhyaya and Ortiz 2001) collections, representing divergence in the total collection of a given species, were observed as a setback to augment implementation of PGR in crop advancement programs.

Despite a diverse spectrum of genetic resources which are accessible nationally and globally, researchers are more expected to rely only on adapted accessions,

Table 2.4 Trait-specific germplasm registered in sorghum from 2020

Sr. no.	National identity	Year	INGR number	Novel unique features
1	IC343586	2020	INGR20084	Drought tolerant
2	IC485033	2020	INGR20085	Low HCN content. High protein yield. High seed yielding single-cut forage genotype (dual-purpose type)
3	IC635700	2020	INGR20086	Resistant to foliar diseases (anthracnose, zonate leaf spot, leaf blight, and gray leaf spot). High per day productivity for green fodder. High seed yield
4	IC635025	2020	INGR20087	Tolerance to shoot fly and downy mildew. High protein content (12.2%) and high grain yield with higher nutrient-use efficiency
5	IC637580	2021	INGR20087	Advanced genetic material. High protein content (11.73%)
6	IC635028	2021	INGR2103	Grain mold resistance (3.8)
7	IC338975	2021	INGR21039	More seed weight (3.54 g) and high dry fodder yield (9798 kg/ha)
8	IC415833	2021	INGR21040	More seed weight (3.61 g) and high dry fodder yield of 9720 kg/ha
9	IC585181	2021	INGR21042	Early flowering (<56 days)
10	IC637581	2021	INGR21043	Grain mold-resistant kharif sorghum genotype with field grade grain mold score of 3.10 and threshed grade grain mold score of 3.53
11	IC640695	2021	INGR21130	Grain mold field grade of AKGMR 118 is in the resistant category and the resistant check B 58586 is also in same category. Grain mold thresh grade (1–9 scale) of AKGMR 118 is in the resistant category, while the resistant check B 58586 is in highly resistant category
12	IC640645	2021	INGR21134	Total soluble sugar (9.14% TSS)
13	IC640646	2021	INGR21135	Resistant to anthracnose
14	IC0288432	2022	INGR22024	More leaf stem ratio (0.32). Low HCN content (40.9%)
15	IC643757	2022	INGR22025	Longer leaves (86.5 cm). More plant height (276 cm). More number of leaves per plant (12.9)
16	IC643980	2022	INGR22026	High fresh stalk yield. High biomass. Sorghum x maize cross derivative
17	IC643968	2022	INGR22027	High oil content (3.14% to 4.76%). Hard seed with bigger germ size ($r = 0.484$)
18	IC643969	2022	INGR22028	High oil content (3.14% to 4.76%). Very bold seed with medium hardness and small germ size ($r = 0.484$)
19	IC643970	2022	INGR22029	High in vitro true digestibility of dry matter (TDDM/IVDMD), organic matter digestibility (OMD), and metabolizable energy content (ME). Low acid detergent fiber (ADF) and acid detergent lignin (ADL) content
20	IC643981	2022	INGR22031	Higher brix and total sugars (14.3%), early maturing (116 days), early flowering (80 days)

(continued)

Table 2.4 (continued)

Sr. no.	National identity	Year	INGR number	Novel unique features
21	IC643961	2022	INGR22078	Early maturing (110.3 days) dwarf, yellow-grained sorghum
22	IC646841	2022	INGR22143	High fresh biomass (59.6 t/ha) and high juice yield (19,654 l/ha)
23	IC0546931	2022	INGR22144	High total fresh biomass (53.8 t/ha) and high juice yield (14,318 l/ha)
24	IC618406	2022	INGR22145	High Brix content (17.4%). High fresh stalk yield (4203 t/ha) and high juice content (14,533 l/ha)
25	IC648612	2023	INGR23036	Longer leaf length (83.1 cm) and wider leaf width (8.39 cm)
26	IC648613	2023	INGR23037	Longer leaf length (85.5 cm) and leaf width (7.32 cm)
27	IC648614	2023	INGR23038	Hurda sorghum with free threshability, excellent fragrance, and high hundred tender grain weight (4.12 g)
28	IC647559	2023	INGR23039	Hurda sorghum with earliness (50% flowering in 67 days) in dough stage (89 days), free threshability, and excellent fragrance
29	IC286441	2023	INGR23040	High zinc (30.61 ppm) and iron (37.14 ppm) content
30	EC488403	2023	INGR23041	High zinc (Zn) (29.78 ppm) and high iron (Fe) content (35.51 ppm)
31	IC643984	2023	INGR23042	Early flowering (67 days) and early maturity (110 days) line
32	IC651981	2024	INGR24022	Tolerance to stem borer (9.7% dead hearts at 45 DAE). Tolerance to shoot fly (34.8% dead hearts at 28 DAE)

ICAR-NBPGR, Germplasm Registration Information System (2024)

avoiding wild and weedy relatives, and landraces in their crop improvement program (Marshall 1989; Nass and Paterniani 2000). Plant breeders are well attentive about the risk of including exotic germplasm in breeding program. Use of wild germplasm and landrace accessions in crop breeding programs is minimal due to illiteracy about the genetic value of the materials or the linkage drag correlated with the inheritance of useful traits from such germplasm (Upadhyaya et al. 2014b).

2.4.4 Core and Mini Core Set

In collaboration with the Generation Challenge Programme (GCP), ICRISAT developed universal composite sorghum collections, which include a total of 383 germplasm accessions in the sorghum reference set. Core and mini core collections, or genotype-based reference sets (Glaszmann et al. 2010), have proven significant in identifying germplasm with agronomically valuable traits for use in crop breeding programs. These subsets are typically tested on a large scale for morphological and nutritional traits, as well as resistance to biotic and abiotic stresses. Numerous germplasm lines with superior agronomic performance have been reported in sorghum.

Several accessions have demonstrated resistance to abiotic and biotic stresses (Vadez et al. 2013; Sharma et al. 2010). Additionally, many seed-nutrient-dense accessions (rich in oil, protein, iron, and zinc) have been identified in sorghum (Kumar and Kumar 2009). Recently, sweet sorghum has gained importance as a bioenergy crop, particularly for ethanol production, due to its sugar-rich stalks. Extensive screening of mini core sorghum collections for stalk sugar content, measured in Brix (%) over multiple seasons, has identified several lines with high Brix (%) and desirable agronomic traits. For instance, accessions such as IS13294, 23684, 13549, 23216, 24139, 24953, and 24939 exhibited significantly higher average Brix values (14.0% to 15.2%) compared to the best check, IS33844 (12.4%), across different environments. However, these accessions were lower yielding and had lower test weights. In contrast, accessions IS4698, 23891, 1004, and 28141 demonstrated superior performance over IS33844 (Upadhyaya et al. 2014b). These accessions are valuable genetic resources for improvement programs aimed at developing sweet sorghum as a bioenergy crop.

2.4.5 Genotyping-Based Reference Set and Composite Collection

Universally 3384 cultivated and wild lines of sorghum are maintained and held by the Global Composite Germplasm Collection (GCGC). This consisted diverse collections from different institutes such as ICRISAT, India, CAAS, China, and CIRAD, France, having 280 breeding accessions and elite cultivars, 68 weedy and wild relatives, and above 3000 traditional lines collected by ICRISAT especially selected from formerly authenticated mini core and core stocks (Grenier et al. 2001; Upadhyaya et al. 2009). This has significant role in protection to abiotic and biotic stresses along with variations in morphological and qualitative traits. Additionally, these sorghum accessions included in GCGC were assessed for molecular characterization using 41 simple sequence repeat (SSR) markers and created a testimonial set of 383 lines on the basis of genotyping which represents 28 breeding cultivars, 332 landraces, and 23 undomesticated relatives, describing entire 5 races along with 10 hybrid races from divergent geographic origins (Billot et al. 2012). Diverse experimental set of sorghum having 107 accessions was created (Shehzad et al. 2009) on the basis of markers' (SSR) diversity ealuation of 320 lines.

2.5 Pre-breeding for Germplasm Enhancement

Sorghum germplasm can be further enhanced by utilizing species from wild relatives of cultivated sorghum. Domesticated sorghum is known to be sexually compatible with its wild-weedy counterparts in the section *Sorghum* (Andersson and Carmen 2010), including *S. bicolor* ssp. *drummondii*, *S. halepense* (Johnsongrass), *S. propinquum*, *S. bicolor* ssp. *arundinaceum*, and its four races (*virgatum*, *aethiopicum*, *arundinaceum*, and *verticilliflorum*). Germplasm valuable for crop improvement programs can be derived through artificial hybridization between Johnsongrass and sorghum. Hybrid plants generally fall into two categories: (1) sterile plants with 30 chromosomes and vigorous rhizomes and (2) fertile plants with 40 chromosomes and weaker rhizomes (Dweikat 2005).

According to Sangduen and Hanna (1984), hybrids between Johnsongrass and sorghum exhibited traits such as increased leafiness, vigor, and a growth habit that resembled Johnsongrass more closely than *Sorghum bicolor*. These hybrids displayed open panicles, altered seed color and shape, and a tendency for seed shattering. Traits like rhizome expression, stem thickness, and seed size were intermediate between the parent species. Piper and Kulakow (1994) performed a similar experiment, incorporating two backcross generations, to develop perennial sorghum varieties. Molecular markers were used in hybrids between cultivated sorghum and Johnsongrass (Dweikat 2005) to confirm normal meiotic behavior, evidenced by a 1:2:1 segregation ratio of 34 markers. These hybrids also exhibited beneficial traits, such as cold tolerance and resistance to greenbug and chinch bug.

The weed *S. bicolor* ssp. *drummondii* (commonly known as shattercane) closely resembles cultivated sorghum (Defelice 2006). It is a natural hybrid between *S. bicolor* ssp. *bicolor* and *verticilliflorum* (De Wet 1978). Genetic marker validation has primarily grouped *S. bicolor* ssp. *drummondii* with *bicolor* race sorghums (Casa et al. 2005). Resistance to nematodes such as *Meloidogyne chitwoodi* and *Meloidogyne hapla* has been observed in Sudan grass (Mojtahedi et al. 1993; Viaene and Abawi 1998). Sudan grass cultivars produce cyanogenic glucoside dhurrin, which decomposes into HCN and p-hydroxybenzaldehyde (pHBA). This property reduces nematode populations, as shown by a 55% decrease in *M. hapla* juvenile penetration into lettuce roots when eggs were incubated in Sudan grass extract (Widmer and Abawi 2000). Sudan grass-sorghum hybrids, which are crosses between Sudan grass and forage-type sorghums, are vigorous, tall annual grasses that can suppress nematodes and weeds (Clark 2007). These hybrids account for approximately three-quarters of the hybrid sorghum seed market (Smith and Frederiksen 2000).

In cultivated sorghum, economically significant traits can be improved by leveraging wild relatives. Wild sorghum is vital for introducing resistance to *Striga* spp., a parasitic weed affecting sorghum (Rich et al. 2004). Johnsongrass has been a source of resistance to greenbug, chinch bug, and cold temperatures (Dweikat 2005). Additionally, Johnsongrass (accession IS 14212) and *Sorghum dimidiatum* (accession IS 18945) have shown resistance to sorghum shoot fly (Kamala et al. 2009; Nwanze et al. 1995). *Sorghum bicolor* ssp. *arundinaceum* has been shown to increase grain yield in hybrid grain production (Jordan et al. 2004). Hybrids derived from *Sorghum purpureosericeum* and *S. bicolor* ssp. *drummondii* also performed better than the cultivated sorghum parent in single backcross generations (Jordan et al. 2011). Texas A&M University has utilized *S. propinquum* to develop cultivars with early maturity and high yields (Hajjar and Hodgkin 2007). Cultivated sorghum has also been successfully hybridized with *Sorghum macrospermum* ($2n = 40$), which provides resistance to sorghum midge, downy mildew, and shoot fly (Sharma et al. 2005). Both *S. propinquum* (Washburn et al. 2013) and Johnsongrass (Sangduen and Hanna 1984) have been utilized in crossing program to introduce rhizomatous traits, facilitating the development of perennial sorghum varieties.

2.6 Genomic Resources

2.6.1 Structural Genomics

The marker system known as restriction fragment length polymorphism (RFLP) was initially widely adopted for sorghum (Tao et al. 1998). Subsequently, various marker systems were developed for this crop, including randomly amplified polymorphic DNA (RAPD), diversity array technology (DArT), amplified fragment length polymorphism (AFLP), and single sequence repeats (SSRs) (Haussmann et al. 2002; Bowers et al. 2003; Wu et al. 2006; Mace et al. 2008) (Table 2.5). These marker systems have been used in sorghum for multiple purposes, such as genetic diversity assessment, fingerprinting, and QTL mapping. The utility of SSR molecular markers has expanded due to their availability through sorghum genome sequencing, with approximately 71,000 SSRs predicted based on genome sequence data. Advancements in next-generation sequencing (NGS) techniques, such as genotyping-by-sequencing (GBS) and the complete sorghum genome sequence, have made it possible to develop cost-effective and efficient marker systems.

The availability of reference genome accessions for sorghum (Paterson et al. 2009), combined with innovations in NGS and computational tools, has facilitated the generation of genomic DNA markers suitable for high-throughput genotyping (HTG). Recent advances, including whole-genome marker arrays (WGMAs) and NGS, have enhanced the aggregation of comprehensive genomic resources for sorghum (Morris et al. 2013a; McCormick et al. 2018). Re-sequencing efforts on sorghum's genomic structure have identified numerous structural variants and SNPs using NGS approaches such as whole-genome shotgun sequencing (WGSS) and Illumina Genome Analyzer sequencing technology (IGAST) (Nelson et al. 2011; Morris et al. 2013b; Bekele et al. 2013). Structural variations in the genome include insertions-deletions (indels), translocations, duplications, inversions, SNPs, copy number variations, and presence/absence variations (Saxena et al. 2014).

In genome-wide association studies (GWAS), genotyping techniques like GBS are frequently employed. GBS combines the use of restriction enzymes to reduce genome complexity with the ability to detect high-density SNPs (Elshire et al. 2011). This technique has been widely applied to sorghum germplasm collections to estimate genetic diversity and structure (Morris et al. 2013a). Analyzing genetic variation within sorghum germplasm helps identify valuable genes and their alternative forms linked to important traits, which can be incorporated into breeding programs.

Sorghum breeders have access to a variety of genomic resources, including DNA markers like SNPs, SSRs, and DArT (Billot et al. 2012; Evans et al. 2013), high-density genetic maps (Kong et al. 2013), and sequenced genomes (Mace et al. 2013). SNP identification and high-resolution genotyping (HRG) rely on reducing

Table 2.5 Constructed genetic linkage map in sorghum using different markers

Parents	Population	Marker(s) deployed	Linkage group	Map length (cM)	References
Tx7000 × B35	98 RIL(F7)	3 SSRs, 214 RFLPs, and 7 RAPD	10	1200	Subudhi and Nguyen (2000)
BTx 623 × IS3620C	138F6–8RIL	114 RFLPs, 31 SSRs	10	1287	Kong et al. (2000)
QL41 × QL39	152RIL(F5)	281 RFLPs, 5 morphological markers, 25 SSRs	10	>2750	Tao et al. (2000)
Tx7000 × B35	98 RIL(F7)	162 RFLPs	10	837	Xu et al. (2000)
BTx 623 × IS3620C	139F6–8RIL	323 RFLPs, 143 SSRs	10	1406	Bhattramakki et al. (2000)
Mixedmap		154 RFLPs, 10 morphological markers, 34 SSRs	10	1450	Bennetzen et al. (2001)
BTx 623 × IS3620C	137F6–8RIL	Morphological markers, 259 RFLPs, 226 SSRs, 303 DArTs	10	1528	Mace et al. (2008)
B2 V4 × 1383–2	150F2	122 MSAP, 22 SSRs	11	483.6	Duan et al. (2009)
296B × IS18551	168RIL	38 EST-SSRs, 107 SSRs, 3 morphological markers, 10 unigene SSRs	16	1143	Satish et al. (2009)
N13 × E36–1	94RIL	14 RFLPs, 45 genomic SSRs, 3 RAPD markers, 125 AFLPs, 55 EST-SSRs	10	2838	Ramu et al. (2009)
296B × IS18551	168RIL	100 SSRs, 28 EST-SSRs	15	1074.50	Srinivas et al. (2009)
M71 × SS79	188RIL	6 EST-SSR, 49 SSR, 102 AFLP	11	1029	Shiringani et al. (2010)
IS2122 × 27B	210RIL	127 SSRs, one morphological marker, and 21 genic-SSRs	10	700	Aruna et al. (2011)

(continued)

Table 2.5 (continued)

Parents	Population	Marker(s) deployed	Linkage group	Map length (cM)	References
Shihong137 × L-Tian	186F2	118 SSRs	15	1884.60	Guan et al. (2011)
SS79 × M71	188RIL	102 AFLP, 6 EST-SSR, and 49 SSRs	11	1029	Shiringani and Friedt (2011)
B 923296 × SC 170–6-8	141RIL	377 DArTs	10	2259	Mace and Jordan (2011)
BR007 × SC283	90RIL	255 DArTs, one RFLP marker, 83 SSRs, and 5 sequence-tagged site (STS)	10	2034.90	Sabadin et al. (2012)
M35–1 × B35	245RIL	228 SSRs and 3 morphological markers	10	1235.50	Nagaraja Reddy et al. (2012)
74LH3213 × MS138B	133RIL	247 SSRs	10	1697.20	Takai et al. (2012)
654 × LTR108	244RIL	3418 bin markers, SNPs	10	1591.40	Zou et al. (2012)
BTx 623 × *Sorghum propinquum*	163RIL	141 SSRs	10	773.1	Kong et al. (2013)
ICSV745 × R890562	119RIL	234 DArTs and 244 RFLPs	12	1487.96	Alam et al. (2014)
296B × PVK801	342RIL	10 SSRs, 1148 DArTs, 927 DArTSeqs, and 13 SSRs		1355.52	Kotla et al. (2019)
BT X 623 and NOG	213RILs	3710 SNPs		1299.70	Kajiya-Kanegae et al. (2020)
Tx623Aandsudangrasssa	RIL	1065 SLAF	10	1191.7	Jin et al. (2021)
Scattered ear sorghum × redhull Sudan grass	150 F2	8009 SNPs	10	2192.84	Lu et al. (2022)
Erdurmus × Ogretmenoglu	175 F2	1438 SNPs	10	1349.72	Guden et al. (2023)

genome complexity. Using the RAD-seq technique, researchers identified approximately 283,000 SNPs across 8 sorghum lines (Nelson et al. 2011). Structural changes in the genome are fundamental to biological systems and crop development, and further research should aim to explore the relationships between genetic variations and improved crop performance.

2.6.2 Functional Genomics

The field of functional genomics aims to describe the functions and interactions of genes and proteins by utilizing genome-wide approaches, in contrast to classical molecular biology techniques. Several genome sequencing projects have been conducted in sorghum (Paterson et al. 2009; McCormick et al. 2018; Deschamps et al. 2018; Cooper et al. 2019), and large-scale data are accessible in public databases such as Phytozome at the Joint Genome Institute (JGI). All the sequencing projects revealed the existence of extensive diversity in sorghum. Breakthroughs in sequencing technologies over the last decade have also led to the identification of QTLs for various target traits, major genes, and marker-assisted breeding approaches, paving the way for understanding the genetic basis of economically valuable traits. Whole-genome sequencing of 44 sorghum accessions revealed racial structure and a complex domestication history involving two distinct domestication events (Mace et al. 2013).

Structural and functional studies of the pan-genome revealed approximately two million SNPs and about 4.0 million indels from the genome sequence data of 354 diverse sorghum lines (Ruperao et al. 2021). Progress in protein function studies has improved understanding of altered protein patterns underlying sorghum's responses to various stresses. Data generated in sorghum indicate an increase in extracellular matrix proteins under drought stress (Ngara et al. 2018) and the identification of 31 heat-responsive proteins involved in heat tolerance (Ngcala et al. 2020). Studies on post-flowering drought stress revealed nine candidate proteins across six different functional categories and three subcellular localizations (Woldesemayat et al. 2018).

Comparative proteomics studies have shed light on the role of kafirin as an energy source under salt stress (Punia et al. 2020), and proteomics in sorghum roots under osmotic stress has focused on the regulation of reactive oxygen species (Li et al. 2020). Collectively, these studies have paved the way for a better understanding of the molecular mechanisms controlling various abiotic stresses in sorghum, which is a key aspect of sorghum improvement.

The flowering time in sorghum, which significantly impacts its cultivation across a wide range of environments, is believed to be controlled by photoperiodic responses and circadian clock mechanisms. Six loci (Ma1 to Ma6) are known to control sorghum flowering in response to daylight length, with natural mutations in these loci aiding its adaptation from tropical to temperate regions. Regarding grain yield improvement, jasmonic acid pathway genes are key targets for high-yield breeding in sorghum. A set of multi-seeded (MSD) genes involved in the JA biosynthesis pathway were identified to regulate grain number by restoring the sterility of pedicellate spikelets (Jiao et al. 2018; Dampanaboina et al. 2019; Gladman et al. 2019; Ge et al. 2023). Sorghum dwarfing genes are not part of the gibberellin (GA) biosynthesis and signaling pathways, as in rice and wheat. Instead, they are related to brassinosteroid or auxin signaling pathway genes, potentially leading to abnormal culm bending due to gibberellic acid deficiency. Collectively, these point mutations in dw-series genes remain vital for designing sorghum dwarf breeding programs. Overexpression of SbSNF4–2 in sorghum significantly enhances biomass by increasing plant height, promoting tillers, and delaying flowering (Upadhyaya et al. 2022).

Functional genomics has played a significant role in advancing our understanding of plant genotype-phenotype correlations. Despite these advances, which have led to the discovery of numerous potential QTLs, the functional characterization of genes responsible for vital agronomic traits remains a major challenge. Future improvements should focus on creating new sorghum varieties by incorporating all the information available in the public domain, primarily to enhance key target traits through the formulation of promising, well-structured intelligent breeding strategies combined with gene editing and genomic selection approaches.

2.7 Future Prospects

Sorghum is recognized as one of the most differentiated species. It exhibits enormous morphological diversity, including variations in plant type and grain characteristics, and is well adapted to a wide range of environments often considered too adverse for other cultivated crops. This diversity is also reflected in seed size and luster. The comprehensive coexistence of sorghum with its wild relatives in its center of origin may be responsible for the divergence in seed size. Conserving and utilizing such genetic variations in a productive manner is a challenging task, but it is essential.

Core and mini core collections, which create representative subsets of genetic material, enable plant breeders to identify novel genetic variations associated with agronomically beneficial traits. These subsets are instrumental for breeding and genomics research in sorghum. Sorghum has a rich repository of genomic resources, and the expanding utilization of these resources will assist researchers in developing trait-specific cultivars or populations with distinct adaptations.

Globally, in terms of sorghum germplasm collection, ICRISAT and US collections are the largest and most effective. The ICAR-National Bureau of Plant Genetic Resources (NBPGR), the nodal agency for managing plant germplasm in India, ranks second in terms of sorghum collection. All three institutes/organizations are consistently augmenting their collections through further evaluations and characterizations of their respective germplasm. They will continue to explore genetic diversity by integrating new genomic and high-throughput evaluation technologies.

References

Abreha KB, Enyew M, Carlsson AS, Vetukuri RR, Feyissa T, Motlhaodi T et al (2021) Sorghum in dryland: morphological, physiological, and molecular responses of sorghum under drought stress. Planta 255(20)

Alam MM, Mace ES, Van Oosterom EJ, Cruickshank A, Hunt CH, Hammer GL et al (2014) QTL analysis in multiple sorghum populations facilitates the dissection of the genetic and physiological control of tillering. Theor Appl Genet 127(10):2253–2266

Ananda GKS, Myrans H, Norton SL, Gleadow R, Furtado A and Henry RJ (2020) Wild Sorghum as a Promising Resource for Crop Improvement. Front. Plant Sci. 11:1108. https://doi.org/10.3389/fpls.2020.01108

Andersson MS, Carmen M (2010) Gene flow between crops and their wild relatives. Johns Hopkins University Press, Baltimore, MD, pp 443–464

Arriola PE, Ellstrand NC (1997) Fitness of interspecific hybrids in the genus Sorghum: persistence of crop genes in wild populations. Ecol Appl 7(2):512–518

Aruna C, Bhagwat VR, Madhusudhana R, Sharma V, Hussain T, Ghorade RB et al (2011) Identification and validation of genomic regions that affect shoot fly resistance in sorghum (*Sorghum bicolor* (L) Moench). Theor Appl Genet 122(8):1617–1630

Aruna C, Visarada KBRS, Bhat BV, Tonapi VA (2018) Breeding sorghum for diverse end uses. Woodhead Publishing

Baker HG (1972) Human influences on plant evolution. Econ Bot 26(1):32–43

Barnaud A, Trigueros G, Mckey D, Joly HI (2008) High outcrossing rates in fields mixed with sorghum landraces: how are landraces maintained? Heredity 101:445–452

Bekele WA, Wieckhorst S, Friedt W, Snowdon RJ (2013) High-throughput genomics in sorghum: from whole-genome resequencing to a SNP screening array. Plant Biotechnol J 11(9):1112–1125

Bennetzen JL, Subramanian V, Xu J, Salimath SS, Subramanian S, Bhattramakki D (2001) A framework genetic map of sorghum containing RFLP, SSR, and morphological markers. DNA-Based Markers Plants:347–355

Bhandari HR, Bhanu AN, Srivastava K, Singh MN, Shreya HA (2017) Assessment of genetic diversity in crop plants—an overview. Adv Plants Agric Res 7:279–286

Bhattramakki D, Dong J, Chhabra AK, Hart GE (2000) An integrated SSR and RFLP linkage map of *Sorghum bicolor* (L) Moench. Genome 43(6):988–1002

Billot C, Rivallan R, Sall MN, Fonceka D, Deu M, Glaszmann JC (2012) A reference microsatellite kit to assess for genetic diversity of *Sorghum bicolor* (Poaceae). Am J Bot 99(6):e245–e250

Bowers JE, Abbey C, Anderson S, Chang C, Draye X, Hoppe AH (2003) A high-density genetic recombination map of sequence-tagged sites for sorghum, as a framework for comparative structural and evolutionary genomics of tropical grains and grasses. Genetics 165(1):367–386

Brotero A (1804) FLORA LUSITANIOA, SEU, PLANTABUM, QUAE IN LUSITANIA Vel, SPONTE CRESOUNT, VEL FUEQUENTIUS COLUNTUE, EX FLORUM PRAESERTUR Sexullus SYSTEMATICE DISTRIBUTARUM, SYNOPSIS. 2 V. Ollsslpone

Casa AM, Mitchell SE, Hamblin MT, Sun H, Bowers JE, Paterson AH (2005) Diversity and selection in sorghum: simultaneous analyses using simple sequence repeats. Theor Appl Genet 111:23–30

Chandel KPS, Paroda RS (2000) Status of plant genetic resources conservation and utilization in Asia-Pacific Region, Regional Synthesis Report 32. Asia-Pacific Association of Agricultural Institutions, FAO Regional Office for Asia and the Pacific, Bangkok

Chiovenda E (1912) Genere Sageretia Brongn. in Africa. Annali di Botanic 10:430–455

Clark A (2007) Managing cover crops profitably, 3rd edn. Sustainable Agriculture Network, Beltsville, MD, p 244

Clayton WD, Renvoize SA (1986) Genera Graminum Grasses of the World, Kew Bulletin Addition Series XIII. Royal Botanic Gardens, Kew, London, pp 338–345

Cooper EA, Brenton ZW, Flinn BS, Jenkins J, Shu S, Flowers D (2019) A new reference genome for *Sorghum bicolor* reveals high levels of sequence similarity between sweet and grain genotypes: implications for the genetics of sugar metabolism. BMC Genomics 20(1):420

Cui YX, Xu GW, Magill CW, Schertz KF, Hart GE (1995) RFLP-based assay of *Sorghum bicolor* (L.) Moench genetic diversity. Theor Appl Genet 90:787–796

Dahlberg JA, Spinks MS (1995) Current status of the US sorghum germplasm collection. International Sorghum and Millets Newsletter., 36, pp 4–12

Dahlberg JA, Hash ET, Kresovich S, Maunder B, Gilbert M (1997) Sorghum and pearl millet genetic resources utilization. In: Proceedings of the International Conference on Genetic Improvement of Sorghum and Pearl Millet, September 22–27, 1996, Lubbock, TX, pp 42–54

Dahlberg JA (2000) Classification and characterization of sorghum. In: Smith CW, Frederiksen RA (eds) Sorghum: origin, history, technology, and production. Wiley, New York, pp 99–130
Damania AB (2002) The Hindustan centre of origin of important plants. Asian Agri-Hist 6:333–341
Dampanaboina L, Jiao Y, Chen J, Gladman N, Chopra R, Burow G, Hayes C, Christensen SA, Burke J, Ware D, Xin Z (2019) Sorghum MSD3 encodes an ω-3 fatty acid desaturase that increases grain number by reducing jasmonic acid levels. Int J Mol Sci 2019
De Wet JMJ (1978) Systematics and evolution of sorghum sect. *Sorghum* (Gramineae). Am J Bot 65:477–484
De Wet JMJ (1977) Domestication of African cereals. Afr Econ Hist 3:15
De Jonge (2009) Plants, genes and justice: an enquiry into fair and equitable benefit sharing. Unpublished PhD dissertation. Wageningen University, Wageningen
Defelice MS (2006) Shattercane, *Sorghum bicolor* (L.) Moench ssp. drummondii (Nees ex Steud.) de Wet ex Davidse—black sheep of the family. Weed Technol 20:1076–1083
Deschamps S, Zhang Y, Llaca V, Ye L, Sanyal A, King M et al (2018) A chromosome-scale assembly of the sorghum genome using nanopore sequencing and optical mapping. Bioinformatics 9(1):2460–2461
Deu M, Gonzalez-de-Leon D, Glazsmann JC, Degremont I, Chantereau J, Lanaud C (1994) RFLP diversity in cultivated sorghum in relation to racial differentiation. Appl Genet 88:838–844
Deu M, Rattunde F, Chantereau J (2006) A global view of genetic diversity in cultivated sorghums using a core collection. Genome 49:168–180
Doggett H (1988) Sorghum. Longman Scientific & Technical, Essex
Doggett H, Majisu BN (1968) Disruptive selection in crop development. Heredity 23:1–26
Duan Y, Qian J, Sun Y, Yi Z, Yan M (2009) Construction of methylation linkage map based on MSAP and SSR markers in *Sorghum bicolor* (L). IUBMB Life 61(6):663–669
Duvick DN (1986) Plant breeding: past achievements and expectations for the future. Econ Bot 40:289–297
Dweikat I (2005) A diploid, interspecific, fertile hybrid from cultivated sorghum, *Sorghum bicolor*, and the common Johnsongrass weed *Sorghum halepense*. Mol Breed 16:93–101
Dwivedi SL, Upadhyaya HD, Gowda CLL (2009) Approaches to enhance the value of genetic resources in crop improvement
Eberhart SA, Bramel-Cox PJ, Prasada Rao KE (1997) Preserving genetic resources. In: The Proceedings of the International Conference on Genetic Improvement of Sorghum and Pearl Millet, Lubbock, TX, September 22–27, 1996
Ejeta G, Grenier C (2005) Sorghum and its weedy hybrids. In: Gressel J (ed) Crop ferality and volunteerism. CRC Press Inc., Florida, Boca Raton, pp 123–135
Elangovan M, Karnam V, Amasiddha B, Sushil P, Chitra Devi P (2022) Assessment of variability in sorghum [*Sorghum bicolor* (L.) Moench] germplasm for agro-morphological traits. Electron J Plant Breed 13(2):488–497
Elshire RJ, Glaubitz JC, Sun Q, Poland JA, Kawamoto K, Buckler ES (2011) A robust, simple genotyping-by-sequencing (GBS) approach for high diversity species. PLoS One 6(5):e19379
Evans J, McCormick RF, Morishige D, Olson SN, Weers B, Hilley J (2013) Extensive variation in the density and distribution of DNA polymorphism in sorghum genomes. PLoS One 8(11):e79192
FAO (1983) International undertaking on plant genetic resources. FAO of the United Nations, Rome
FAO (2022) Crops and livestock products. United Nations Food and Agriculture Organization. FAO, Rome
FAO (2009) Commission on Genetic Resources for Food and Agriculture. Draft Second Report on the World's Plant Genetic Resources for Food and Agriculture (Final Version). Rome, 330 p. Available from: FAO Report. Accessed September 2014
FAOSTAT (2019) Production/Yield quantities of Sorghum in World (Rome). Available at: FAOSTAT. Accessed 28th December 2019
Folkertsma RT, Rattunde HFW, Chandra S, Raju GS, Hash CT (2005) The pattern of genetic diversity of Guinea-race *Sorghum bicolor* (L.) Moench landraces as revealed with SSR markers. Theor Appl Genet 111:399–409

Frankel OH (1984) Genetic perspective of germplasm collection. In: Arber W, Limensee K, Peacock WJ, Stralinger P (eds) Genetic manipulations: Impact on man and society. Cambridge University Press, pp 161–170
Ge F, Xie P, Wu Y, Xie Q (2023) Genetic architecture and molecular regulation of sorghum domestication. aBIOTECH 4(1):57–71
Gebrekidan B (1982) Utilization of germplasm in sorghum improvement. In: House LR, Mughogho LK, Peacock JM (eds) Proceedings of the International Symposium on Sorghum. ICRISAT, Patancheru, India., November 2–7, 1981, 1, pp 335–345
Gladman N, Jiao Y, Lee YK, Zhang L, Chopra R, Regulski M, Burow G, Hayes C, Christensen SA, Dampanaboina L (2019) Fertility of pedicellate spikelets in sorghum is controlled by a jasmonic acid regulatory module. Int J Mol Sci 20:4951
Glaszmann JC, Kilian B, Upadhyaya HD, Varshney RK (2010) Accessing genetic diversity for crop improvement. Curr Opin Plant Biol 13:167–173
Grenier C, Bramel-Cox PJ, Harmon P (2001) Core collection of sorghum: I. Stratification based on eco-geographical data. Crop Sci 41:234–240
Guan YA, Wang HL, Qin L, Zhang HW, Yang YB, Gao FJ et al (2011) QTL mapping of bio-energy related traits in Sorghum. Euphytica 182(3):431–440
Guden B, Yol E, Erdurmus C, Lucas SJ, Uzun B (2023) Construction of a high-density linkage map and QTL mapping for bioenergy-related traits in sweet sorghum. Front Plant Sci 14:1081931
Hajjar R, Hodgkin T (2007) The use of wild relatives in crop improvement: a survey of developments over the last 20 years. Euphytica 156:1–13
Hargrove TR, Cabanilla VL, Coffman WR (1988) Twenty years of rice breeding. Bioscience 38:675–681
Harlan JR, De Wet JMJ (1972) A simplified classification of cultivated sorghum. Crop Sci 12:172–176
Haussmann B, Hess D, Seetharama N, Welz H, Geiger H (2002) Construction of a combined sorghum linkage map from two recombinant inbred populations using AFLP, SSR, RFLP, and RAPD markers, and comparison with other sorghum maps. Theor Appl Genet 105(4):629–637
Heal G, Walker B, Levin S, Arrow K, Dasgupta P, Daily G (2004) Genetic diversity and interdependent crop choices in agriculture. Resour Energy Econ 26:175–184
House LR (1985) A Guide to Sorghum Breeding, 2nd edn. International Crops Research Institute for the Semi-Arid Tropics (ICRISAT), Patancheru, Andhra Pradesh
IBPGR/ICRISAT (1980) Sorghum descriptors. IBPGR, Rome
IPGRI (1993) Diversity for Development. IPGRI, Rome
Jiao Y, Burow G, Gladman N, Acosta-Martinez V, Chen J, Burke J, Ware D, Xin Z (2018) Efficient identification of causal mutations through sequencing of bulked F2 from two allelic bloomless mutants of *Sorghum bicolor*. Front Plant Sci 8:2267
Jin P, Wang L, Zhao W, Zheng J, Wang YH, Liu Y (2021) Construction of high-density genetic map and QTL mapping in sorghum × sudangrass. Euphytica 217(8):162–211
Jordan D, Butler D, Henzell B, Drenth J, McIntyre L (2004) Diversification of Australian sorghum using wild relatives. In: New Directions for a Diverse Planet: Proceedings of the Fourth International Crop Science Congress, September 26–October 1, 2004, Brisbane, Australia
Jordan DR, Mace ES, Cruickshank AW, Hunt CH, Henzell RG (2011) Exploring and exploiting genetic variation from unadapted sorghum germplasm in a breeding program. Crop Sci 51:1444–1457
Jordan DR, Tao YZ, Godwin ID, Henzell RG, Cooper M, McIntyre CL (1998) Loss of genetic diversity associated with selection for resistance to sorghum midge in Australian sorghum. Euphytica 102:1–7
Kajiya-Kanegae H, Takanashi H, Fujimoto M, Ishimori M, Ohnishi N, Wacera WF (2020) RAD-seq-based high-density linkage map construction and QTL mapping of biomass-related traits in sorghum using the Japanese landrace Takakibi NOG. Plant Cell Physiol 61(7):1262–1272. https://doi.org/10.1093/pcp/pcaa056

Kamala V, Sharma HC, Manohar Rao D, Varaprasad KS, Bramel PJ (2009) Wild relatives of sorghum as sources of resistance to sorghum shoot fly, Atherigona soccata. Plant Breed 128:137–142

Kamesawara Rao N, Bramel PJ, Reddy VG, Deb UK (2004) Conservation, utilization, and distribution of sorghum germplasm. In: Bantilan MCS, Deb UK, Gowda CLL, Reddy BVS, Obilana AB, Evenson RE (eds) Sorghum genetic enhancement: research process, dissemination and impacts. International Crops Research Institute for the Semi-Arid Tropics, Patancheru, AP, pp 43–62

Kimber CT (2000) Origins of domesticated sorghum and its early diffusion to India and China. In: Smith CW, Frederiksen RA (eds) Sorghum: origin, history, technology and production. Wiley, New York, pp 3–98

Kong L, Dong J, Hart GE (2000) Characteristics, linkage-map positions, and allelic differentiation of *Sorghum bicolor* (L.) Moench DNA simple-sequence repeats (SSRs). Theor Appl Genet 101(3):438–448

Kong W, Jin H, Franks CD, Kim C, Bandopadhyay R, Rana MK (2013) Genetic analysis of recombinant inbred lines for *Sorghum bicolor* × *Sorghum propinquum*. G3 Genes|Genomes|Genet 3(1):101–108

Kotla A, Phuke R, Hariprasanna K, Mehtre SP, Rathore A, Gorthy S (2019) Identification of QTLs and candidate genes for high grain Fe and Zn concentration in sorghum (*Sorghum bicolor* (L.) Moench). J Cereal Sci 90:102850

Kuhlman LC, Burson BL, Stelly DM, Klein PE, Klein RR, Price HJ (2010) Early-generation germplasm introgression from *Sorghum macrospermum* into sorghum (*S. bicolor*). Genome 53:419–429

Kumar B, Kumar J (2009) Evaluation of small millet genotypes against endemic diseases in mid-western Himalayas. Indian Phytopathol 62(4):518–521

Lazarides M, Hacker JB, Andrew MH (1991) Taxonomy, cytology, and ecology of indigenous Australian sorghums (*Sorghum* Moench: Andropogoneae: Poaceae). Aust Syst Bot 4:591–635

Li H, Li Y, Ke Q, Kwak SS, Zhang S, Deng X (2020) Physiological and differential proteomic analyses of imitation drought stress response in *Sorghum bicolor* root at the seedling stage. Int J Mol Sci 21:9174

Lu Q, Yu X, Wang H, Yu Z, Zhang X, Zhao Y (2022) Construction of ultra-high-density genetic linkage map of a sorghum-Sudan grass hybrid using whole genome sequencing. PLoS One 17(11):e0278153

Mace ES, Jordan DR (2011) Integrating sorghum whole genome sequence information with a compendium of sorghum QTL studies reveals uneven distribution of QTL and of gene-rich regions with significant implications for crop improvement. Theoretical and Applied Genetics 123(1):169–191. https://doi.org/10.1007/s00122-011-1575-y

Mace ES, Xia L, Jordan DR, Halloran K, Parh DK, Huttner E (2008) DArT markers: diversity analyses and mapping in *Sorghum bicolor*. BMC Genomics 9(1):26–11

Mace E, Tai S, Gilding E et al (2013) Whole-genome sequencing reveals untapped genetic potential in Africa's indigenous cereal crop sorghum. Nat Commun 4:2320

Mann JA, Kimber CT, Miller PR (1983) The origin and early cultivation of sorghums in Africa. Tex Agric Exp Stn Bull:1454

Maredia MK, Byerlee D, Pee P (2000) Impacts of food crop improvement research: evidence from sub-Saharan Africa. Food Policy 25:531–559

Marshall DR (1989) Limitations to the use of germplasm collections. In: Brown AHD, Marshall DR, Frankel OH, Williams JT (eds) The use of plant genetic resources. Cambridge University Press, New York, pp 105–120

McCormick RF, Truong SK, Sreedasyam A, Jenkins J, Shu S, Sims D (2018) The *Sorghum bicolor* reference genome: improved assembly, gene annotations, a transcriptome atlas, and signatures of genome organization. Plant J 93(2):338–354

Mekbib F (2008) Genetic erosion of sorghum (*Sorghum bicolor* (L.) Moench) in the centre of diversity, Ethiopia. Genet Resour Crop Evol 55:351–364

Miller PR, Kebede Y (1984) Genetic contributions to yield gains in sorghum, 1950 to 1980. In: Fehr WR (ed) Genetic Contributions to Yield Gains of Five Major Crop Plants, CSSA Special Publication 7. CSSA, ASA, 13, pp 1–14

Mojtahedi H, Santo GS, Ingham RE (1993) Suppression of *Meloidogyne chitwoodi* with Sudangrass cultivars as green manure. J Nematol 25:303–311

Morrell PL et al (2005) Crop-to-weed introgression has impacted allelic composition of Johnsongrass populations with and without recent exposure to cultivated sorghum. Mol Ecol 14(7):2143–2154

Morris GP, Ramu P, Deshpande SP, Hash CT, Shah T, Upadhyaya HD (2013b) Population genomic and genome-wide association studies of agroclimatic traits in sorghum. Proc Natl Acad Sci 110(2):453–458

Morris GP, Rhodes DH, Brenton Z, Ramu P, Thayil VM, Deshpande S (2013a) Dissecting genome-wide association signals for loss-of-function phenotypes in sorghum flavonoid pigmentation traits. G3 Genes Genomes Genet 3(11):2085–2094

Nagaraja Reddy R, Madhusudhana R, Murali Mohan S, Chakravarthi DVN, Seetharama N (2012) Characterization, development, and mapping of Unigene-derived microsatellite markers in sorghum (*Sorghum bicolor* (L) Moench). Mol Breed 29(3):543–564

Nass LL, Paterniani E (2000) Pre-breeding: a link between genetic resources and maize breeding. Agric Sci 57:581–587

Nelson JC, Wang S, Wu Y, Li X, Antony G, White FF (2011) Single nucleotide polymorphism discovery by high-throughput sequencing in sorghum. BMC Genomics 12(1):352–415

Ngara R, Ramulifho E, Movahedi M (2018) Identifying differentially expressed proteins in sorghum cell cultures exposed to osmotic stress. Sci Rep 8:8671

Ngcala MG, Goche T, Brown AP, Chivasa S, Ngara R (2020) Heat stress triggers differential protein accumulation in the extracellular matrix of sorghum cell suspension cultures. Proteomes 8

Nwanze KF, Seetharama N, Sharma HC, Stenhouse JW (1995) Biotechnology in pest management: improving resistance in sorghum to insect pests. Afr Crop Sci J 3:209–215

Ortiz R, Ruiz-Tapia EN, Mujica-Sanchez A (1998) Sampling strategy for a core collection of Peruvian quinoa germplasm. Theory Appl Genet 96:475–483

Paterson AH, Bowers JE, Bruggmann R, Dubchak I, Grimwood J, Gundlach H (2009) The *Sorghum bicolor* genome and the diversification of grasses. Nature 457(7229):551–556

Pedersen JF, Toy JJ, Johnson B (1998) Natural outcrossing of sorghum and Sudangrass in the Central Great Plains. Crop Sci 38:937–939

Perumal R, Krishnaramanujam R, Menz MA, Katilé S, Dahlberg J, Magill CW (2007) Genetic diversity among sorghum races and working groups based on AFLPs and SSRs. Crop Sci 47:1375–1383

Piper. (1915) Proc Biol Soc Wash 28:42

Piper JK, Kulakow PA (1994) Seed yield and biomass allocation in *Sorghum bicolor* and F1 and backcross generations of *S. bicolor* × *S. halpense* hybrids. Can J Bot 72:468

Popescu A, Dinu TA, Stoian E (2018) Sorghum—an important cereal in the world, in the European Union and Romania. Sci Pap Ser Manag Econ Eng Agric Rural Dev 18(4):271–284

Prasada Rao KE, Gopal Reddy V, Stenhouse JW (1995) Sorghum genetic resources at ICRISAT Asia Center. Int Sorghum Millets Newsl 36:15–19

Prasada Rao KE, Mengesha MH, Reddy YG (1989) International use of sorghum germplasm collection. In: Brown AHD, Frankel OH, Marshall DR, Williams JT (eds) The use of plant genetic resources. Cambridge University Press, Cambridge, UK, pp 150–167

Price HJ (2005) A *Sorghum bicolor* × *S. macrospermum* hybrid recovered by embryo rescue and culture. Aust J Bot 53:579–582

Punia H, Tokas J, Bhadu S, Mohanty AK, Rawat P, Malik A, Satpal. (2020) Proteome dynamics and transcriptome profiling in sorghum (*Sorghum bicolor* (L.) Moench) under salt stress. 3 Biotech 10(9):412

Qingshan L, Dahlberg JA (2001) Chinese sorghum genetic resources. Econ Bot 55(3):401–425

Rai M (2002) Genetic resources and intellectual property rights in agriculture perspective. Indian J Pulses Res 15:1–18

Ramu P, Kassahun B, Senthilvel S, Ashok Kumar C, Jayashree B, Folkertsma RT (2009) Exploiting rice–sorghum synteny for targeted development of EST-SSRs to enrich the sorghum genetic linkage map. Theor Appl Genet 119(7):1193–1204

Reddy GN, Kameshwara Rao N, Reddy BVS, Prasada Rao KE (2002) Geographic distribution of basic and intermediate races in the world collection of sorghum germplasm. Int Sorghum Millet Newsl 43:15–17

Reddy VG, Upadhyaya HD, Gowda CLL (2006) Current status of sorghum genetic resources at ICRISAT: their sharing and impacts. Int Sorghum Millets Newsl 47:9–13

Rhoades RE (1991) The world's food supply at risk. Natl Geogr 179(4):74–103

Rich PJ, Grenier C, Ejeta G (2004) Striga resistance in the wild relatives of sorghum. Crop Sci 44:2221–2229

Ritter KB, McIntyre CL, Godwin ID, Jordan DR, Chapman SC (2007) An assessment of the genetic relationship between sweet and grain sorghums, within *Sorghum bicolor* ssp. bicolor (L.) Moench, using AFLP markers. Euphytica 157:161–176

Rooney WL, Waniska RD (2000) Sorghum food and industrial utilization. In: Smith CW, Frederiksen RA (eds) Sorghum: origin, history, technology, and production. Wiley, New York, pp 689–729

Ruperao P, Thirunavukkarasu N, Gandham P, Selvanayagam S, Govindaraj M, Nebie B et al (2021) Sorghum pan-genome explores the functional utility for genomic-assisted breeding to accelerate the genetic gain. Front Plant Sci 12:666342

Sabadin PK, Malosetti M, Boer MP, Tardin FD, Santos FG, Guimaraes CT (2012) Studying the genetic basis of drought tolerance in sorghum by managed stress trials and adjustments for phenological and plant height differences. Theor Appl Genet 124(8):1389–1402

Salgotra RK, Gupta BB, Ahmed MI (2012) Characterization of thermo-sensitive genic male sterility (TGMS) rice genotypes (*Oryza sativa* L.) at different altitudes. Aus J Crop Sci 6:957–962

Sangduen N, Hanna WW (1984) Chromosome and fertility studies on reciprocal crosses between two species of autotetraploid *Sorghum bicolor* (L.) Moench and *S. halepense* (L.) Pers. J Hered 75:293–296

Sasaki T, Antonio BA (2009) Plant genomics: sorghum in sequence. Nature 457:547

Satish K, Srinivas G, Madhusudhana R, Padmaja PG, Nagaraja Reddy R, Murali Mohan S (2009) Identification of quantitative trait loci for resistance to shoot fly in sorghum (*Sorghum bicolor* (L) Moench). Theor Appl Genet 119(8):1425–1439

Saxena RK, Edwards D, Varshney RK (2014) Structural variations in plant genomes. Brief Funct Genomics 13(4):296–307

Sharma HC, Reddy BVS, Dhillon MK, Venkateswaran K, Singh BU, Pampapathy G (2005) Host plant resistance to insects in sorghum: present status and need for future research [online]. J Semi-Arid Trop Agric Res 1:1–8

Sharma R, Rao VP, Upadhyaya HD, Reddy VG, Thakur RP (2010) Resistance to grain mold and downy mildew in a mini-core collection of sorghum germplasm. Plant Dis 94:439–444

Shehzad T, Iwata H, Okuno K (2009) Genome-wide association mapping of quantitative traits in sorghum (*Sorghum bicolor* (L.) Moench) by using multiple models. Breed Sci 59:217–227

Shewayrga H, Jordan DR, Godwin ID (2008) Genetic erosion and changes in distribution of sorghum (*Sorghum bicolor* L. (Moench)) landraces in north-eastern Ethiopia. Plant Genet Res 6(1):1–10

Shiringani AL, Friedt W (2011) QTL for fibre-related traits in grain × sweet sorghum as a tool for the enhancement of sorghum as a biomass crop. Theor Appl Genet 123(6):999–1011

Shiringani AL, Frisch M, Friedt W (2010) Genetic mapping of QTLs for sugar-related traits in a RIL population of *Sorghum bicolor* L. Moench. Theor Appl Genet 121(2):323–336

Smith CW, Frederiksen RA (2000) History of cultivar development in the United States: from "Memoirs of A. B. Maunder-Sorghum Breeder". In: Smith CW, Frederiksen RA (eds) Sorghum-Origin, History, Technology, and Production. Wiley, pp 191–224

Snowden JD (1955) The wild fodder sorghums of the section Eu-sorghum. J Linn Soc Lond Bot 55:191–260

Srinivas G, Satish K, Madhusudhana R, Nagaraja Reddy R, Murali Mohan S, Seetharama N (2009) Identification of quantitative trait loci for agronomically important traits and their association with genic-microsatellite markers in sorghum. Theor Appl Genet 118(8):1439–1454

Stapf O (1917) Flora of Tropical Africa, vol 9, p 1

Subudhi PK, Nguyen HT (2000) Linkage group alignment of sorghum RFLP maps using a RIL mapping population. Genome 43(2):240–249

Takai T, Yonemaru JI, Kaidai H, Kasuga S (2012) Quantitative trait locus analysis for days-to-heading and morphological traits in an RIL population derived from an extremely late flowering F1 hybrid of sorghum. Euphytica 187(3):411–420

Tao YZ, Henzell RG, Jordan DR, Butler DG, Kelly AM, McIntyre CL (2000) Identification of genomic regions associated with stay green in sorghum by testing RILs in multiple environments. Theor Appl Genet 100(8):1225–1232

Tao YZ, Jordan DR, Henzell RG, McIntyre CL (1998) Construction of a genetic map in a sorghum recombinant inbred line using probes from different sources and its comparison with other sorghum maps. Aust J Agric Res 49(4):729–736

Tao Y, Manners JM, Ludlow MM, Henzell RG (1993) DNA polymorphisms in grain sorghum (*Sorghum bicolor* (L.) Moench). Theor Appl Genet 86:679–688

Teshome A, Patterson D, Asfew Z, Torrance JK, Arnason JT (2007) Changes of *Sorghum bicolor* landrace diversity and farmers' selection criteria over space and time, Ethiopia. Genet Resour Crop Evol 54:1219–1233

Upadhyaya HD, Wang L, Prakash CS, Liu Y, Gao L, Meng R, Wang YH (2022) Genome-wide association mapping identifies an SNF4 ortholog that impacts biomass and sugar yield in sorghum and sugarcane. J Exp Bot 73(11):3584–3596

Upadhyaya HD, Dwivedi SL, Ramu P, Singh SK, Singh S (2014a) Genetic variability and effect of post flowering drought on stalk sugar content in sorghum mini core collection. Crop Sci 54:2120–2130

Upadhyaya HD, Ortiz R (2001) Mini core subset for capturing diversity and promoting utilization of chickpea genetic resources in crop improvement. Theor Appl Genet 102:1292–1298

Upadhyaya HD, Pundir RPS, Dwivedi SL, Gowda CLL, Reddy VG, Singh S (2009) Developing a mini core collection of sorghum for diversified utilization of germplasm. Crop Sci 49:1769–1780

Upadhyaya HD, Sharma S, Dwivedi SL, Singh SK (2014b) Sorghum genetic resources: conservation and diversity assessment for enhanced utilization in sorghum improvement. In: Wang YH, Upadhyaya HD, Kole C (eds) Genetics, genomics, and breeding of sorghum. CRC Press, pp 28–55

Vadez V, Deshpande S, Kholova J, Ramu P, Hash CT (2013) Molecular breeding for stay-green: progress and challenges in sorghum. In: Varshney RK, Tuberosa R (eds) Translational genomics for crop breeding, vol 2. Wiley-Blackwell, pp 125–141

Van de Wouw M, Kik C, van Hintum T, van Treuren R, Visser B (2010) Genetic erosion in crops: concept, research results, and challenges. Plant Genet Resour 8:1–15

Vavilov NI (1992) In: Dorofeev VP (ed) Origin and geography of cultivated plants. Cambridge University Press

Viaene NM, Abawi GS (1998) Management of *Meloidogyne hapla* on lettuce in organic soil with Sudangrass as a cover crop. Plant Dis 82:945–952

Wang YH, Upadhyaya HD, Burrell AM, Sahraeian SM, Klein RR, Klein PE (2013) Genetic structure and linkage disequilibrium in a diverse, representative collection of the C4 model plant, *Sorghum bicolor*. G3 Genes Genomes Genet 3:783–793

Washburn JD, Murray SC, Burson BL, Klein RR, Jessup RW (2013) Targeted mapping of quantitative trait locus regions for rhizomatousness in chromosome SBI-01 and analysis of overwintering in a *Sorghum bicolor* × *S. propinquum* population. Mol Breed 31:153–162

Wendorf F, Close AE, Schild R, Wasylikowa K, Housley RA, Harlan JR (1992) Saharan exploitation of plants 8,000 years BP. Nature 359:721–724

Widmer TL, Abawi GS (2000) Mechanism of suppression of *Meloidogyne hapla* and its damage by a green manure of Sudan grass. Plant Dis 84:562–568

Wilkes G (1991) In situ conservation of agricultural systems. In: Oldfield ML, Alcorn JB (eds) Biodiversity: culture, conservation, and eco-development. Westview Press, pp 86–101

Woldesemayat AA, Modise DM, Ndimba BK (2018) Identification of proteins in response to terminal drought stress in sorghum (*Sorghum bicolor* (L.) Moench) using two-dimensional gel-electrophoresis and MALDI-TOF-TOF MS/MS. Indian J Plant Physiol 23(1):24–39

Wu YQ, Huang Y, Tauer CG, Porter DR (2006) Genetic diversity of sorghum accessions resistant to greenbugs as assessed with AFLP markers. Genome 49(2):143–149

Xu W, Subudhi PK, Crasta OR, Rosenow DT, Mullet JE, Nguyen HT (2000) Molecular mapping of QTLs conferring stay-green in grain sorghum (*Sorghum bicolor* L. Moench). Genome 43(3):461–469

Zou G, Zhai G, Feng Q, Yan S, Wang A, Zhao Q (2012) Identification of QTLs for eight agronomically important traits using an ultrahigh-density map based on SNPs generated from high-throughput sequencing in sorghum under contrasting photoperiods. J Exp Bot 63(15):5451–5462

Finger Millet

3

Mani Vetriventhan, C. Nandini, Ankita Roy, D. Naresh, Sobhan Sajja, Himabindu Kudapa, Sweta Mishra, and Kuldeep Singh

Abstract

The conservation and sustainable use of plant genetic resources are imperative for breeding cultivars with increased genetic gains. Finger millet (*Eleusine coracana* (L.) Gaertn.) is an important staple food crop that is cultivated and consumed largely in Africa and Asia, contributing significantly to food, fodder, and nutritional security. It is grown largely in marginal lands, where productivity is constrained by biotic and abiotic stresses. Globally, over 43,800 germplasm accessions have been conserved ex-situ in the genebanks, particularly in Asian and African countries. Phenotypic and genetic diversity assessment of germplasm indicated considerable diversity, and the diversity of the finger millet is structured based on geographical origin, while racial classification helped breeders choose appropriate panicle types suitable for various growing regions. To support the use of diverse germplasm in crop improvement, germplasm diversity subsets such as "core" and "mini core" collections have been developed. Evaluation of germplasm and diversity subsets has resulted in the identification of promising trait-specific germplasm for use in crop improvement. This chapter describes the status of finger millet germplasm conservation, diversity, promising trait-specific sources, key traits, genomic resources, and their utilization in crop improvement.

M. Vetriventhan (✉) · A. Roy · D. Naresh · S. Sajja · H. Kudapa · K. Singh
International Crops Research Institute for the Semi-Arid Tropics (ICRISAT), Hyderabad, India
e-mail: Vetriventhan.Mani@icrisat.org

C. Nandini
Zonal Agricultural and Horticultural Research Station, Hiriyur, KSNUAHS, Shivamogga, Karnataka, India

S. Mishra
Dr Rajendra Prasad Central Agricultural University, Pusa, Samastipur, Bihar, India

K. Tripathi et al. (eds.), *Plant Genebank Utilization for Trait Discovery in Millets*, https://doi.org/10.1007/978-981-96-4043-0_3

Keywords

Finger millet · Plant genetic resources · Germplasm · Breeding · Utilization · Pre-breeding · Trait-specific sources

3.1 Introduction

Finger millet (*Eleusine coracana* (L.) Gaertn.), an allotetraploid self-pollinated crop with a chromosome count of $2n = 4x = 36$ (AABB) and a genome size of 1.5Gb, traces its evolutionary lineage from its wild progenitor *E. coracana* subsp. *africana* (=*E. africana*) (Devos et al. 2023). Finger millet is widely cultivated and consumed as a staple food in Africa and Asia. The crop thrives under both rainfed and irrigated conditions, including in marginal lands with limited agricultural inputs. It is a small-seeded minor cereal with a light brown to brick red color or dark brown seed coat, rich in phytochemicals, such as polyphenols and dietary fibers (Abioye et al. 2022). Finger millet grains are processed into various edible forms and brewed products through malting, fermentation, milling, cooking, roasting, popping, or puffing (Mubiru et al. 2020). Although finger millet is grown in over 50 countries, largely in Africa and Asia, its actual total area and production of finger millet are not known because production statistics are usually reported together with other millets. However, in some countries, such as India, Ethiopia, Nepal, Uganda, Rwanda, and Tanzania, finger millet production statistics are available because they are grown in considerable acreage. The area and production of finger millet on a global scale are estimated to be approximately 2.12 million ha with an annual production of 3.7 million tons. The major finger millet-producing countries include India (1.8 million tons), Ethiopia (1.2 million tons), Nepal (0.31 million tons), Uganda (0.20 million tons), and Tanzania (0.10 million tons) (Gebreyohannes et al. 2024). Important finger millet-growing countries in Eastern and Southern Africa are the subhumid regions of Ethiopia, Kenya, Tanzania, Zambia, Malawi, the Democratic Republic of Congo, Uganda, and Zimbabwe. In South Asia, this crop is widely cultivated in India, Nepal, Bhutan, and Sri Lanka (Ganapathy 2017; Guarino 2012). The resilience of finger millet to diverse climatic conditions, coupled with its nutritional richness, cultural heritage, and versatile usability, underscores its importance in addressing food security challenges, especially in marginalized communities and regions prone to climatic variability.

Beyond its significance as a staple food crop, finger millet offers additional benefits, such as straw fodder and grains with excellent storage qualities, providing sustenance during periods of food scarcity. Notably, its high calcium (Ca) content (364 ± 58 mg/100 g) surpasses that of other staples such as rice (raw, brown, 10.91 ± 1.79 mg/10 g), wheat (whole, 39.36 ± 5.65 mg/100 g), maize (8.91 ± 0.61 mg/100 g), pearl millet (27.6 ± 2.16 mg/100 g), and even milk (121 ± 3 mg/100 g in buffalo milk, 118 ± 2.9 mg/100 g in cow milk) (Shobana et al. 2013; Kumar et al. 2016; Gupta et al. 2017; Puranik et al. 2017; Antony Ceasar et al. 2018; Vetriventhan et al. 2020). A finger millet-based diet provides 4.4 times more

Ca than a rice-based diet. Upon cooking, the bioavailability of Ca in finger millet reaches approximately 28.6%, comparable to 32.1% in milk, and can be further enhanced through various processing methods (Anitha et al. 2021). Moreover, finger millet is rich in dietary fiber (10–18%), balanced protein (6–13%), minerals (2.5–3.5%), phytates (0.48%), tannins (0.61%), and phenolic compounds (0.3–3%), making it a beneficial dietary choice for all age groups (Chandra et al. 2016; Devi et al. 2014). The conservation of genetic resources is paramount for safeguarding the biodiversity of agricultural crops, ensuring their resilience to environmental challenges, and sustainable utilization for future generations. The conservation and sustainable use of finger millet genetic resources are imperative for breeding cultivars with increased genetic gains. This chapter explores the status of finger millet germplasm conservation, diversity, promising trait-specific sources, genomic resources, and their utilization in crop improvement.

3.2 The Crop

3.2.1 History, Origin, Taxonomy, Domestication

The evolutionary history of cultivated finger millet (*Eleusine coracana* (L.) Gaertn.) traces back to its wild progenitor *E. africana* (=*E. coracana* subsp. *africana*). Domestication of cultivated finger millet, *E. coracana*, started around 5000 years ago in Western Uganda and Ethiopian highlands, and the crop has extended to the Western Ghats of India around 3000 BC, marking India as the secondary center of diversity (Hilu et al. 1979; Hilu and Wet 1976). This has resulted in geographically stratified cultivated germplasm pools (Backiyalakshmi et al. 2021b; Devos et al. 2023). A single nucleotide polymorphism (SNP)-based genetic diversity and population structure assessment of finger millet germplasm indicated two major populations in accordance with their place of origin, Asia and Africa, and the significant population differentiation is mainly due to within individuals between regions or between populations, while races had a negligible impact on population structure (Backiyalakshmi et al. 2021b). This study also indicated a similar average gene diversity in both regions (0.132 in Asia, 0.129 in Africa), which shows the introduction and extensive use of African landraces in the Asian finger millet breeding program, especially for introgression of blast resistance and for new sources of diversity (Backiyalakshmi et al. 2021a, b).

Eleusine is a small genus that includes 9–12 species with 3 basic chromosome numbers, 8, 9, and 10 (Table 3.1) (Dwivedi et al. 2012; Hiremath and Chennaveeraiah 1982; Neves et al. 2005), and is distributed in the tropical and subtropical regions of Africa, Asia, and South America. Both *E. coracana* and *E. africana* have the same genome and chromosome number ($2n = 4x = 36$, AABB). Cytological, isozyme, and random amplified polymorphic DNA (RAPD) studies, chloroplast DNA restriction analysis, and in situ hybridization of diploid and polyploidy species showed that *E. indica* is the "A" genome donor while *E. floccifolia* is the "B" genome donor of cultivated *E. coracana* (Bisht and Mukai 2001; Hilu 1988; Hiremath and Salimath

Table 3.1 Chromosome number, genome, growth habits, and distribution of *Eleusine* species

Name of the species	Genome	Chromosome number	Gene pool	Growth habit	Geographical distribution
E. coracana	AABB	$2n = 4x = 36$	Primary	Annual	India, Nepal, Uganda, Kenya, Ethiopia
E. africana	AABB	$2n = 4x = 36$	Primary	Annual	Malawi, Kenya, Rhodesia, Tanzania
E. indica	AA	$2n = 2x = 18$	Secondary	Annual or perennial	Japan, India, North America, Tanzania, Belgium, Hungary
E. tristachya	AA	$2n = 2x = 18$	Secondary	Annual	South America
E. floccifolia	BB	$2n = 2x = 18$	Secondary	Perennial	Ethiopia, Eritrea, Somalia, Kenya, Yemen
E. kigeziensis	AADD	$2n = 4x = 36 / 38$	Tertiary	Perennial	Uganda, Rwanda, Burundi, and adjacent parts of the Congo
E. multiflora	CC	$2n = 2x = 16$	Tertiary	Annual	Ethiopia, Eritrea, Kenya, Tanzania
E. intermedia	AB	$2n = 2x = 18$	Tertiary	Perennial	Ethiopia
E. jaegeri	DD	$2n = 2x = 20$	Tertiary	Perennial	East Africa

Source: Bisht and Mukai (2002), Philips (1972)

1992; Zhang et al. 2019). In contrary to this, Liu et al. (2014) on the basis of waxy sequences together with multicolor genomic in situ hybridization (McGISH) suggested *E. indica* as the primary A genome parent while *E. tristachya* or its extinct sister or ancestor as the secondary A genome parent for derivation of *E. coracana* and B genome donor as most likely extinct. This is also supported by the close phylogenetic relationships of diploids *E. indica* and *E. tristachya* with *E. africana*, *E coracana*, and *E. kigeziensis* for both cpDNA and nrDNA *Pepc4* (Liu et al. 2011; Neves et al. 2005).

3.2.2 The *Eleusine* Gene Pool

Cultivated finger millet is highly variable within its primary center of origin in Africa and secondary center of diversity in the Indian subcontinent. *E. africana* is completely cross-compatible with the cultivated finger millet, *E. coracana*, and produces fertile hybrids that form the primary gene pool (Hiremath and Salimath 1992; Mehra 1962). The diploid wild species *E. indica*, *E. floccifolia*, and *E. tristachya* form the secondary gene pool, and the tertiary gene pool comprises *E. intermedia*, *E. jaegeri*, *E. kigeziensis*, *E. multiflora*, and *E. semisterilis* (Table 3.1). Based on inflorescence morphology, finger millet can be grouped into races and subraces (Table 3.2) (Rao et al. 1993). *E. africana* has two wild races, *africana* and *spontanea*, whereas *E. coracana* has four cultivated races: *elongata*, *plana*, *compacta*, and *vulgaris*. These cultivated races are further divided into subraces, namely, *laxa*,

Table 3.2 Characteristics of races and subraces in the cultivated gene pool of the finger millet

Race	Prominent inflorescence features	Subraces
Elongata	Characterized by long slender inflorescence branches, digitately arranged, spreading and curved outward at the time of maturity	*Laxa*: long open fingers with spikelets arranged in narrow rows on inflorescence branches, closely resembles wild race *africana* *Reclusa*: short open fingers without any curving out of finger branches *Sparsa*: short open fingers but spikelets are arranged in cluster on the inflorescence branch with naked space
Plana	Large spikelets that are arranged in two, almost even rows along the rachis, giving the inflorescence branch a flat ribbon-like appearance	*Seriata*: the spikelets are serially arranged giving a typical ribbon-like appearance *Confundere*: the fertile florets are numerous and almost surround the rachis at maturity, thus giving a compact appearance to the panicle *Grandigluma*: large, pointed glumes, which are several times longer than the spikelets
Compacta	Spikelets are composed of nine or more florets, sometimes with the inflorescence axis divided at the base, ascending and incurved at the tip to form a fist-like inflorescence	No subraces
Vulgaris	Most common type with 4–8 spikelets giving an appearance of semi-compactness with tip of finger incurved at maturity	*Liliacea*: inflorescence branches are reflexed *Stellata:* inflorescence branches twisted *Incurvata:* inflorescence branches incurved giving fist-like appearance *Digitata:* inflorescence branches top curved

reclusa, and *sparsa* in race *elongata*; *seriata*, *confundere*, and *grandigluma* in race *plana*; and *liliacea*, *stellata*, *incurvata*, and *digitata* in race *vulgaris*. The race *compacta* has no subraces (De Wet et al. 1984; Prasada Rao and de Wet 1997).

3.2.3 Production Constraints

Finger millet production is affected by various biotic and abiotic factors. Various fungal and bacterial diseases cause considerable yield loss when effective management is not implemented. Blast caused by *Pyricularia grisea* is a very prominent disease in finger millet, which affects productivity, utilization, and trade in Eastern and Southern Africa and South Asia. The average loss due to blast has been reported to be approximately 28–36% (Nagaraja and Das 2016), and in certain areas, yield losses could be as high as 80–90% (Vishwanath et al. 1986). Symptoms vary with intermittent rainfall, when leaves are continuously wet for more than 10 h, which promotes rapid development and spread of the pathogen. Rainfall that persists throughout the heading stage causes a finger blast. This pathogen infects the stem, leaves, collar, neck, and fingers and causes significant yield loss. Another important disease in finger millet is blight caused by *Helminthosporium nodulosum* (Kalaisekar et al. 2017). Higher temperatures and humidity are conducive for the emergence of this disease, causing serious threat to crop production. In addition to these two diseases, finger millet is also affected by leaf spot caused by *Cercospora eleusinis* and downy mildew disease caused by *Sclerophthora macrospora* (Nagaraja and Das 2016; Patro et al. 2018). The adoption of appropriate management practices in the initial stages of the crop growth period will aid in reducing the invasion of these diseases. Additionally, *Striga*, a parasitic weed found naturally in Africa, Asia, and Australia, has a significant negative impact on finger millet production, causing yield losses of up to 80% (Atera et al. 2011). In addition to diseases and parasites, finger millet is affected by pests such as pink stem borers (*Sesamia inferens*) (Baladhiya et al. 2018), white borers (*Saluria inficita*) (www.aicrpsm.res.in), earhead caterpillar (*Euproctis subnotata*) (Kalaisekar et al. 2017), and defoliators like cutworm (*Spodoptera exigua* Hübner), armyworm (*Spodoptera* sp.), and black hairy caterpillar (*Estigmene lactinea*) (Kalaisekar et al. 2017). The estimated yield loss due to insect pests in finger millet ranges from 23% to 56% (Mulualem and Melak 2013).

The important abiotic stresses that limit finger millet productivity are drought, salinity, and heat stress. Periodic drought and low, irregular rainfall affect crop productivity in semi-arid and arid regions. Severe drought at flowering and grain filling stages results in 40–70% yield loss in finger millet (Maqsood and Ali 2007; Numan et al. 2021; Tadele 2018). Drought stress in finger millet reduces leaf area, dry matter accumulation, seed weight, biomass, and yield (Maqsood and Ali 2007). Drought stress also increases the activity of superoxide dismutase, ascorbate peroxidase, and glutathione reductase in tolerant varieties (PR 202 and VL 315) but lowers the activity in susceptible varieties (PES 400 and VR 708) (Bhatt et al. 2011). The ascorbate peroxidase/superoxide dismutase ratio, which is a crucial factor in alleviating drought stress, was higher in drought-tolerant varieties than in drought-susceptible varieties under stress. The susceptible varieties recorded maximum stress-induced damage, wherein higher accumulation of malondialdehyde and hydrogen peroxide was observed (Bhatt et al. 2011). Salinity affects almost 20% of the world's farmed area and up to half of the irrigated fields (Rhoades and Loveday 1990). In arid and semi-arid regions, finger millet production is threatened by salinity stress, which affects plant development throughout the growth cycle. Salinity prevents germination in finger millet seedlings, resulting in chlorosis, scorched leaves, slowed growth, and stunted leaves (Mbinda and Masaki 2021; Mukami et al. 2020). Salinity inhibits root and shoot lengths, fresh and dry weights, leaf water potential, and chlorophyll content of finger millet during the vegetative stage (Divya et al. 2022; Mukami et al. 2020). Salinity also interferes with vital metabolic and physiological processes. High-temperature stress has a significant influence on the growth, development, and yield of finger millet. Traits that influence yield, such as panicle number, finger number, finger length, seed number, and seed weight, were significantly reduced by high-temperature stress (Opole et al. 2010; Babu et al. 2013). Babu et al. (2013) established a temperature induction response (TIR) technique for screening high temperature tolerance at the seedling stage in finger millet, wherein they standardized the sublethal, that is, challenging temperatures of 38–54 °C (for 5 h) and lethal temperatures of 57 °C (for 2 h), and found 5 (out of 100) most thermotolerant genotypes with intrinsic heat tolerance.

In addition to biotic and abiotic challenges, finger millet production patterns vary across regions due to several other factors such as climate, soil fertility, water availability, farming methods, economic incentives, cultural preferences, research and development, government policies, awareness, processing, value addition, and market trends. Addressing these issues is crucial for closing the yield gap and achieving crop potential, which could greatly benefit food security and nutrition. Efforts to improve finger millet production must consider these constraints and focus on developing strategies to enhance resilience to these challenges.

3.3 Germplasm Resources Conservation Status

Globally, over 43,800 germplasm accessions of finger millet have been conserved in various national and international genebanks (Bramel et al. 2022). Major collections of finger millet accessions are conserved in India, Kenya, Ethiopia, Uganda, and Zambia. The ICAR-National Bureau of Plant Genetic Resources, New Delhi, India, has the largest germplasm collection, maintaining 11,809 accessions, most of which are indigenous in nature (http://genebank.nbpgr.ernet.in/). The ICRISAT's genebank in Patancheru, India, conserves 7519 germplasm accessions originating from 26 countries. These accessions are conserved in medium-term storage (MTS, temperature 4 °C, relative humidity 20%) and long-term storage (LTS, temperature −20 °C) conditions. Accessions are regenerated when the germination % of the given accessions goes below 85% for cultivated and below 50% for wild species accessions in the MTS and below 90% for cultivated and below 50% for wild species accessions in the LTS. The complete passport and characterization data are available in the ICRISAT Genebank website https://genebank.icrisat.org/. Figure 3.1 shows the panicle and seed color diversity of the finger millet germplasm.

The global status of finger millet germplasm conserved ex situ, as presented in the FAO WIEWS database, is presented in Table 3.3. The global PGR database called Genesys contains about 14,000 accessions that can be accessed to view country- and institute-wise germplasm availability (https://www.genesys-pgr.org/a/overview/v2EYGa9DDlg). Germplasm maintained by local farmers and national and international research institutions could be an excellent source of novel genetic variation for finger millet improvement.

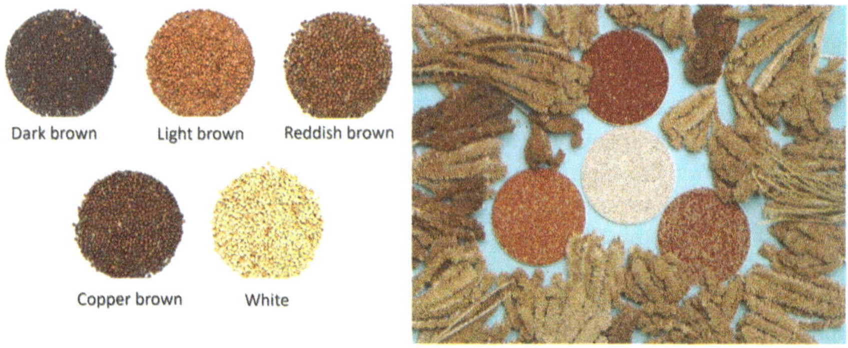

Fig. 3.1 Panicle and seed color diversity of finger millet germplasm

Table 3.3 Status of the finger millet germplasm conserved ex situ in the genebanks globally

Regions	Accessions	Genera	Species	Genebanks	Countries
Sub-Saharan Africa	7,184	1	7	14	13
Northern America	1458	1	6	3	2
Latin America and the Caribbean	44	1	4	6	5
Asia	14,041	1	2	9	9
Europe	122	1	6	10	9
Oceania	44	1	3	2	2
Regional	674	1	2	1	–
International	7557	1	7	3	–
Total	31,124	1	8	47	41

https://www.fao.org/wiews/data/ex-situ-sdg-251/overview/en/

3.3.1 Germplasm Diversity Representative Subsets

The development of diversity representative subsets such as "core" and "mini core" collections is a significant advancement in agricultural biodiversity conservation and for their use in crop improvement. By selecting a representative subset from a larger collection, researchers can streamline the study of genetic diversity and enhance the efficiency of breeding programs. ICRISAT has developed manageable yet diverse subsets of accessions for focused research and development, which include a core collection of 622 accessions (~10% of the total collection) and a mini core collection with 80 accessions (10% of the core collection or 1% of the total collection) (Upadhyaya et al. 2006, 2010). Furthermore, as part of the Generation Challenge Program, a composite collection of germplasm, including 1000 accessions, was created, which included the ICRISAT core collection, Indian NARS core collection, and promising sources identified for biotic and abiotic stresses, grain nutrients, and agronomic traits (Upadhyaya et al. 2006). Simple sequence repeat (SSR)-based genotyping of the composite collection resulted in the development of a reference set (300 accessions) for finger millet. These strategies not only conserve valuable genetic resources but also facilitate the discovery of desirable traits for crop improvement. In addition, the mini core collection, the trait-specific sources identified in the core and mini core collection, new diversity, and breeding lines were included to constitute a diversity subset of 310 accessions at ICRISAT, which was further evaluated for several traits and genotyped using DArTSeq approach for trait discovery (Backiyalakshmi et al. 2021b).

3.4 Germplasm Evaluation

The genetic variability present in the germplasm of finger millet for various important traits could be harnessed only when these resources are evaluated and used in breeding programs. To enhance the use of diversity in varietal development, diverse representative subsets, such as core and mini core collections, have been developed in finger millet. The global collection of finger millet conserved at the ICRISAT Genebank has large variability for several morpho-agronomic, adaptation, quality, and stress tolerance traits. For example, in the core collection of finger millet, the days to flowering varied from 51 to 96 days, plant height from 71 cm to 134 cm, inflorescence length from 51 mm to 166 mm, longest finger length from 46 mm to 130 mm, and fingers per ear from 6 to 9 (Upadhyaya et al. 2010). Core and/or mini core collections have undergone extensive evaluations of various traits, including agronomic performance, grain nutrients (Backiyalakshmi et al. 2023; Upadhyaya et al. 2011), blast (Babu et al. 2013), salinity tolerance (Krishnamurthy et al. 2014), drought resistance (Krishnamurthy et al. 2016), and fodder quality (Backiyalakshmi et al., 2021a, b) (Table 3.4). The assessment of finger millet core collection for grain nutrient revealed substantial variability for Fe (21.71–65.23 mg/kg), Zn (16.58–25.33 mg/kg), Ca (1.84–4.89 g/kg), and protein (6.00–11.09%) content (Upadhyaya et al. 2011) and identified the top 15 promising accessions for each for grain Fe (37.66–65.23 mg/kg), Zn (22.46–25.33 mg/kg), Ca (3.86–4.89 g/kg), and protein (8.66–11.09%) contents and 24 accessions for 2 or more grain nutrients (Upadhyaya et al. 2011). Screening of the finger millet mini core collection under both field and glasshouse conditions resulted in the identification of 66 accessions with combined resistance to leaf, neck, and finger blast. Nine accessions exhibited resistance and favorable agronomic features, such as early flowering, medium plant height, and semi-compact to compact inflorescence (Babu et al. 2013). Similarly, Krishnamurthy et al. (2014) assessed the grain yield performance of finger millet mini core collection in soil saturated with 100 or 125 mM NaCl solution. Salinity delayed phenology and marginally reduced shoot biomass and grain yield in finger millet, and the finger millet mini core (80) accessions were grouped into salinity-tolerant (22), moderately tolerant (20), sensitive (21), and sensitive and late ones (5) based on yield under saline conditions (Krishnamurthy et al. 2014). In another study, the mini core collection was assessed for drought tolerance under the ICRISAT lysimeter facility for the variation in total water use (*T*) and transpiration efficiency (*TE*) (Krishnamurthy et al. 2016). This study revealed a wide range of genotypic variations in drought response, and finger millet uses relatively more

Table 3.4 Finger millet trait-specific genetic sources identified from the ICRISAT collection

Finger millet	Accession ID (IE numbers)	Reference
Early flowering	IE# 49, 120, 189, 196, 234, 501, 509, 581, 588, 600, 641, 694, 847, 2030, 2093, 2158, 2275, 2293, 2322, 2323, 2957, 3104, 3537, 3543, 4425, 4431, 4432, 4442, 4711, 4734, 4755, 4759, 6013, 6550	Bharathi (2011)
Basal tillers	IE# 2296, 2034, 4711, 2293, 2299, 2608, 2619, 5145, 6553, 847, 2408, 2534, 3987, 1013, 120, 2042, 2091, 2106, 2139, 2146, 2233, 2288, 2367, 2410, 2504, 2645, 2657, 2674	Bharathi (2011)
Finger number	IE# 6033, 3790, 4586, 6059, 3111, 4476,3106, 2914, 4677,5733, 5875, 5877, 4257,5105, 5563, 6510, 4297,2957, 5689, 5956, 4563,3120, 2816, 6013, 2303, 2591, 6252, 6241, 4866	Bharathi (2011)
Head length	IE# 2223, 2621, 2789, 6553, 3581, 3431, 3722, 6512, 2108, 2781, 3046, 2486, 5321, 3704, 798, 3489, 5022, 2591, 2608, 4476, 2611, 3531, 2336,4125, 4658, 6546	Bharathi (2011)
Grain yield	IE# 94,2340, 2498,2578,2587,2683,2773, 2903, 2983, 2992, 3194, 3790, 3802, 4600, 4974, 5198, 5472, 3663, 3693, 3744,4121, 4310, 4679, 5862, 6142, 6236, 667, 1010, 2299, 2590, 2678, 2684, 2698, 2712, 2756, 2827, 2872, 3135, 136,3270	Bharathi (2011)
Iron (37.66–65.23 mg/kg)	IE# 4708, 2921, 4709, 588, 5736, 4476, 942, 4734, 5794, 4107, 7338, 2093, 5870, 4443, 817, 9, 956, 2008, 2055, 3935, 4115, 6473, 6645, 7386	Backiyalakshmi et al. (2023) and Upadhyaya et al. (2011)
Zinc (22.46–34.39 mg/kg)	IE# 3120, 7508, 6546, 3025, 7386, 7407, 615, 712, 5788, 633, 2008, 1023, 886, 4817, 510, 2872, 2875, 3443, 3723, 3935, 4115, 4192, 6473, 7386	Backiyalakshmi et al. (2023) and Upadhyaya et al. (2011)
Calcium (3.86–4.89 mg/kg)	IE# 4476, 2030, 6546, 4708, 2568, 2957, 6537, 2608, 2572, 2921, 4443, 2780, 4866, 7386, 4709, 2756, 2869, 2872, 2875, 3814, 4134, 6537, 8599	Backiyalakshmi et al. (2023), Upadhyaya et al. (2011)
Protein (8.66–11.09%)	IE# 6537, 9, 4709, 4708, 6541, 2921, 6546, 4476, 4443, 588, 6013, 2093, 4817, 3120, 3101, 501, 510, 817, 2008, 2030, 2608, 3935, 4698, 4734, 6013	Backiyalakshmi et al. (2023) and Upadhyaya et al. (2011)
Magnesium (1790.01–1908.54 mg/kg)	IE# 954, 2689, 2760, 2872, 2875, 3443, 3973, 4218, 5433, 6533	Backiyalakshmi et al. (2023)
Copper (7.49–8.68 mg/kg)	IE# 2760, 2875, 2911, 2957, 3132, 3157, 3428, 4700, 5367, 6957	Backiyalakshmi et al. (2023)
Sulfur (1421.32–1489.58 mg/kg)	IE# 2008, 2030, 2416, 2430, 2872, 3973, 4073, 4192, 4866, 5433	Backiyalakshmi et al. (2023)

(continued)

Table 3.4 (continued)

Finger millet	Accession ID (IE numbers)	Reference
Phosphorus (3278.40–3457.79 mg/kg)	IE# 817, 2416, 2872, 2957, 3663, 3788, 4218, 6655, 6964, 7508	Backiyalakshmi et al. (2023)
Potassium (5222.10–5551.20 mg/kg)	IE# 2581, 2875, 3758, 3788, 4110, 4497, 4656, 4817, 5433, 6082	Backiyalakshmi et al. (2023)
Sources for multiple grain nutrients	IE# 817, 2008, 2872, 2875, 3935, 3973, 5433, 6473, 7386, 2760	Backiyalakshmi et al. (2023)
Blast resistance	IE# 501, 518, 1055, 2034, 2042, 2217, 2296, 2312, 2430, 2437, 2457, 2572, 2589, 2606, 2619, 2710, 2790, 2821, 2871, 2872, 2911, 2957, 3045, 3077, 3104, 3317, 3391, 3392, 3470, 3475, 3614, 3721, 3945, 3952, 3973, 4028, 4057, 4073, 4121, 4329, 4491, 4497, 4545, 4565, 4570, 4622, 4646, 4671, 4709, 4734, 4757, 4795, 4797, 4816, 5066, 5091, 5106, 5201, 5306, 5367, 5537, 5817, 5870, 6059, 6082, 6154, 6165, 6221, 6240, 6294, 6326, 6337, 6350, 6421, 6473, 6514, 6537, 7018, 7079, 7320	Babu et al. (2013)
Forage yield and quality	Stover yield (SY): IE# 2039, 2045, 2939, 3473, 4699, 4700, 4701, 4707, 5435, 2117¸24, 2568, 2651¸2753, 2796, 2811¸ 2880, 2942, 2979, 3789, 50, 672, 715, 860, 908, 916, 96, 99 In vitro organic matter digestibility: IE# 2217, 2341, 2760, 3129, 3614, 3723, 3821, 4646, 5435, 8790 Multiple quality traits: IE# 2066, 2437, 3129, 3614, 3723, 3821, 3952, 4700, 5435, 8790	Backiyalakshmi et al. (2021a) and Bharathi (2011)
High transpiration efficiency	IE# 5066, 2296, 5165, 2606, 5057, 5106, 3614, 2957, 6421	Krishnamurthy et al. (2016)
Salinity tolerance	IE# 518, 2034, 2217, 2790, 2872, 3045, 3077, 3391, 3470, 3973, 4073, 4329, 4671, 4673, 4757, 4789, 4795, 4797, 5066, 6154, 6165, 6326	Krishnamurthy et al. (2014)

water and has a high *TE*. Both *T* and *TE* positively contributed to shoot biomass under drought stress conditions. Eleven accessions showed high transpiration efficiency, while the remaining were moderate to low (Krishnamurthy et al. 2016). In addition to characterization of the germplasm for biotic/abiotic responses, the collection was also evaluated for grain color. In general, there are two major grain colors in finger millet, brown and white, while brown is predominantly present with varying intensity from light brown to dark brown. White-colored finger millet grains are highly preferred in the baking and food industries. A set of 118 white-colored finger millet accessions available in the ICRISAT Genebank has been made as a subset for researchers to use in breeding white-seeded cultivars. The list/subset of

Table 3.5 Finger millet genetic stock registered and conserved at the National Genebank, ICAR-NBPGR, New Delhi, India

Sr. no.	National identity	INGR no.	Novel unique features
1	IC0598201	INGR14015	Partial sterility, useful in hybridization, easy maintenance
2	IC0614156	INGR16014	High calcium content 452.8 mg/100 g
3	IC635027	INGR20022	Finger blast resistance
4	IC635026	INGR21044	Resistance to neck and finger blast
5	IC0403065	INGR21045	Early flowering (65 days) and high grain yielding
6	IC637583	INGR21046	White grain, blast resistant, medium maturity, and high grain yield
7	IC473958	INGR21047	Resistant to neck blast (3.71 score in 1–9 scale)
8	IC0473970	INGR21048	Resistant to finger blast (2.9 score in 1–9 scale)
9	IC640692	INGR21128	Resistant to foot rot, leaf blast, neck blast, and finger blast. High harvest index and high grain yield
10	IC640693	INGR21129	Broad resistance to finger blast and neck blast
11	IC640697	INGR21132	Neck and finger blast resistance
12	IC640698	INGR21133	Neck and finger blast resistance
13	IC624599	INGR21211	Early flowering and early maturity. Red grain and high yielding
14	IC644006	INGR22022	Banded blight resistance
15	IC644005	INGR22030	Neck blast resistance
16	IC618375	INGR22032	Bold white grains (3.14 g), longer finger length (12.27 cm), multi-fingers/earhead (10.13 cm)
17	IC643979	INGR22033	Long finger length (10.7 cm)
18	IC640694	INGR22096	Early maturity (100 days) and white grain
19	IC646842	INGR22134	Neck and finger blast resistance
20	IC0642429	INGR22135	Early flowering (65 days) and early maturing (99 days)
21	IC0308859	INGR22136	Early flowering (68 days) and early maturing (103 days)
22	IC0331688	INGR22137	High productive tillers (4.7 cm) and tall plant (119 cm)
23	IC647589	INGR23030	Early duration (98.8 days), neck and finger blast resistance
24	IC647590	INGR23031	Banded blight resistance
25	IC647592	INGR23032	Neck and finger blast resistance
26	IC595249	INGR23033	Non-lodging and having uniform maturity
27	IC648610	INGR23034	High number of fingers/earhead (>10 numbers). Long earhead length (13.4 cm)
28	IC648611	INGR23035	Long finger length (13.3 cm) as per DUS test

Source: http://www.nbpgr.ernet.in:8080/registration/InventoryofGermplasm.aspx

accessions with white-seeded finger millet is available at ICRISAT (https://genebank.icrisat.org/) and the Genesys PGR database (https://www.genesys-pgr.org/). Furthermore, assessment of the finger millet diversity panel (310 accessions) conserved at the ICRISAT Genebank for forage quality traits resulted in the identification of promising germplasm for different fodder quality traits, including accessions with multiple forage quality traits (Backiyalakshmi et al. 2021a). Trait-specific genetic stocks registered with ICAR-NBPGR are presented in Table 3.5.

3.4.1 Germplasm Use in Crop Improvement

The productivity of finger millet is low and varies across different growing regions. Various breeding methods, such as pure line selection, pedigree selection, mass selection, and mutation breeding, which are applicable to self-pollinating crops, have also been used in finger millet. Globally, approximately 235 improved finger millet varieties have been released, with India having the highest number (149), followed by Africa (72) (Gebreyohannes et al. 2024; Vetriventhan et al. 2020). Of these, 62% of improved finger millet varieties have been developed following pure line selection and 32% from the recurrent selection approach (Gebreyohannes et al. 2024). In India, 52% of the varieties were developed through hybridization (52%), 41% through selection from landraces, and 6% through mutation breeding approaches. The ICRISAT Genebank distributed over 53,000 seed samples of finger millet germplasm to researchers in 58 countries. A few of them were directly selected and released as cultivars, indicating the direct use of germplasm for variety release (Table 3.6).

The floral morphology, size, and anthesis behavior of finger millet make it the most difficult species for hybridization. Emasculation and crossing techniques have also been reported. Manual emasculation of florets is very difficult in finger millet; therefore, hot water treatment for emasculation and contact methods for pollination are mainly followed by breeders for hybridization (Nagaraja et al. 2023). Following this approach, recombination breeding has been widely used in finger millet since the 1950s in India, whereas in Africa, it started in the early 2000s (Devos et al. 2023). Reports indicate that Asian and African germplasm pools remained separated until the 1950s, while intercrossing between these two pools led to the development and release of popular Indaf varieties in India and their subsequent use as parents in crop improvement (Backiyalakshmi et al. 2021b; Devos et al. 2023).

Although considerable variability exists in finger millet germplasm collections, they have not yet been used for their full potential. To use diversity in crop improvement, germplasm diversity representative subsets called core and mini core collections have been established in finger millet. Extensive evaluation of these subsets resulted in the identification of promising sources for various traits, including yield, maturity duration, grain quality, biomass yield, and stress tolerance traits (Table 3.4).

Table 3.6 List of varieties released in finger millet through direct selection from ICRISAT-supplied germplasm

Accession number	Country of origin	Country released	Release name	Year released
IE 723	Unknown	India	BM 1	1985
IE 524	Unknown	India	VL 101	1978
IE 4115	Uganda	Kenya	KAK-WIMBI 2	2016
IE 1010	Unknown	Uganda	NAROMIL 1	2018
IE 882	Unknown	Uganda	NAROMIL 2	2018
IE 2440	Kenya	Uganda	NAROMIL 5	2018
IE 2947	Malawi	Zambia	FMV 287	1987
IE 2929	Malawi	Zambia	Lima	1987

In India, exotic germplasms from Africa have been largely used in recombination breeding to introduce new variability and disease resistance and contributed to the release of many superior high-yielding varieties (Seetharam 2011). Genetic male sterility (GMS) has also been reported in finger millet. The GMS line, called "INFM 95001," was created in collaboration with the Division of Agronomy at the University of Nebraska in the United States and the ICRISAT in Matopos, Zimbabwe, and Kano, Nigeria (Gowda et al. 2014). However, a maintenance issue with the GMS prevents its use for hybridization in finger millet. Partial male sterile mutant PS 1 (Gowda et al. 2014) is easy to maintain by selfing or growing in isolation. This mutant sets approximately 10% of seeds upon selfing and approximately 20% during open pollination. When the PS 1 seed set was assessed over several seasons, it was found to be consistent (Gowda et al. 2014).

Finger millet germplasms are a vital resource for breeding programs aimed at developing improved varieties. These germplasms contain diverse genetic traits such as disease resistance, drought tolerance, high yield potential, and nutritional quality, which can be incorporated into new cultivars through hybridization and selection. Some finger millet germplasms exhibit superior nutritional qualities such as higher calcium, protein, essential amino acids, vitamins, and minerals. Researchers can use these germplasms to develop varieties with enhanced nutritional profiles, address malnutrition, and improve food security in regions where finger millet is a staple crop. Finger millet germplasm offers valuable genetic resources for the development of climate-resilient varieties. Traits such as heat tolerance, water use efficiency, and pest resistance present in certain germplasms can be crucial for ensuring the sustainability of crops in a changing climate. Germplasms contain many variations in these traits; these germplasms could be effectively utilized in breeding programs to create variation in these traits. Furthermore, implementing a selection germplasm will provide better lines that could be used in breeding programs.

3.5 Pre-breeding

Wild and weedy relatives harbor valuable genes for adaptation, productivity, quality, and stress tolerance traits. The self-pollinating nature and floral morphology of finger millet make it one of the difficult crops for hybridization; however, emasculation and crossing methods have been documented in finger millet (Nagaraja et al. 2023). In finger millet, interspecific hybridization studies have mostly focused on determining genome relationships; however, these studies provide basic information on crossability barriers, genome relationships, etc. (Devarumath et al. 2005, 2010; Mallikharjun et al. 2005; Salimath et al. 1995). The evaluation and utilization of wild gene pools for crop improvement have received limited research attention. *E. africana* is a close relative of cultivated species of *E. coracana*. Earlier reports have indicated that gene transfer from *E. africana* to *E. coracana* is feasible and useful for breeding. *E. africana* has more tillering ability (15–20), has high drought tolerance capacity, matures early (95–100 days), and has more fingers per ear and long finger length (Shet et al. 2010). Interspecific hybridization between *E. africana* and cultivated finger millet varieties (HR 911, PR 201, Indaf 8, HR911, and PR 202) was attempted to transfer some of the desirable characteristics from wild species to popular cultivars. The hybrids were intermediate in productive tillers, finger length, finger number, and days to 50% flowering and exhibited reduced pollen fertility (Shet et al. 2009; Shet et al. 2010). The interspecific hybridization attempted between *E. intermedia* × *E. coracana*, *E. tristachya* × *E. coracana*, *E. africana* × *E. indica*, *E. africana* × *E. floccifolia*, and *E. intermedia* × *E. africana* showed crossability of 3.2–8.3%, and the resultant triploid hybrids showed normal growth and substantial flowering, but the anthers were mostly shriveled, resulting in drastic reduction in pollen stainability (2–8%) in F_1 hybrids, and all these plants were completely sterile (Mallikharjun et al. 2005). Devarumath et al. (2005) produced successful interspecific hybrids, when *E. intermedia* was used as a female parent in two crosses involving *E. indica* and *E. floccifolia* as male parent. However, all F_1s were completely sterile, with E. tristachya as the male parent. Reciprocal crosses of all three combinations (*E. intermedia* × *E. indica*, *E. intermedia* × *E. floccifolia*, and *E. tristachya* × *E. intermedia*) did not yield F_1 hybrids. The F_1 hybrids exhibited characteristics that were intermediate in nature or similar to those of their parents (Devarumath et al. 2005).

3.6 Genomic Resources

The application of genomic tools and molecular marker approaches enhances the understanding and utilization of genetic diversity, leading to improvements in yield, stress resistance, and nutritional quality. Genomic resources in finger millet are considerably less developed when compared with other cereal crops, delaying crop improvement. Although there are some studies on finger millet on genomic resource information, such as molecular markers and physical/genetic maps, these studies are inadequate. Various molecular markers, such as RAPD (Bezaweletaw 2011), amplified fragment length polymorphism (AFLP) (Dida et al. 2007), inter-simple sequence repeat (ISSR) (Brhane et al. 2017), simple sequence repeat (SSR) (Pandian et al. 2018; Gimode et al. 2016; Lee et al. 2017), single nucleotide polymorphism (SNP) (Brhane et al. 2022; Backiyalakshmi et al. 2021b), and expressed sequence tag-SSR (EST-SSR) (Bwalya et al. 2020; Brhane et al. 2021), have been employed to characterize the genetic diversity and population structure of finger millet. The first set of 82 SSR markers was developed in finger millet to facilitate the use of marker-assisted selection. In addition to these identified SSR markers, RFLP, AFLP, and EST markers were used to generate a genetic map of tetraploid finger millet. Covering all 18 chromosomes, the maps span 721 cM in the A genome and 787 cM in the B genome (Dida et al. 2007).

Advancements in next-generation sequencing (NGS) have highlighted the need for more informative genetic and genomic resources. Two draft reference genomes have been published for finger millet (Hatakeyama and Aluri 2018; Hittalmani et al. 2017), and a chromosome-level complete genome assembly of finger millet has recently been made available (Devos et al. 2023). Furthermore, a population structure analysis using the SNPs mined from whole-genome resequencing (~3.5 m SNPs) or genotyping-by-sequencing (GBS, ~10,700 SNPs) data corresponding to 308 *E. coracana* subsp. *coracana* and 22 *E. coracana* subsp. *africana* accessions resulted in two subpopulations separating wild from cultivated accessions. These accessions represent the major finger millet-growing regions in Africa and South Asia (Devos et al. 2023). Other sequencing techniques such as whole-genome resequencing (WGRS), restriction site-associated DNA (RAD) sequencing, genotyping-by-sequencing (GBS), etc. have accelerated the development of genomic resources. SNPs in genomes and populations can be identified and genotyped using GBS, which is a unique NGS application. These genomic resources provide an opportunity for the genomic assessment of finger millet.

Over the last 10 years, several reports have been published detailing the successful application of GBS in large-scale breeding programs for finger millet. These include genomic diversity studies, genetic linkage analysis, molecular marker discovery, and GWAS. Backiyalakshmi et al. (2021b) assessed the genetic diversity and population structure of 306 accessions of the global finger millet collection conserved at the ICRISAT Genebank. This study showed considerable genetic diversity, and the average genetic distance of the studied accessions was 0.301 (range from 0.040 to 0.450); accessions of the race *elongata* showed the highest average genetic distance (0.326) and the least in the race *plana* (0.275). This study also indicated higher genetic divergence between the *elongata* and *vulgaris* races (0.320), while the least was between the compacta and plana races (0.281). Landraces had higher gene diversity (0.144) and genetic distance (0.299) than the breeding lines (0.117 and 0.267, respectively), whereas similar average gene diversity was observed in the accessions in Asia (132) and Africa (129), and the distance between these two regions was 0.327. Geographical origin is a key determinant of population structure in the finger millet germplasm, structured mainly within individuals between regions (Asia and Africa) and between populations, while races showed a negligible impact on population structure (Backiyalakshmi et al. 2021b). A few SNP-based genetic diversity assessments and GWAS were conducted in finger millet for agronomic and grain nutrient traits, and several marker-trait associations were identified (Table 3.7). Puranik et al. (2020) identified a marker-trait association for nutritional traits using 190 finger millet accessions. SNP markers associated with yield, yield-attributing traits, and disease resistance have been assessed in 186 germplasm accessions (Sood et al. 2023). Similarly, Sharma et al. (2022) performed a marker-trait association for calcium content in 202 accessions of finger millet. Integration of marker-trait association studies and genomic techniques into finger millet research greatly improves our understanding and utilization of the genetic diversity of the crop.

Integration of genomic technologies in finger millet breeding programs would facilitate the identification of potential genomic regions, candidate genes, and other traits of interest. Eight QTLs for blast resistance (leaf and panicle blast) and agronomic traits (flowering date, pH, and panicle number) were identified using a linkage mapping approach (Pendergast et al. 2022). Another study identified 55 and 37

Table 3.7 Genome-wide association studies in finger millet

Trait	Marker type	Number of marker-trait association	Population size	References
Days to 50% flowering, basal tiller number, flag leaf width, and plant height	SSRs	5	190	Babu et al. (2014)
Finger blast and neck blast resistance	SSRs	7	190	Babu et al. (2014)
Protein and tryptophan contents	SSRs	3	190	Babu et al. (2014)
Days to maturity, finger number, grain yield per plant, number of grains per spikelet, productive tiller number, and thousand grain weight	SSRs	16	138	Lule et al. (2018)
Days to 50% flowering, days to maturity, and grain yield	SNPs	111	131	Sharma et al. (2018)
Grain yield, days to maturity, and seed protein content	SNPs	81	113	Tiwari et al. (2020)
Iron, zinc, magnesium, sodium, and potassium contents	SNPs	418	190	Puranik et al. (2020)
Calcium content	SNPs	44	202	Sharma et al. (2022)
Days to 50% flowering, days to maturity, plant height, number of basal tillers, flag leaf length, flag leaf width, ear length, ear width, number of fingers, green fodder weight, dry fodder weight, and neck blast	SNPs	132	186	Sood et al. (2023)
Zinc and iron contents	SNPs	45	202	Chandra et al. (2024)

QTLs under low and high phosphorous (P) conditions, respectively, for agro-morphological (primary root length, shoot fresh weight, root fresh weight, shoot dry weight, root dry weight, root hair length, and root hair density) and P (phosphorous) content (inorganic phosphate content in shoots, inorganic phosphate content in roots, and total P content in shoots (TPS)) traits (Maharajan et al. 2023). These developments have opened the door to the creation of improved cultivars that are more productive, resilient, and nutrient-dense, enhancing food security and sustainability in areas reliant on this main crop. Combining genomic tools with traditional breeding methods would be more efficient for targeted crop improvement.

3.7 Future Prospects

In the context of climate change, the world is looking at millet crops to tap into their climate resilience in addition to food and nutritional security. Finger millet is an ideal candidate for this situation, being a rich source of desirable physiological and nutritional traits. The stage is set with numerous genetic resources of finger millet coupled with modern tools and techniques related to breeding and genomics. Finger millet germplasms contain many valuable genes related to productivity traits, nutritional traits, and biotic and abiotic stresses. However, this has not yet been utilized in crop improvement. The availability of advanced crossing techniques and male sterile lines coupled with modern tools and techniques has made it easier to incorporate these genes into our cultivated varieties and create variability for these traits. Numerous varieties have been released worldwide, and all breeding programs have centered on traditional breeding methods such as selection, germplasm introduction, mass selection, and pure line selection. There is an ample scope for extensive research to produce a male sterile system like those seen in other major food crops. This will help in making controlled crosses in finger millet and facilitate a more efficient use of available germplasm resources. Furthermore, high priority should be given to the development of biotic and abiotic resistant cultivars. Blast resistance is the most desirable traits in all future genotypes that are to be developed. A concerted effort by all stakeholders involved may result in the high-yielding and stress-tolerant genotypes that farmers have been looking for. The food industry around millet has also evolved. Thorough research on food processing and product diversification is therefore required. Mechanization is currently minimal in finger millet. Weed management is another important area of concern. If these two persistent problems are solved, many farmers will look forward to cultivating finger millet. Research on different food products, such as "ready-to-cook" or "ready-to-eat," is increasing and needs to be intensified to cater to the needs of the population. Together, these steps contribute to crop success and horizontal expansion.

References

Abioye VF, Babarinde GO, Ogunlakin GO, Adejuyitan JA, Olatunde SJ, Abioye AO (2022) Varietal and processing influence on nutritional and phytochemical properties of finger millet: a review. Heliyon 8(12):e12310. https://doi.org/10.1016/j.heliyon.2022.e12310

Anitha S, Givens DI, Botha R, Kane-Potaka J, Sulaiman NLB, Tsusaka TW, Subramaniam K, Rajendran A, Parasannanavar DJ, Bhandari RK (2021) Calcium from finger millet—a systematic review and meta-analysis on calcium retention, bone resorption, and in vitro bioavailability. Sustain For 13(16):1–14. https://doi.org/10.3390/su13168677

Antony Ceasar S, Maharajan T, Ajeesh Krishna TP, Ramakrishnan M, Victor Roch G, Satish L, Ignacimuthu S (2018) Finger millet [*Eleusine coracana* (L.) Gaertn.] improvement: current status and future interventions of whole genome sequence. Front Plant Sci 9:1054. https://doi.org/10.3389/fpls.2018.01054

Atera AE, Itoh K, Onyango JC (2011) Evaluation of ecologies and severity of Striga weed on rice in sub-Saharan Africa. Agric Biol J N Am 2(5):752–760. https://doi.org/10.5251/abjna.2011.2.5.752.760

Babu TK, Thakur RP, Upadhyaya HD, Reddy PN, Sharma R, Girish AG, Sarma NDRK (2013) Resistance to blast (*Magnaporthe grisea*) in a mini-core collection of finger millet germplasm. Eur J Plant Pathol 135(2):299–311. https://doi.org/10.1007/s10658-012-0086-2

Babu BK, Dinesh P, Agrawal PK, Sood S, Chandrashekara C, Bhatt JC, Kumar A (2014) Comparative genomics and association mapping approaches for blast resistant genes in finger millet using SSRs. PLoS One 9:e99182

Backiyalakshmi C, Babu C, Reddy DN, Padmakumar VP, Prasad KVSV, Azevedo VCR, Vetriventhan M (2021a) Assessing forage potential of the global collection of finger millet (*Eleusine coracana* (L.) Gaertn.) conserved at the ICRISAT genebank. Agronomy 11(9). https://doi.org/10.3390/agronomy11091706

Backiyalakshmi C, Vetriventhan M, Deshpande S, Babu C, Allan V, Naresh D, Gupta R, Azevedo VCR (2021b) Genome-wide assessment of population structure and genetic diversity of the global finger millet germplasm panel conserved at the ICRISAT genebank. Front Plant Sci 12:692463. https://doi.org/10.3389/fpls.2021.692463

Backiyalakshmi C, Babu C, Deshpande S, Govindaraj M, Gupta R, Sudhagar R, Naresh D, Anitha S, Peerzada O, Sajja S, Singh K, Vetriventhan M (2023) Characterization of finger millet global germplasm diversity panel for grain nutrients content for utilization in biofortification breeding. Crop Sci. https://doi.org/10.1002/csc2.21085

Baladhiya HC, Sisodiya DB, Pathan NP (2018) A review on pink stem borer, *Sesamia inferens* Walker: a threat to cereals. J Entomol Zool Stud 6(3):1235–1239

Bezaweletaw K (2011) Genetic diversity of finger millet [*Eleusine coracana* (L.) Gaertn.] landraces characterized by random amplified polymorphic DNA analysis. Innov Syst Design Eng 2(4):207–217

Bharathi A (2011) Phenotypic and genotypic diversity of global finger millet (*Eleusine coracana* (L.) Gaertn.) composite collection, vol Issue 04. Tamil Nadu Agricultural University, Coimbatore, Tamil Nadu

Bhatt D, Negi M, Sharma P, Saxena SC, Dobriyal AK, Arora S (2011) Responses to drought induced oxidative stress in five finger millet varieties differing in their geographical distribution. Physiol Plant 17:347–353

Bisht MS, Mukai Y (2001) Identification of genome donors to the wild species of finger millet, *Eleusine africana* by genomic in situ hybridization. Breed Sci 51(4):263–269. https://doi.org/10.1270/jsbbs.51.263

Bisht MS, Mukai Y (2002) Genome organization and polyploid evolution in the genus *Eleusine* (Poaceae). Plant Syst Evol 233:243–258. https://doi.org/10.1007/s00606-002-0201-5

Bramel P, Giovannini P, Dulloo ME (2022) Global strategy for the conservation and use of genetic resources of selected millets. Global Crop Diversity Trust, Bonn

Brhane H, Haileselassie T, Tesfaye K (2017) Genetic diversity and population structure of Ethiopian finger millet (*Eleusine coracana* (L.) Gaertn) genotypes using inter-simple sequence repeat (ISSR) markers. Afr J Biotechnol 16(21):1203–1209

Brhane H, Haileselassie T, Tesfaye K, Hammenhag C, Ortiz R, Abreha KB, Geleta M (2021) Novel expressed sequence tag-derived and other genomic simple sequence repeat markers revealed genetic diversity in Ethiopian finger millet landrace populations and cultivars. Front Plant Sci 12:735610

Brhane H, Haileselassie T, Tesfaye K, Ortiz R, Hammenhag C, Abreha KB, Geleta M (2022) Novel GBS-based SNP markers for finger millet and their use in genetic diversity analyses. Front Genet 13:848627. https://doi.org/10.3389/fgene.2022.848627

Bwalya EC, Marzougui S, Mwangi E, Wooseon C, Lee M (2020) Genetic diversity analysis and population structure of some African and Asian finger millet (*Eleusine coracana* L.) accessions using expressed sequence tags—simple sequence repeat (EST-SSR) markers. Int J Environ Agric Biotechnol 5(3):643–646

Chandra D, Chandra S, Pallavi, & Sharma, A. K. (2016) Review of Finger millet (*Eleusine coracana* (L.) Gaertn): A powerhouse of health benefiting nutrients. Food Sci Human Wellness 5(3):149–155. https://doi.org/10.1016/j.fshw.2016.05.004

Chandra AK, Pandey D, Sood S, Joshi DC, Tiwari A, Sharma D, Gururani K, Kumar A (2024) Uncovering the genomic regions underlying grain iron and zinc content using genome-wide association mapping in finger millet. 3 Biotech 14:47. https://doi.org/10.1007/s13205-023-03889-1

De Wet JMJ, Rao KEP, Brink DE, Mengesha MH (1984) Systematics and evolution of *Eleusine coracana* (Gramineae). Am J Bot 71(4):550–557

Devarumath RM, Hiremath SC, Rao SR, Kumar A, Bewal S (2005) Genome analysis of finger millet *E. coracana* by interspecific hybridization among diploid wild species of *Eleusine* (Poaceae). Cytologia 70(4):427–434

Devarumath RM, Sheelavantmath SS, Hiremath SC (2010) Chromosome pairing analysis in interspecific hybrids among tetraploid species of *Eleusine* (Poaceae). Cytologia 70(3):299–303

Devi PB, Vijayabharathi R, Sathyabama S, Malleshi NG, Priyadarisini VB (2014) Health benefits of finger millet (*Eleusine coracana* L.) polyphenols and dietary fiber: A review. J Food Sci Technol 51(6):1021–1040. https://doi.org/10.1007/s13197-011-0584-9

Devos KM, Qi P, Bahri BA, Gimode DM, Jenike K, Manthi SJ, Lule D, Lux T, Martinez-Bello L, Pendergast TH, Plott C, Saha D, Sidhu GS, Sreedasyam A, Wang X, Wang H, Wright H, Zhao J, Deshpande S et al (2023) Genome analyses reveal population structure and a purple stigma color gene candidate in finger millet. Nat Commun 14(1). https://doi.org/10.1038/s41467-023-38915-6

Dida MM, Srinivasachary, Ramakrishnan S, Bennetzen JL, Gale MD, Devos KM (2007) The genetic map of finger millet, *Eleusine coracana*. Theor Appl Genet 114:321–332

Divya S, Geetha K, Kumar RS, Prabu PC, Parasuraman P, Kumari NA, Sharavanan PT (2022) Early identification of salt tolerant genotypes in finger millet (*Eleusine coracana* L.) at germination stage by observing the morphological characters. Ind J Agric Res 56:631–637. https://doi.org/10.18805/IJARe.A-6008

Dwivedi S, Upadhyaya H, Senthilvel S, Hash C, Fukunaga K, Xia Min D, Santra D, Baltensperger D, Prasad M (2012) Millets: genetic and genomic resources. Plant Breeding Rev 35:247–375

Ganapathy KN (2017) Improvement in finger millet: status and future prospects. In: Patil JV (ed) Millets and sorghum: biology and genetic improvement. Wiley, Chichester, pp 87–111. https://doi.org/10.1002/9781119130765.ch6

Gebreyohannes A, Shimelis H, Mashilo J, Odeny DA, Tadesse T, Ojiewo CO (2024) Finger millet (Eleusine coracana) improvement: Challenges and prospects—A review. Plant Breeding 143(3):350–374. https://doi.org/10.1111/pbr.13169

Gimode D, Odeny DA, de Villiers EP, Wanyonyi S, Dida MM, Mneney EE, Muchugi A, Machuka J, de Villiers SM (2016) Identification of SNP and SSR markers in finger millet using next generation sequencing technologies. PLoS One 11:e0159437

Gowda MVC, Pushpalatha N, Jadhav SS, Satish RG, Boranayaka MB, Sujay V, Pramila CK, Manasa KG, Gowda J, Ravishankar P, Krishnappa M, Murthy DN (2014) PS-1 (IC0598201; INGR14015), a finger millet (*Eleusine coracana*) germplasm with partial sterility, useful in hybridization and easy maintenance. Ind J Plant Genetic Res 27(2):192

Guarino, L. (2012). *Global strategy for the ex situ conservation of finger millet and its wild relatives*. https://www.croptrust.org/wp/wp-content/uploads/2014/12/Finger-Millet-Strategy-FINAL-14May2012.pdf

Gupta SM, Arora S, Mirza N, Pande A, Lata C, Puranik S, Kumar J, Kumar A (2017) Finger millet: A "certain" crop for an "uncertain" future and a solution to food insecurity and hidden hunger under stressful environments. Front Plant Sci 8:643. https://doi.org/10.3389/fpls.2017.00643

Hatakeyama M, Aluri S, Balachadran MT, Sivarajan SR, Poveda L, Shimizu-Inatsugi R, Patrignani A, Gru S, Baeten J, Francoijs K, Nataraja KN, Reddy YAN, Phadnis S, Ravikumar RL, Schlapbach R, Sreeman SM, Shimizu KK (2018) Multiple hybrid de novo genome assembly of finger millet, an orphan allotetraploid crop. DNA Res 25:39–47. https://doi.org/10.1093/dnares/dsx036

Hilu KW (1988) Identification of the "A" genome of finger millet using chloroplast DNA. Genetics 118(1):163–167

Hilu KW, De Wet JMJ (1976) Racial evolution in *Eleusine coracana* ssp. *coracana* (Finger Millet). Am J Bot 63(10):1311. https://doi.org/10.2307/2441839

Hilu KW, De Wet JMJ, Harlan JR (1979) Archaeobotanical studies of *Eleusine coracana* (finger millet). Am J Bot 63(3):330–333. https://doi.org/10.2307/2442610

Hiremath SC, Chennaveeraiah MS (1982) Cytogenetical studies in wild and cultivated species of *Eleusine* (Gramineae). Caryologia 35(1):57–69. https://doi.org/10.1080/00087114.1982.10796921

Hiremath SC, Salimath SS (1992) The 'A' genome donor of *Eleusine coracana* (L.) Gaertn. (Gramineae). Theor Appl Genet 84(5):747–754. https://doi.org/10.1007/BF00224180

Hittalmani S, Mahesh HB, Shirke MD, Biradar H, Uday G, Aruna YR, Lohithaswa HC, Mohanrao A (2017) Genome and transcriptome sequence of finger millet (*Eleusine coracana* (L.) Gaertn.) provides insights into drought tolerance and nutraceutical properties. BMC Genomics 18:465. https://doi.org/10.1186/s12864-017-3850-z

Kalaisekar PG, Padmaja VR, Bhagwat JV (2017) Insect pests of millets: systematics, bionomics and management. In: Indian Institute of Millets Research (ICAR), Hyderabad

Krishnamurthy L, Upadhyaya HD, Purushothaman R, Gowda CLL, Kashiwagi J, Dwivedi SL, Singh S, Vadez V (2014) The extent of variation in salinity tolerance of the minicore collection of finger millet (*Eleusine coracana* L. Gaertn.) germplasm. Plant Sci 227:51–59. https://doi.org/10.1016/j.plantsci.2014.07.001

Krishnamurthy L, Upadhyaya HD, Kashiwagi J, Purushothaman R, Dwivedi SL, Vadez V (2016) Variation in drought-tolerance components and their interrelationships in the minicore collection of finger millet germplasm. Crop Sci 56(4):1914–1926. https://doi.org/10.2135/cropsci2016.03.0191

Kumar A, Metwal M, Kaur S, Gupta AK, Puranik S, Singh S, Singh M, Gupta S, Babu BK, Sood S, Yadav R (2016) Nutraceutical value of finger millet [*Eleusine coracana* (L.) Gaertn.], and their improvement using omics approaches. *Frontiers*. Plant Sci 7:934. https://doi.org/10.3389/fpls.2016.00934

Lee KJ, Yoon M, Shin M, Lee J, Cho Y, Lee H, Ma K, Lee G (2017) Development of SSR markers and their use in studying genetic diversity and population of finger millet (*Eleusine coracana* L. Gaertn.). Plant Breed Biotechnol 5:183–191

Liu Q, Triplett JK, Wen J, Peterson PM (2011) Allotetraploid origin and divergence in *Eleusine* (Chloridoideae, Poaceae): evidence from low-copy nuclear gene phylogenies and a plastid gene chronogram. Ann Bot 108(7):1287–1298. https://doi.org/10.1093/aob/mcr231

Liu Q, Iang B, Wen J, Peterson PM (2014) Low-copy nuclear gene and McGISH resolves polyploid history of *Eleusine coracana* and morphological character evolution in *Eleusine*. Turk J Bot 38:1–12. https://doi.org/10.3906/bot-1305-12

Lule D, de Villiers S, Fetene M, Odeny DA, Rathore A, Das RR, Tesfaye K (2018) Genetic diversity and association mapping of Ethiopian and exotic finger millet accessions. Crop Pasture Sci 69(9):879–891

Maharajan T, Ajeesh Krishna TP, Rakkammal K, Ramakrishnan M, Ceasar SA, Ramesh M, Ignacimuthu S (2023) Identification of QTL associated with agro-morphological and phosphorus content traits in finger millet under differential phosphorus supply via linkage mapping. Agriculture 13(2):262

Mallikharjun DR, Hiremath SC, Rao SR, Kumar A, Sheelavanthmath SS (2005) Genome interrelationship in the genus *Eleusine* (Poaceae) as revealed through heteroploid crosses. Caryologia 58(4):300–307

Maqsood M, Ali AA (2007) Effects of drought on growth, development, radiation use efficiency and yield of finger millet (*Eleusine coracana*). Pak J Bot 39:123–134

Mbinda W, Masaki H (2021) Breeding strategies and challenges in the improvement of blast disease resistance in finger millet: a current review. Front Plant Sci 11:602882. https://doi.org/10.3389/fpls.2020.602882

Mehra KL (1962) Natural hybridization between *Eleusine coracana* and *E. africana* in Uganda. J Indian Bot Soc 41:531–539

Mubiru D, Recha T, Otieno G (2020) Sensory evaluation of finger millet and bean products in Hoima, Uganda. Report of field work conducted 6–12 September 2019. Bioversity International, Rome, p 25. ISBN: 978-92-9255-200-8

Mukami A, Ng'etich A, Syombua E, Oduor R, Mbinda W (2020) Varietal differences in physiological and biochemical responses to salinity stress in six finger millet plants. Physiol Mol Biol Plants 26:1569–1582. https://doi.org/10.1007/s12298-020-00853-8

Mulualem T, Melak A (2013) A survey on the status and constraints of finger millet (*Eleusine coracana* L.) production in Metekel Zone, Northwestern Ethiopia. Science 1(5):67–72

Nagaraja A, Das IK (2016) Disease resistance in pearl millet and small millets. In: Das IK, Padmaja PG (eds) Biotic stress resistance in millets. Academic Press, pp 69–104. https://doi.org/10.1016/B978-0-12-804549-7.00003-2

Nagaraja TE, Nandini C, Bhat S, Parveen SG (2023) Artificial hybridization techniques in small millets—a review. Front Plant Sci 14:1112117. https://doi.org/10.3389/fpls.2023.1112117

Neves SS, Swire-Clark G, Hilu KW, Baird WV (2005) Phylogeny of *Eleusine* (Poaceae: Chloridoideae) based on nuclear ITS and plastid trnT-TrnF sequences. Mol Phylogenet Evol 35(2):395–419. https://doi.org/10.1016/j.ympev.2004.12.005

Numan M, Serba DD, Ligaba-Osena A (2021) Alternative strategies for multi-stress tolerance and yield improvement in millets. Genes 12(5):739. https://doi.org/10.3390/genes12050739

Opole R, Prasad PVV, Kirkham MB (2010) Effect of high temperature stress on growth, development, and yield of finger millet. ASA, CSSA, and SSSA 2010 International Annual Meetings

Pandian S, Satish L, Rameshkumar R, Muthuramalingam P, Rency AS, Rathinapriya P, Ramesh M (2018) Analysis of population structure and genetic diversity in an exotic germplasm collection of *Eleusine coracana* (L.) Gaertn. using genic-SSR markers. Gene 653:80–90. https://doi.org/10.1016/j.gene.2018.02.018

Patro TSSK, Meena A, Divya M, Anuradha N (2018) Screening of finger millet for major diseases and identification of resistant varieties. J Pharmacogn Phytochem 7(3):2681–2683

Pendergast IVTH, Qi P, Odeny DA, Dida MM, Devos KM (2022) A high-density linkage map of finger millet provides QTL for blast resistance and other agronomic traits. Plant Genome 15:e20175

Philips SM (1972) A survey of the genus *Eleusine* Gaertn. (Gramineae) in Africa. Kew Bull 27(2):251–270

Prasada Rao KE, de Wet JMG (1997) In: Fuccillo D, Sears L, Stableton P (eds) Small millets. Biodiversity in trust: conservation and use of plant genetic resources in CGIAR centers, pp 259–272

Puranik S, Kam J, Sahu PP, Yadav R, Srivastava RK, Ojulong H, Yadav R (2017) Harnessing finger millet to combat calcium deficiency in humans: Challenges and prospects. Front Plant Sci 8:1311. https://doi.org/10.3389/fpls.2017.01311

Puranik S, Sahu PP, Beynon S, Srivastava RK, Sehgal D, Ojulong H, Yadev R (2020) Genome-wide association mapping and comparative genomics identifies genomic regions governing grain nutritional traits in finger millet (*Eleusine coracana* L. [Gaertn]). Plants People Planet 2(6):649–662

Rao KEP, De Wet JMJ, Reddy VG, Mengesha MH (1993) Diversity of the small millets collection at ICRISAT. In: R. K. W., G. S. C., S. A., & M. J. N. (ed) Advances in small millets. Oxford and IBH Publishing Co. Pvt. Ltd, New Delhi, p 557

Rhoades JD, Loveday J (1990) Salinity in irrigated agriculture. In: American Society of Civil Engineers, Irrigation of Agricultural Crops (Monograph 30). American Society of Agronomists, Madison, pp 1089–1142

Salimath SS, Hiremath SC, Murthy HN (1995) Genome differentiation patterns in diploid species of *Eleusine*. Caryologia 195:189–195

Seetharam, A. (2011). Genetic improvement of small millets in India during Pre and Post Crop Coordinated Project era

Sharma D, Tiwari A, Sood S, Jamra G, Singh NK, Meher PK, Kumar A (2018) Genome wide association mapping of agro-morphological traits among a diverse collection of finger millet (*Eleusine coracana* L.) genotypes using SNP markers. PLoS One 13:e0199444. https://doi.org/10.1371/journal.pone.0199444

Sharma D, Tiwari A, Sood S, Meher PK, Kumar A (2022) Identification and validation of candidate genes for high calcium content in finger millet (*Eleusine coracana* (L.) Gaertn.) through genome-wide association study. J Cereal Sci 107:103517

Shet RM, Gireesh C, Jagadeesha N, Lokesh GY, Gowda J (2009) Genetic variability in segregating generation of interspecific hybrids of finger millet (*Eleusine coracana* (L.) Gaertn.). Environ Ecol 27(3):1013–1016

Shet RM, Jagadeesha N, Lokesh GY, Gireesh C, Gowda J (2010) Genetic variability, association and path coefficient studies in two interspecific crosses of finger millet (*Eleusine coracana* (L.) Gaertn.). International Journal of Plant Sciences (Muzaffarnagar) 5(1):24–29

Shobana S, Krishnaswamy K, Sudha V, Malleshi NG, Anjana RM, Palaniappan L, Mohan V (2013) Finger Millet (Ragi, *Eleusine coracana* L.). A review of its nutritional properties, processing, and plausible health benefits. In: Advances in Food and Nutrition Research, vol 69. Academic Press Inc., pp 1–39. https://doi.org/10.1016/B978-0-12-410540-9.00001-6

Sood S, Joshi DC, Rajashekara H, Tiwari A, Bhinda MS, Kumar A, Kant L, Pattanayak A (2023) Deciphering the genomic regions governing major agronomic traits and blast resistance using genome-wide association mapping in finger millet. Gene 854:147115

Tadele Z (2018) African orphan crops under abiotic stresses: challenges and opportunities. Scientifica 2018:1451894. https://doi.org/10.1155/2018/1451894

Tiwari A, Sharma D, Sood S, Jaiswal JP, Pachauri SP, Ramteke PW, Kumar A (2020) Genome-wide association mapping for seed protein content in finger millet (*Eleusine coracana*) global collection through genotyping by sequencing. J Cereal Sci 91:102888

Upadhyaya HD, Gowda CLL, Pundir RPS, Reddy VG, Singh S (2006) Development of core subset of finger millet germplasm using geographical origin and data on 14 quantitative traits. Genet Resour Crop Evol 53(4):679–685. https://doi.org/10.1007/s10722-004-3228-3

Upadhyaya HD, Sarma NDRK, Ravishankar CR, Albrecht T, Narasimhudu Y, Singh SK, Varshney SK, Reddy VG, Singh S, Dwivedi SL, Wanyera N, Oduori COA, Mgonja MA, Kisandu DB, Parzies HK, Gowda CLL (2010) Developing a mini-core collection in finger millet using multilocation data. Crop Sci 50(5):1924–1931. https://doi.org/10.2135/cropsci2009.11.0689

Upadhyaya HD, Ramesh S, Sharma S, Singh SK, Varshney SK, Sarma NDRK, Ravishankar CR, Narasimhudu Y, Reddy VG, Sahrawat KL, Dhanalakshmi TN, Mgonja MA, Parzies HK, Gowda CLL, Singh S (2011) Genetic diversity for grain nutrients contents in a core collection of finger millet (*Eleusine coracana* (L.) Gaertn.) germplasm. Field Crop Res 121(1):42–52. https://doi.org/10.1016/j.fcr.2010.11.017

Vetriventhan M, Azevedo VCR, Upadhyaya HD, Nirmalakumari A, Kane-Potaka J, Anitha S, Ceasar SA, Muthamilarasan M, Bhat BV, Hariprasanna K, Bellundagi A, Cheruku D, Backiyalakshmi C, Santra D, Vanniarajan C, Tonapi VA (2020) Genetic and genomic resources, and breeding for accelerating improvement of small millets: current status and future interventions. Nucleus (India) 1:13. https://doi.org/10.1007/s13237-020-00322-3

Vishwanath S, Sanne Gowda S, Seetharam A, Shankare Gowda BT (1986) Reaction to blast disease of released and pre-released varieties of finger millet from different states. Millet Newsletter 5:31

Zhang H, Hall N, Goertzen LR, Chen CY, Peatman E, Patel J, McElroy S (2019) Transcriptome analysis reveals unique relationships among Eleusine species and heritage of Eleusine coracana. G3: Genes, Genomes. Genetics 9(6):2029–2036. https://doi.org/10.1534/g3.119.400214

Foxtail Millet

4

K. Hariprasanna, P. Rajendrakumar, M. Vetriventhan, R. Narasimhulu, and P. Parashuram

Abstract

Foxtail millet, a self-pollinating short-duration cereal, is one of the oldest cultivated millets in the world, which has originated in China, and its main centre of diversity is East Asia, including China and Japan. From Central Asia, it spread to India and European countries and currently is adapted to a wide range of elevations and environmental conditions. China and India are two major foxtail millet-producing countries for grain purpose, though cultivated area has come down drastically in the recent years. Considerable genetic diversity is present within the cultivated germplasm of foxtail millet. Globally, over 46,000 accessions of foxtail millet have been conserved in the genebanks. China has the largest collection of more than 26,000 accessions, followed by India with over 4600 accessions. ICRISAT genebank conserves a global collection of 1542 foxtail millet accessions from 26 countries. The Indian germplasm exploration has covered almost all the states, but further collections have to be planned from western and eastern states where rich diversity exists. In spite of availability of good genetic variability, the use of germplasm in crop improvement is limited due to difficulties in crossing and obtaining the hybrids. Germplasm representative subsets

K. Hariprasanna (✉) · P. Rajendrakumar
ICAR-Indian Institute of Millets Research, Hyderabad, Telangana, India
e-mail: k.hariprasanna@icar.gov.in

M. Vetriventhan
International Crops Research Institute for the Semi-Arid Tropics, Patancheru, Telangana, India

R. Narasimhulu
Regional Agricultural Research Station, ANGRAU, Nandyal, Andhra Pradesh, India

P. Parashuram
Centre on Rabi Sorghum, ICAR-Indian Institute of Millets Research, Solapur, Maharashtra, India

K. Tripathi et al. (eds.), *Plant Genebank Utilization for Trait Discovery in Millets*, https://doi.org/10.1007/978-981-96-4043-0_4

have been developed along with identification of trait-specific genotypes for agronomic traits, tolerance to biotic and abiotic stress and nutritional parameters for enhanced use in breeding programme. Availability of genomic sequence and large quantum of genomic resources is an advantage in foxtail millet compared to other millets. Use of trait-specific germplasm in recombination breeding with refined crossing techniques and deployment of genomic resources in crop improvement will lead to enhanced genetic gains in foxtail millet in future.

Keywords

Chinese millet · Foxtail millet · Genebank · Germplasm · Grain yield · Protein · *Setaria*

4.1 Introduction

Foxtail millet [*Setaria italica* (L.) Beauv.], known globally as Italian millet, German millet, Chinese millet or Hungarian millet, is one of the oldest grown millets in the world (Oelke et al. 1990) cultivated in about 23 countries in Asia, Africa and America in the dry lands. It is a self-pollinating, C_4 cereal crop and nutritious food for human consumption, feed for poultry and cage birds and fodder for cattle. The foxtail millet is valued as a crop of short life cycle maturing in 60–100 days, providing a grain harvest even under adverse weather conditions. Its cultivation is successful even in low annual rainfall areas and altitudes from mean sea level up to 2000 m above mean sea level (Oduori 1993; Brink 2006). It is mainly grown as a rain-fed crop but yields well under irrigated conditions and good management. It is tolerant to drought and its water-use efficiency (WUE) is found to be better than maize, sorghum and wheat (Zhang et al. 2007).

Regarding its early domestication, foxtail millet was found in large quantities at the early Yangshao site of Banpo in Shaanxi (c. 5000), as well as in several other northern Neolithic centres in China in the warmer postglacial period (Pernes 1983; Jiang et al. 2008), and formed the staple diet of early farming communities in the north (Chang 1977). The very early presence of foxtail millet in central North China was confirmed by a few archaeological remains dating from 7000 to 6500 BC. Throughout the historical period, foxtail millet has been considered as northern rather than southern crop in China (Bray 1981). It was cultivated in Gansu, Shaanxi, Shanxi, Hebei and Henan where foxtail millet was the principal crop for at least four millennia (Yang 1997). In China, it continues to have an important role providing food to millions of people thriving mainly on poor or marginal soils. Currently, it is primarily grown in China, India and the USA, though there is minor presence in Afghanistan, Japan, North Korea, South Korea and Georgia. It is known by different names around the world: in Mandarin it is usually called *Xiǎomǐ*, meaning 'little rice'; in Japanese *Awa*; in Chinese *Ji*, *Su*, etc.; in Korean *Jo* while grains obtained *Jopsal*; in Indonesian *Jewawut*; in Malay *Sekoi*; in Nepali *Kaguno*; in Russian *Mogara* or *Chumiza*; and in Sri Lanka *Thana Haal* in Sinhala. In India it is known by different names in different languages as *Kangni* or *Kakum* (Hindi), *Navane* (Kannada), *Tenai* (Tamil), *Korralu* (Telugu), *Kang* or *Rala* (Marathi), *Kang* (Gujarati), *Kaon* (Bengali), *Kanghu* (Odiya), etc.

4.2 Foxtail Millet: A Brief Background

4.2.1 History, Origin and Domestication

Foxtail millet is believed to have originated in China or Central Asiatic centre of origin. The principal centre of diversity for foxtail millet is East Asia, encompassing China and Japan (Vavilov 1926). From the centre of origin, it probably spread to India and European countries (Oelke et al. 1990). Independent domestication of foxtail millet in China and in Europe was suggested by de Wet and Harlan (1975). Currently, a multiple domestication hypothesis is widely accepted though several hypotheses have been proposed on the origin and domestication of foxtail millet (de Wet et al. 1979). Cytological and genetic studies of the interspecific cross of the cultivated foxtail millet indicated that *Setaria viridis* is the wild ancestor of foxtail millet (Li et al. 1945). It spontaneously hybridizes with cultivated *S. italica* (Li et al. 1945). Phylogenetic analyses involving nuclear and chloroplast genes showed that foxtail and green millet (*S. viridis*) are close relatives (Giussani et al. 2001; Doust et al. 2007). However, the green foxtail is found commonly in various areas of Europe and Asia, and hence, the geographical origin of domesticated foxtail millet could not be determined based on the distribution of green foxtail (Li et al. 2012).

It is believed that domestication of foxtail millet had taken place about 16,000 years ago from its wild progenitor, green foxtail (*Setaria viridis* [L.] Beauv.), based on the evidence collected at an archaeological site in Xiachuan (Wei 1994; Yang et al. 2012). The planting of the crop has been mentioned in Chinese records as early as 2700 BC (Vinall 1924). The earliest archaeological remains of foxtail millet were found in the Cishan ruins of northern China, approximately 7400 years before present, and Peiligang ruins in the Yellow River Valley approximately 7935 years before present (Li and Wu 1996). Recent evidence indicated that the domestication of foxtail millet dates back to 10,000 years before present (Lu et al. 2009; Zhao 2014).

Foxtail millet served as the first staple food in northern parts of China for more than 7000 years until the introduction of maize about 400 years ago (Diao 2017). From China or Central Asia, where foxtail millet originated, it spread into different parts of Eurasia as early as 5000 years ago and contributed greatly to the development of human civilization (He and Hui 2015). In Russia, there is evidence that foxtail millet was grown since ancient times as long as 1500 years ago (Movsisyants 1950). In Europe, presence of foxtail millet cultivation about 4000 years ago itself supports the hypothesis of a second centre of domestication in Europe, since no archaeological evidence of migration or exchange between Europe and China during the Neolithic period has ever been reported.

In Africa, foxtail millet can be found in the upland areas in East Africa, Cameroon and Southern Africa (Brink 2006). In Europe, where it used to be cultivated as a summer crop until the seventeenth century, it has become marginal, and Central Europe is now the main area of production (Panaud 2006). In the USA, the crop was introduced from the old world around the middle of the nineteenth century, mainly as a forage crop, and cultivated as a catch crop, especially for fodder and pet food for cage and wild birds (Baltensperger 2002). A third centre of domestication in an area ranging from Afghanistan to Lebanon was suggested by Li et al. (1995), using morphological traits for studying the diversity in an international collection (Le Thierry d'Ennequin et al. 1999). Thus, three main groups of cultivated foxtail millet gene pool, namely, Chinese (from China, Japan and Korea), tropical (from Taiwan, India and Kenya) and European (Jusuf and Pernes 1985; Panaud 2006), were suggested in the accessions of *S. italica* and *S. viridis* on the basis of isozyme studies.

In China, foxtail millet was used as an important crop as rice, providing about 15–17% of the total food consumed (Baltensperger 2002). Before 1950, foxtail millet was one of the leading crops in many provinces of northern China. It was the third most important food crop with a cultivated area of nearly 10 m ha. In the early 1990s, it was estimated that foxtail millet represented about 18% of grain millet production with about 5 mt of grains (ICRISAT and FAO 1996). However, in the recent years, the area under foxtail millet cultivation has declined drastically due to the expansion of other high-yielding and early-maturing crops like hybrid maize. In the USA, it used to be mainly grown in the northern and western Great Plains, Midwest, the Dakotas, Colorado, Kansas, Nebraska and Wyoming (Oelke et al. 1990; Baltensperger 1996).

In India, foxtail millet used to be an important crop in the rain-fed ecosystem in the semi-arid regions because of its tolerance to drought and short growth duration. It was consumed as a staple diet by rural people in the central and southern states till the introduction of subsidized rice through Public Distribution Scheme in the early 1980s. When the people switched over to rice as their staple due to increased availability, the demand for foxtail millet started declining. Currently foxtail millet is cultivated in a very limited area mostly to meet the domestic needs of the rural people in the states of Andhra Pradesh, Karnataka and Tamil Nadu. It is also found sparsely in the states of Telangana, Maharashtra, Rajasthan, Madhya Pradesh and north-eastern region, where mostly some low-yielding local cultivars are grown. It is still used as an energy source for pregnant and lactating women, for sick people and children and especially for diabetics (Sema and Sarita 2002). Because of high contents of dietary fibre and protein, it finds place as a diabetic food. Foxtail millet releases glucose steadily to the blood stream without affecting the metabolism of the body. The incidence of diabetes is found to be less among the people who regularly consume foxtail millet (Hariprasanna 2017).

4.2.2 Taxonomy and Gene Pool

Foxtail millet belongs to the family Poaceae and tribe Paniceae in the subfamily Panicoideae with diploid genetic constitution ($2n = 2x = 18$) and a small genome size (515 Mb). The genus *Setaria* consists of about 125 species extensively dispersed in warm and temperate regions of the world (Dwivedi et al. 2012). The species comprises crops, wild and weed species with different breeding systems, life cycles and ploidy levels (Li et al. 2012). There are many wild and cultivated types; the cultivated types differ in height, habit, structure of inflorescences, number and colour of grain, etc. (Brink 2006; Prasada Rao et al. 1987). *Setaria italica* is the most economically important genus (Baltensperger 1996). Green foxtail, *S. viridis*, the closest wild relative (de Wet et al. 1979) is a weedy annual. Yellow foxtail [*S. pumila* (Poiret) Roemer & Schultes or *S. glauca* (L.) P. Beauv.] is occasionally cultivated as a cereal crop but yield levels are lower. Bristly foxtail [*S. verticillata* (L.) P. Beauv.] has been harvested as a wild cereal by local people in Africa, Asia, South America and Australia (de Wet et al. 1979; Austin 2006) though not cultivated. Among the species, bristly foxtail, giant foxtail (*S. faberii* Herrm), green foxtail, knot root foxtail [*S. geniculata* (Lamarck) Beauv.] and yellow foxtail are the most widely distributed weedy types worldwide (Rominger 1962).

Based on feasibility of interspecific hybridization and hybrid pollen fertility, the genus *Setaria* is organized into three gene pools. Primary gene pool comprises cultivated foxtail millet (*S. italica*) and its putative wild ancestor green foxtail (*S. viridis*) (Harlan and de Wet 1971). The secondary gene pool contains diploid *S. adhaerens* ($2n = 2x = 18$) and two allotetraploid species, *S. verticillata* and *S. faberii* ($2n = 4x = 36$) (Li et al. 1942; Benabdelmouna et al. 2001). The tertiary gene pool contains *S. pumila* (or *S. glauca*, 4× to 8×) and many other wild species.

The *S. italica* has abundant within-species diversity, mostly for inflorescence. Based on the comparative morphology of the foxtail millet, Prasada Rao et al. (1987) suggested three races of foxtail millet: (1) race *moharia* (common in Europe, southeast Russia, Afghanistan and Pakistan), (2) race *maxima* (common in eastern China, Georgia (Eurasia), Japan, Korea, Nepal and northern India) (also introduced in the USA) and (3) race *indica* (found in the remaining parts of India and Sri Lanka). These races were further divided into ten subraces (*moharia* → *aristata*, *fusiformis* and *glabra*; *maxima* → *compacta*, *spongiosa* and *assamense*; and *indica* → *erecta*, *glabra*, *nana* and *profusa*). Later on, a separate race *nana* was added by Li et al. (1995) along with *moharia*, *maxima* and *indica*. The plants resembling the wild green foxtail with very short and slender stature, many tillers and very short inflorescence and having poor yield and early maturity were grouped under race *nana*. Two phenotypically distinct forms, the waxy (lowered levels of amylose in the grain endosperm, which gives a sticky texture to the grain when cooked) and non-waxy grain type, are distributed globally (Van et al. 2008). In East and Southeast Asia, where local communities prefer sticky cereals like glutinous rice, the waxy-type foxtail millet is more commonly found, while the non-waxy grain type is more widespread and cultivated throughout Eurasia and parts of Africa (Kawase et al. 2005).

Understanding the taxonomic or phylogenetic relationships in the genus *Setaria* is useful for successful selection of parents for hybridization purpose and to ensure sampling from a broad range of genetic variability (Li et al. 1998). The green foxtail is believed to be one of the world's worst weeds of arable land (Harlan 1992). In the course of domestication of foxtail millet, selection for high grain yield through larger inflorescences with a more complex branching pattern, less tillering and axillary vegetative branching has taken place (Harlan 1992; Doust and Kellogg 2002; Doust et al. 2004).

4.2.3 Area and Production

Foxtail millet is adapted to a wide range of elevations, soils and temperatures. It can grow well in sandy to heavy clay loamy soils with pH ranging between 5.5 and 8.3 (Brink 2006; Sheahan 2014). It is also tolerant to saline conditions to some extent (Krishnamurthy et al. 2014). It grows well in places receiving annual rainfall of 500–700 mm and where rains are received during summer. It can grow even at 300–400 mm annual rainfall in semi-arid areas (Brink 2006), but well-distributed rainfall to ensure moisture availability is essential for good yields. It cannot tolerate water-logged conditions due to shallow root system. Foxtail millet is frost sensitive and grows better between 16 °C and 26 °C but has temperature tolerance up to 35 °C.

China had the largest area (9.207 m ha) under foxtail millet cultivation and topped the grain production at 7.797 m tonnes in 1949. But due to low grain yield potential and other marketing reasons, the area under cultivation drastically decreased to 0.718 m ha in 2014 with a total grain production of 1.809 m tonnes (Diao 2017). Reduction in the cultivated area of foxtail millet and other minor millets in China reflects historical changes in Chinese agriculture over the past decades (Diao 2017). In spite of area reduction, foxtail millet remains a leading crop in many arid and semi-arid regions of northern China (Cheng and Dong 2010). India is the third-largest producer of foxtail millet currently, and the crop has importance in the history of agriculture. However, India's contribution to world production of foxtail millet is only about 2%. The estimated area under cultivation during 2015–2016 was 87,000 ha with a production of 66,000 tonnes and productivity of 762 kg/ha. However, area has further declined and production is concentrated only in the two states of Andhra Pradesh and Karnataka. The world area under foxtail millet is estimated to be 1.057 m ha producing 2.29 m tonnes of grains (IIMR Estimates). The contribution of foxtail millet to the total millet production of the world is very meagre (2.4%).

4.2.4 Use and Nutritional Importance

The foxtail millet grain is similar to paddy rice, but small and ovoid in shape, around 2 mm long and pale yellow to orange or brown in colour. The kernel is enclosed in thin hulls and should undergo dehusking before being stored and processed for food

(Ecoport 2011). Dehusking can be done by using a rice processing machine such as a cone polisher or a centrifugal sheller (Brink 2006; Ravi 2004). De-hulled whole grains are usually cooked like rice or made into meal. It is also consumed as stiff porridge (*sargati*) or as leavened bread (*roti*), prepared using the flour from de-hulled grains. Confectionary items like breads, cakes, chips, rolls, noodles, pudding, etc. are also prepared from foxtail millet grains. Grain is fermented to make vinegar and wine in China and to make beer in Russia and Myanmar. Sprouted seeds are also eaten in some regions (Brink 2006).

Foxtail millet grains have high fibre content (Geervani and Eggum 1989; Pawar and Pawar 1997) and low glycaemic index (GI) of 54.5 (Anitha et al. 2021). The protein content is higher among major cereals as well as millets and also the content of dietary fibre (as β-glucans; 42.6%) (Amadou et al. 2011) (Table 4.1). The metabolism of sugar and cholesterol gets increased due to β-glucans resulting in hypoglycaemic and hypocholesterolaemic effects, which are beneficial for prevention of diabetes and cardiovascular diseases (Krishnakumari and Thayumanvan 1997; Itagi et al. 2012). Foxtail millet is used in the preparation of low-GI foods for treating diabetes, particularly type 2 diabetes (Anju and Sarita 2010; Thathola et al. 2010; Itagi et al. 2012; Chhavi and Sarita 2012; Jali et al. 2012), and cardiovascular diseases (Choi et al. 2005). A blood glucose reduction by 14–18%, triglyceride and cholesterol reduction by 8–10%, LDL cholesterol reduction by 5% and HDL cholesterol increase by 2–3% have been observed in the clinical trials of foxtail millet diabetic food on diabetic volunteers. As the foxtail millet grain is highly priced, its use as livestock feed is non-existent in India, though studies on its value in poultry diets have been undertaken (Rao et al. 2000). In developed countries, the grain and by-products from processing like the bran are used as bird feed or in animal feeds (Brink 2006; Heuzé et al. 2015).

4.3 Status of Germplasm Collections

Considerable genetic variability is present within the cultivated germplasm of foxtail millet. Globally, over 46,000 germplasm accessions of foxtail millets have been conserved ex situ in the genebank (Bramel et al. 2022). Collection of landraces of foxtail millet in China started in the 1920s, and at present the Institute of Crop Science, Chinese Academy of Agricultural Sciences (ICS-CAAS) conserves the largest collection of foxtail millet (>26,000 accessions) (Doust et al. 2009; Vetriventhan et al. 2015), 90% of which are landraces. The second largest collection is maintained at ICAR-National Bureau of Plant Genetic Resources (NBPGR), New Delhi (4685 accessions) (http://genebank.nbpgr.ernet.in). At the CGIAR genebanks, 9494 accessions of foxtail millet have been conserved, while at Svalbard genebank, Norway, 1442 accessions have been conserved (Padulosi et al. 2009). The ICRISAT, Patancheru genebank conserves a global collection of foxtail millet (1542 accessions originating from 26 countries). The major collections of foxtail millet germplasm conserved at the different genebanks are presented in Table 4.2. Phenotypic

Table 4.1 Nutritive value of foxtail millet vs. rice and wheat (per 100 g)

Parameter	Foxtail millet	Wheat (whole)	Rice (raw, milled)
Carbohydrates (g)	60.9	71.2	78.2
Protein (g)	12.3	11.8	6.8
Fat (g)	4.3	1.5	0.5
Energy (kcal)	331	346	345
Crude fibre (g)	8	1.2	0.2
Mineral matter (g)	3.3	1.5	0.6
Amylose (%)	17.5	25	12–19
Amylopectin (%)	82.5	75	88–81
Ca (mg)	31	41	10
P (mg)	290	306	160
Essential amino acid (mg/g of N)			
Arginine	220	290	480
Histidine	130	130	130
Lysine	140	170	230
Tryptophan	60	70	80
Phenyl alanine	420	280	280
Tyrosine	–	180	290
Methionine	180	90	150
Cystine	100	140	90
Threonine	190	180	230
Leucine	1040	410	500
Isoleucine	480	220	300
Valine	430	280	380
Vitamins (mg/100 g)			
Thiamine	0.59	0.41	0.41
Niacin	3.2	5.1	4.3
Riboflavin	0.11	0.1	0.04
Vitamin A (carotene) (μg/100 g)	32	64	0
Vitamin B6	–	0.57	–
Folic acid	15	36.6	8
Vitamin B5	0.82	–	–
Vitamin E	31	–	–
Micronutrients (mg/100 g)			
Fe	2.8	5.3	0.7
Zn	2.4	2.7	1.4
Mg	81	138	90
Na	4.6	17.1	–
K	250	284	–
Cu	1.4	0.68	0.14
Mn	0.6	2.29	0.59
Mb	0.07	0.051	0.058
Cr	0.03	0.012	0.004
Su	171	128	–
Cl	37	47	–

Source: Nutritive value of Indian foods (2007), National Institute of Nutrition, India

Table 4.2 Status of global foxtail millet germplasm collections

Organization	Location	No. of accessions	Resource
China National GeneBank (CNGB)	Beijing, China	26,670	Wang et al. (2012) http://icgr.caas.net.cn/cgris_english.html
National Bureau of Plant Genetic Resources (NBPGR)	New Delhi, India	4685	Umesh Chandra et al. (1987) http://genebank.nbpgr.ernet.in
ORSTOM-MONTP	Montpellier, France	3500	Dwivedi et al. (2012)
Svalbard Global Seed Vault	Norway	2505	http://www.nordgen.org/sgsv/
University of Agricultural Sciences (UAS)	Bangalore, India	1750	http://www.smallmillets.res.in/
International Crops Research Institute for the Semi-Arid Tropics (ICRISAT)	Patancheru, India	1542	www.icrisat.org/crop-foxtailmillet.htm
National Institute of Agrobiological Sciences (NIAS)	Kannondai, Japan	1299	http://www.gene.affrc.go.jp/index_en.php
North Central Regional Plant Introduction Station, USDA-ARS	Ames, USA	1000	Dwivedi et al. (2012)
Biologie Végétale Appliquée, Institut Louis Pasteur (IUT)	l'Argonne, Strasbourg, France	850	Dwivedi et al. (2012)
Tamil Nadu Agricultural University (TNAU)	Coimbatore, India	741	www.tnau.ac.in
Kenya Agricultural Research Institute (KARI)	Muguga, Kenya	772	Dwivedi et al. (2012)
Plant Genetic Resources Conservation Unit (PGRCU) of the USDA-ARS	Griffin, USA	762	http://www.ars-grin.gov/npgs/index.html
Estación de Iguala, Instituto Nacional de Investigaciones Agrícolas (INIA)	Iguala, Mexico	350	Dwivedi et al. (2012)

characterization of the collected materials has indicated that foxtail millet germplasm is highly diverse (Li and Wu 1996; Reddy et al. 2006).

Good genetic variability, though not comparable to other major cereals, has been captured in the genebank collection maintained at the ICAR-National Bureau of Plant Genetic Resources, New Delhi. The collection represents a large portion of indigenous germplasm from India belonging to four species under genus *Setaria* (Table 4.3) and 97 exotic accessions of *Setaria italica* (http://genebank.nbpgr.ernet.in). Catalogues have been published for most of these accessions (Umesh Chandra et al. 1987). The accessions of cultivated species, *S. italica*, have been collected from 29 states/union territories across India, and majority of them are from Tamil

Table 4.3 Status of foxtail millet germplasm collection at NBPGR, New Delhi

Species	IC	No. of states	EC	No. of countries
Setaria italica	4528	29	97	15
Setaria italica subsp. *viridis*	51	10	0	–
Setaria pumila	7	6	0	–
Setaria sativum	2	1	0	–
Total	**4588**		**97**	

IC indigenous collection, *EC* exotic collection

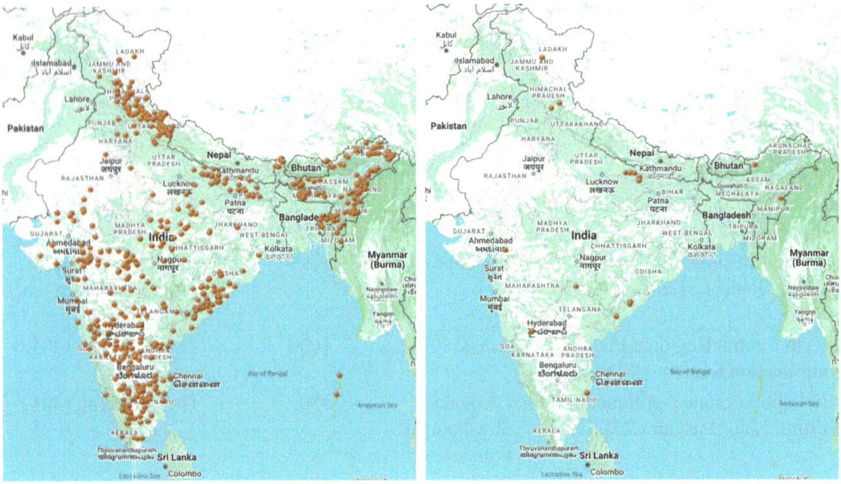

(a). *Setaria italica* (b). *Setaria italica* subsp. *viridis*

Fig. 4.1 Foxtail millet germplasm collection sites in India (**a**) *Setaria italica* (**b**) *Setaria italica* subsp. *viridis*

Nadu followed by Andhra Pradesh and Maharashtra (Table 4.3). The germplasm collection has covered almost all the states but more concentrated in the southern states followed by central region. Collection from western and eastern states, except Bihar, is relatively low (Fig. 4.1). However, most of the states from where germplasm have been collected do not have any reported area under foxtail millet cultivation. Further collections can be planned from these states/regions to identify potential accessions that have good adaptation to regional environment, which may be used in breeding programme to develop cultivars to meet the regional requirements and extend the area to non-traditional areas.

In India, the importance of germplasm resources for crop improvement was recognized even prior to initiation of coordinated project on millets. As a result, the first attempt to collect the available germplasm of millets was made in 1961 under PL 480 Project on ‘Storage, Maintenance and Distribution of Millets Germplasm’. This resulted in the collection of nearly 3000 accessions of different small millets, of which 615 were that of foxtail millet. For long-term conservation of genetic resources of small millets, a germplasm unit came into being at the University of

Agricultural Sciences (UAS), Bangalore, in 1980 (Seetharam 1997; Seetharam and Prasada Rao 1988). During the VII Five Year Plan period of India, a separate All India Coordinated Small Millets Improvement Project (AICSMIP) was launched in 1986 by the Indian Council of Agricultural Research, New Delhi, with its headquarters at UAS, Bangalore, for concerted research and development efforts on small millets. Following this, the germplasm unit became part of AICSMIP. Also, a National Active Germplasm Site (NAGS) for small millets was established in 1992 at UAS, Bangalore, which conserves a total of 1750 accessions of foxtail millet, comprising landraces and cultivars from 13 states in India and other countries like Afghanistan, the USA, USSR, Turkey, Taiwan and African countries. Among these, catalogue on evaluation of 1482 accessions for qualitative (14) and quantitative (11) descriptors has been published. A core collection comprising 95 accessions has been constituted based on the morpho-agronomic diversity in the accessions.

Different State Agricultural Universities which are engaged in foxtail millet crop improvement also maintain working collection as well as active collection in medium-term storage. For example, the Tamil Nadu Agricultural University (TNAU), Coimbatore, India, is conserving germplasm of foxtail millet in its Ramaiah Genebank, which houses both medium-term and long-term cold storage units. Characterization efforts have revealed considerable diversity among 741 accessions maintained for all the agronomic characters (Nirmalakumari and Vetriventhan 2010). The AICRP on Small millets centre at the Regional Agricultural Research Station, ANGRAU, Nandyal, maintains about 1350 accessions of foxtail millet, and these accessions have been characterized and evaluated for yield and related traits.

The ICAR-Indian Institute of Millets Research, Hyderabad, the premier agricultural research institute for the country engaged in basic and strategic research on sorghum and other millets, acts as a NAGS for all the millets. The institute in its genebank conserves 4575 accessions of foxtail millet including 34 wild accessions. All these accessions have been characterized for the standard descriptors, and selected accessions are being evaluated under multilocations to identify trait-specific genotypes for crop improvement.

4.3.1 Germplasm Use in Crop Improvement

Foxtail millet is a highly self-pollinated crop with very low extent of cross-pollination, which may go up to 4% (Li et al. 1935). Natural outcrossing between cultivated foxtail millet and its wild progenitor green foxtail (*S. viridis*) was up to 2.2–4.0% upon bagging different panicles together (Boudry et al. 1993). Interspecific hybridization occurred in the same magnitude depending on the distance between the male and female lines in the experimental plots (Boudry et al. 1993). The crop improvement efforts in foxtail millet started in the early 1940s. Before the launch of coordinated project, the breeding work on small millets was confined to Tamil Nadu, Andhra Pradesh, Karnataka, Uttar Pradesh and Bihar (Seetharam 1998). The focus was on selection of superior genotypes from the local germplasm. In Tamil

Nadu (erstwhile Madras Presidency), the Millet Research Station was established as early as 1923 at Coimbatore, and later research work extended to Anakapalle (presently in Andhra Pradesh) and Hagari (presently in Karnataka). As a result, a few improved varieties (Co 1, Co 2 and Co 3; 1942) were developed mostly through pure line selection in Tamil Nadu. Selection efforts undertaken at Hagari resulted in development of two varieties (H1 and H2; 1948). During the same period, the work carried out in Uttar Pradesh resulted in the release of variety Type 4 in foxtail millet (Paroda 1989).

Though the small millet crop improvement work under the coordinated programme in India started in 1964, the start of a separate coordinated millet improvement project in 1969 helped in giving greater attention to these crops. However, only finger millet among small millets received more attention. Five crop-specific lead centres were established in 1978–1979 with financial assistance from the International Development Research Centre (IDRC), Canada; out of which Nandyal in Andhra Pradesh was established for improvement of foxtail millet. With the launch of AICSMIP, this centre became part of the coordinated project. Under the AICSMIP (currently AICRP on Small millets), the crop improvement efforts in foxtail millet got a new boost. From pure line selection, the breeding method slowly transformed to recombination breeding. Till date, a total of 41 varieties of foxtail millet have been released, of which 12 were released before the launch of AICRP on Small millets and the rest after 1986. Out of these, only a dozen varieties have been developed through hybridization followed by pedigree method.

The success of hybridization in foxtail millet is limited because of difficulty in crossing and obtaining the hybrids. Contact method of crossing or hot water emasculation of male parent is used for hybridization, but the frequency of obtaining true hybrids is very low. Siles et al. (2004) reported about 60–70% higher yield in hybrid than parents. However, proper crossing technique is largely lacking due to difficulties in hand emasculation because of very small spikelet size, clustered nature of spikelets, delicate florets, etc. Also, anthesis behaviour is strange with flower opening taking place late night to early in the morning, small anthers with poor pollen load and high influence of temperature and humidity on flower opening. Because of difficulty in hybridization, the germplasm utilization in crop improvement remains very low. More efficient crossing techniques and deployment of molecular markers for identification of true hybrids among the probable crossed seeds are essential to successfully utilize more germplasm in crop improvement and to develop more productive cultivars.

4.3.2 Developing Germplasm Representative Subsets for Enhanced Use

Difficulties in emasculation and crossing have resulted in limited use of diverse germplasm in foxtail millet improvement. Enhanced use of diverse germplasm in crop improvement is a means to improve the productivity further. Though a good amount of germplasm has been conserved in the genebanks globally, use of such diversity in crop improvement requires a systematic evaluation and identification of trait-specific germplasm subsets. Upadhyaya et al. (2009) developed a core collection (10% of the entire collection) out of total accessions conserved in ICRISAT using passport information and data on 23 morphological descriptors, in order to enhance the utilization of diversity. For establishing core collection in foxtail millet, the total collection of foxtail millet germplasm was stratified into three taxonomic races (*indica*, *maxima* and *moharia*). Through principal coordinate analysis using 12 qualitative traits in each of the biological races separately, 29 clusters were formed. From each cluster, 10% of the accessions were selected to constitute a core collection comprising 155 accessions (Upadhyaya et al. 2009).

In the core collection developed, race *indica* was predominant (102 accessions, 65.8%), followed by race *maxima* (24 accessions, 15.5%) and *moharia* (29 accessions, 18.7%). The major subraces were *nana* (81 accessions, 52.2%) and *glabra* (16 accessions, 10.3%) in race *indica*; *compacta* (20 accessions, 12.9%) and *spongiosa* (3 accessions, 1.9%) in race *maxima*; and *glabra* (17 accessions, 10.9%) and *aristata* (11 accessions, 7.1%) in race *moharia*. Accessions from other subraces were minimal (0.6–2.0%). Accessions from South and Southeast Asia constituted 66.5% (103 accessions), while those from West Asia and East Asia were 12.9% (20 accessions) and 7.1% (11 accessions), respectively. Other regions constituted less than 5% only (Upadhyaya et al. 2009). The variation in the foxtail millet core collection acts as reservoir of genetic diversity and potential sources of beneficial alleles for breeding (Upadhyaya et al. 2009). A core of core collection called mini core collection (Upadhyaya et al. 2011) was developed following evaluation of core collection for morpho-agronomic traits under multiple environments. Twenty-three accessions flowered significantly early (25–40 days), while 19 produced greater grain yields (990–1760 kg/ha) than the check SiA 326 (48 days; 980 kg/ha). The selected accessions were diverse than the control cultivars used in the study.

4.4 Characterization and Evaluation

4.4.1 Yield and Agro-Morphological Traits

Though many varieties in foxtail millet have been developed for different agro-ecologies with high yielding potential, the average yield levels remain low at national level. Strengthening of the hybridization techniques to augment use of diverse germplasm accessions in recombination breeding is essential. Systematic characterization of available germplasm lines for various agro-morphological traits and identification of trait-specific genotypes needs to be strengthened for isolation of better accessions adapted to low inputs, and moisture and other biotic stresses needed for future breeding programmes. Therefore, characterization of diverse foxtail millet germplasm accessions conserved in different global genebanks including India for multiple agro-morphological traits holds immense potential for their utilization in breeding.

Fig. 4.2 Variability for inflorescence size and compactness in foxtail millet accessions

In several isolated efforts, millets researchers across the country evaluated hundreds of germplasm lines collected and conserved in their working collection. Good amount of genetic variability has been found for different agro-morphological traits like plant height, flowering and maturity duration, inflorescence size and shape, number of lobes and compactness, seed colour, etc. (Figs. 4.2 and 4.3). Phenotypic characterization and classification of foxtail millet accessions conserved at the ICRISAT genebank indicated that these collections (1535 accessions from 26 countries) are highly diverse (Reddy et al. 2006; Upadhyaya et al. 2009) for different agronomic traits as well as seed colour (Fig. 4.4). In these collections, days to flowering varied from 32 to 135 days, plant height from 20 cm to 215 cm, basal tiller number from 1 to 52 and inflorescence length from as short as 1 cm to as long as 39 cm (Vetriventhan et al. 2015). Also, evaluation of the core collection resulted in identification of accessions for higher grain yield and early flowering (Upadhyaya et al. 2011). From the collection, a large number of seed samples have been distributed to 45 countries. Some accessions like ISe 376, ISe 375 and ISe 865 have been requested very frequently indicating their genetic worth in breeding programmes.

Evaluation of 741 germplasm accessions including some varieties at Tamil Nadu Agricultural University, Coimbatore, during 2008–2009 indicated that the use of germplasm with more productive tillers, medium panicle length and medium duration can result in good yield advantage (Nirmalakumari and Vetriventhan 2010). A cluster analysis of these accessions resulted in nine clusters which show wide range of variability in the population. At NAGS, UAS, Bengaluru, 1312 germplasm

Fig. 4.3 Variability for seed colour and size in foxtail millet accessions

accessions collected from NBPGR, New Delhi, were characterized using multivariate traits. Characters like days to flowering, plant height, peduncle length, flag leaf length and flag leaf width were the most important traits that contributed to the overall variability. K means clustering method based on 10 quantitative traits among these accessions was able to separate and group them into 37 clusters (Nandini et al. 2018). Effective selection of appropriate donor lines from this collection can result in good genetic improvement. Zonal Agricultural Research Station, V. C. Farm, Mandya, Karnataka, evaluated 30 indigenous collections of foxtail millet for yield and contributing traits and identified diverse accessions (Neethu Mohan et al. 2019).

Characterization of 50 germplasm accessions for DUS descriptors revealed significant variation among the germplasm accessions for all traits studied (Amarnath et al. 2019). In another study, analysis of variance revealed the presence of

Fig. 4.4 Variability for seed colour in foxtail millet accessions at ICRISAT

significant differences among the 50 genotypes for days to flowering and maturity; plant height; contents of fat, carbohydrate, iron, phosphorus and calcium; and grain yield indicating that the collection under study was genetically diverse (Ayesha et al. 2019). Association and path analysis at phenotypic level in 100 elite foxtail millet accessions for 18 yield-related and nutritional traits revealed a significant positive association of grain yield with chlorophyll content, plant height, panicle length, number of grains per panicle and test weight implying that these traits may be chosen as selection criterion for developing high-yielding cultivars (Pavan Kumar et al. 2019). Similarly, other studies on germplasm accessions also indicated a high and positive effect of panicle weight, test weight and straw weight on grain yield in both rainy and summer seasons (Singh et al. 2023). Thus, direct selection for these traits can help in improving grain yield. Single plant stover weight had relatively high heritability with high genetic advance indicating additive gene action in a set of 70 foxtail millet genotypes (Dharnishkumar et al. 2023). Also, the traits plant height, number of productive tillers, panicle length, bristle length, test weight and grain yield per plant had showed additive gene action in their control with very low environmental effects. Six groups were constructed based on genetic distances in 52 accessions of foxtail millet (Karvar et al. 2021). Donor parents selected from diverse clusters for a hybridization programme will help in increasing grain yield.

Table 4.4 Germplasm identified as donor sources for important traits in foxtail millet

Parameter	Germplasm line	References
Grain yield	ISe# 710, 969, 1820, 388, 842, 49, 1888, 90, 364, 1767, 362, 1808, 846, 869, 1511, 909, 1846, 1610, 795, 1458, 1704	Upadhyaya et al. (2011)
Early flowering	ISe# 1312, 1151, 1227, 1201, 1234, 1335, 1286, 1161, 1320, 1647, 1638, 1037, 1181, 1563, 1254, 1204, 1574, 1187, 403, 1118, 1163	Upadhyaya et al. (2011)
Calcium (171.2–288.7 mg/kg)	ISe# 1227, 1181, 1059, 1419, 827, 751, 1474, 1685, 900, 840, 1629, 1851, 769, 1581, 1286, 1136, 1161, 1773, 931, 869, 663	Upadhyaya et al. (2011)
Iron (58.2–68.0 mg/kg)	ISe# 1151, 1286, 1400, 1305, 1332, 1059, 1581, 1320, 1312, 144, 1163, 1460, 160, 1037, 1597, 1009, 1161, 1704, 1187, 1745, 838	Upadhyaya et al. (2011)
Zinc (54.5–74.2 mg/kg)	ISe# 1286, 748, 1387, 195, 1134, 1408, 1419, 1161, 900, 1820, 1320, 1654, 1704, 1605, 403, 1808, 751, 1674, 144, 1234, 985	Upadhyaya et al. (2011)
Protein (15.6–18.5%)	ISe# 1312, 1227, 1789, 1254, 1541, 827, 748, 1305, 1647, 1335, 751, 1118, 1134, 1151, 195, 1234, 1067, 1419, 144, 735, 1161	Upadhyaya et al. (2011)
Blast (resistance to sheath, neck and head blast)	ISe #748, 751, 769, 1059, 1067, 1181, 1201, 1204, 1320, 1387, 1419, 1547, 1575, 1593, 1685	Sharma et al. (2014)
Sheath blight resistance	SiA 3121, RAU 2, SiA 2723, SiA 2863, SiA 3132, SiA 3155, SiA 3156	Patro and Madhuri (2014)
Blast resistance	IC0547319, EC0529830, EC0529829, IC0404112	Das et al. (unpublished data)
Salinity tolerance	ISe #254, 869, 1851, 96, 388, 480, 995, 1629, 969, 1888	Krishnamurthy et al. (2014)
High TE (drought tolerant)	ISe# 200, 238, 289, 480, 525, 783, 796, 840, 869, 1251, 1454, 1511, 1664, 1805, 1881, 1892	Krishnamurthy et al. (2016)
High biomass, grain yield, low DSI, high WE and TE	ISe 426, GS 147, ISe 1059	Talwar et al. (unpublished data)
Low P stress	ISe 1181, ISe 1655, ISe 783 and ISe 1892 showed tolerance to low P	Ceasar et al. (2020)
Shoot fly tolerant	SiA 1538, SiA 1533, SiA 1507, SIA 1581, SiA1566, SiA 1549	IIMR (unpublished data)

Some of the promising as well as trait-specific accessions of foxtail millet identified in different studies are given in Table 4.4.

The Plant Genetic Resources Centre (PGRC), Bangladesh Agricultural Research Institute (BARI), took up evaluation of 246 foxtail millet germplasm lines collected from different stations and traditional farming villages of Bangladesh. Wide variation was observed for both qualitative and quantitative characters like growth habit, plant pigmentation, inflorescence colour and shape, compactness and seed colour.

Plant height, number of tillers, inflorescence length, peduncle length, leaf blade length and leaf blade width also expressed good variation among the accessions. Phenotypic variation of the foxtail millet accessions was found to be related to the diverse geographic origins (Iftekhar Ahmed et al. 2017).

4.4.2 Genetic Studies on Important Yield-Related Traits

Presence of a wider spectrum of variability in the germplasm will enhance the chances of selecting a desired genotype. Besides genetic variability, knowledge on heritability and genetic advance helps the breeder to employ appropriate breeding strategy to achieve the objective. Significant positive association of grain yield per plant with productive tillers and peduncle length has been reported by many researchers (Santhakumar 1999; Murugan and Nirmala Kumari 2006; Nirmalakumari and Vetriventhan 2010; Upadhyaya et al. 2011; Brunda et al. 2014). Similarly, flowering duration had significant positive association with plant height, panicle length and flag leaf length. Panicle length was positively related with panicle width, peduncle length, flag leaf length and flag leaf width. The positive association of yield with other related characters indicates possibility of simultaneous improvement.

4.4.3 Abiotic Stress Tolerance

Foxtail millet, because of its cultivation mostly on poor and marginal soils in arid and semi-arid region, frequently gets affected by a few abiotic and biotic stress factors. Though millets are tolerant to majority of environmental stresses and adaptable to the impending climate change, further improvement is essential to enhance their resilience to the unexpected consequences of climate change and related environmental stresses. Among abiotic stresses, drought, salinity, water logging, heat (high temperature) and lodging have a negative impact on crop yield and quality. Drought and salinity are the two leading causes of crop loss globally. Due to moisture stress, foxtail millet often has fewer tillers, shorter shoots, lower photosynthesis, reduced panicle length and volume, and seed size leading to lower harvest index, which ultimately causes a loss in yield. Salinity causes mineral imbalances, ion toxicity and osmotic stress, which lower photosynthetic efficiency and eventually lower biomass production. In order to improve the yield of foxtail millet further, there is a need to develop stress-tolerant and climate-resilient varieties.

Foxtail millet has attracted substantial attention from the biological researchers because of its projection as an excellent biological model for studying architecture, evolution, physiology and stress tolerance attributes (Lata et al. 2013). Considerable work has been done towards germplasm characterization and understanding the genetic basis of abiotic stress tolerance, particularly drought, salinity, water logging and lodgings in foxtail millet. Some progress has been made towards understanding the physiology of abiotic stress tolerance and the genomic regions associated with drought stress tolerance.

4.4.3.1 Drought

Drought or moisture stress is one of the most frequent abiotic stresses in many arid and semi-arid regions. Seasonal drought may also occur occasionally in non-arid regions, which can have a negative impact on crop growth and productivity. Foxtail millet is known as a relatively drought-tolerant crop, because of its morphological and anatomical features such as thick cell wall, robust reticulate root system, small leaf area and excellent water-use efficiency (Li 1997). However, the natural tolerance mechanisms to drought can be overwhelmed by severe or prolonged drought. Thus, it is crucial to screen genetic resources in order to identify drought-resistant accessions. The primary challenge in any crop is precisely identifying the drought resistance indicators as it involves looking at several indicators at different stages. Enhancing our knowledge of traits linked to drought resistance might lead to new avenues for foxtail millet crop improvement. Recent investigations have examined the morphology, physiology and biochemistry of drought indicators to identify drought resistance in various crops using a range of identification techniques (Albacete et al. 2014).

The germplasm accessions of foxtail millet must be analysed at different stages like germination, seedling and maturity in order to discover drought resistance (Cattivelli et al. 2008). By examining the drought resistance at various stages, researchers found that they could accurately determine the drought resistance of foxtail millet at its maturity stage (Zhang et al. 2012a). The evaluation of drought resistance has been precisely analysed and identified in recent years by research on drought-resistant genes and proteins (Ghatak et al. 2016; Parvathi et al. 2019). However, due to the high cost of detection, it is challenging to implement the screening on a broad scale. Researchers have evaluated multiple indicators using a combination of principal component, membership function, cluster and stepwise regression analyses in order to avoid the one-sidedness and instability of a single index in the process of evaluating drought resistance (Wang et al. 2017a; Xiao et al. 2021, 2022). Previous assessment techniques did not emphasize much on physiological markers and were unable to accurately depict the absolute yield of foxtail millet under drought conditions. It was thus required to conduct additional assessments of the yield of foxtail millet in dryland conditions using the drought resistance index (DRI) technique and to examine the connection between physiological attributes and drought resistance traits by comprehensive evaluation approach. The grain weight per panicle, grain yield, panicle weight and panicle length of foxtail millet can be used as simple and intuitive drought resistance evaluation indicators. Net photosynthetic and transpiration rates could also be used as reference indicators to further determine the drought resistance in foxtail millet.

Begum et al. (2013) in an experiment under irrigated and rain-fed conditions screened 20 genotypes including the test crop (Titas). It is observed that 17 lines, considering plant height stress index, showed tolerance above the test crop, while 9 lines showed better yield stability ratio. Considering all the parameters, lines BSi-1, EM 15/BSi 467 and EM 8/BSi 467 showed better performance. Krishnamurthy et al. (2016) evaluated 155 accessions of germplasm lines belonging to three races, viz. *moharia*, *maxima* and *indica*, in mini-lysimeters under both terminal drought stress (DS) and well-watered (WW) environments over 2 years. Ten accessions (ISe 2, 49, 96, 289, 1454, 1468, 1704, 1774, 1788 and 1859) were found to be drought tolerant with highest grain yields. Several genes have been identified in foxtail millet including the LEA gene SiLEA14 that plays a crucial role in abiotic stress tolerance (Loni et al. 2023). It is reported that overexpression of SiLEA14 improved the tolerance of foxtail millet to abiotic stresses (Wang et al. 2014a). Similarly, overexpression of SiARDP in foxtail millet induced the expression of drought and salinity-relevant genes and increased the abiotic stress tolerance (Li et al. 2014). Thus, SiARDP and SiLEA14, involved in salt and drought tolerance, have been functionally characterized in foxtail millet (Li et al. 2014; Wang et al. 2014a). Quantitative trait loci (QTLs) in foxtail millet were identified which contribute to a number of traits that regulate drought tolerance including promptness index, radical root length, coleoptile length and lateral root number at germinating stage (Qie et al. 2014). A total of 18 QTLs were detected which included 9 that explained over 10% of the phenotypic variance for a given trait. Several QTLs, which are associated with biomass, plant height, panicle length and diameter and panicle weight, have been identified for drought tolerance in foxtail millet (Fang et al. 2016; Wang et al. 2014a; Zhang et al. 2017).

4.4.3.2 Salinity

Foxtail millet is a promising crop for salt-affected fields because of its great tolerance to salinity (Maas 1985), and its short growing season helps in escaping from salt stress. However, the growth, productivity and quality of foxtail millet are negatively impacted despite its high tolerance to such stress (Panchal et al. 2023). Harvest index, 1000-grain weight and seed and straw yields of foxtail millet were found to decline linearly with saline water irrigations with salinity increasing from 1 to 16 dS/m (Thimmaiah et al. 1989). A decrease in N, P and K uptake by 36%, 56% and 42%, respectively, in crops under salinity stress has been reported by Hussain et al. (2015), which may be the primary cause for crop yield reduction under drought and salt stress.

Considerable variation for salt tolerance has been reported within foxtail millet genotypes (Puranik et al. 2011; Hendawy et al. 2012). The salinity tolerance in foxtail millet genotypes is mostly determined by a single indicator that mostly relies on germination rate. For salinity stress screening at vegetative stage, the shoot Na + concentration may be a viable non-destructive selection criterion (Krishnamurthy et al. 2007). In a rapid salinity screening method, based on the root length, relative seed germination (7-day-old seedlings), root growth and plasma membrane integrity of the root tip under salinity at the seedling stage of foxtail millet, accessions ICERI-5 and ICERI-6 were identified as tolerant, while ICERI-10 was identified as sensitive (Ardie et al. 2015). Akter et al. (2019) reported that genotype BD-878 exhibited a comparatively better tolerance to salt stress, while BD-897 was susceptible in relation to yield. Improved salinity tolerance in BD-878 was linked to higher relative leaf water and chlorophyll content with greater proline accumulation and lower malondialdehyde content accumulation. Krishnamurthy et al. (2014) screened 155 accessions of foxtail millet core collection, along with 5 control accessions (ISe 375, 376, 1468, 1470 and 1541) in a soil saturated once with 100 mM NaCl and in a non-saline control, and a subset of 84 accessions in the next season in a partly controlled environment. Out of them, 82 accessions, common to both, were grouped into 5 sets based on grain yield under saline conditions, and the most highly tolerant group had 13 accessions. Accession ISe 869 was the most reliable and excellent salt-tolerant genotype in the study. Though QTLs for traits related with drought tolerance have been reported, the QTLs related to salt tolerance in foxtail millet are yet unclear due to a paucity of germplasm screening under salt stress (Wu et al. 2023).

4.4.3.3 Other Stresses

Under climate change scenario, changes in rainfall patterns apart from total precipitation are expected, which might lead to either prolonged dry spells or spells of waterlogging. In general, foxtail millet is sensitive to waterlogging, and it has a negative effect on stomatal closure, photosynthesis and root architecture by inhibiting gas diffusion. In China, cultivar Lugu No. 7 was developed that can withstand waterlogging (Chen and Qi 1993). Other abiotic challenges that foxtail millet may face include cold temperatures and lodging. The development of the extremely cold-tolerant cultivar Liggu No 26 made it possible to expand the cultivation of foxtail millet to 54° N in northern China (Chen and Qi 1993; Dwivedi et al. 2012). Studies on cold tolerance in foxtail millet or screening the germplasm are rare. Zhao et al. (2023) studied RNA-Seq data of foxtail millet at different developmental stages and different cold stress periods, which provided new ideas for stress resistance in foxtail millet and the key genes of cold stress tolerance.

Lodging is a limitation, which affects grain yield and quality and poses difficulty in mechanized harvesting of foxtail millet. Because of thin stem, foxtail millet is badly affected by lodging sometimes (Dwivedi et al. 2012). Lodging can result due to climatic conditions like wind and rain, irrigation or combination of these. Stem lodging is characterized by stems which bend or break towards the ground, while root lodging is related to disanchoring due to wind force acting on the stem or panicles especially towards grain filling stage (Van Delden et al. 2010). Stem lodging affects the photosynthetic effectiveness of the leaves, thereby affecting the grain filling. A coefficient based on stem and root traits may serve as a suitable indicator for determining lodging resistance in foxtail millet (Tian et al. 2010). The mechanical strength of the stem, lower plant height and increased weights of the above and underground tissues may impart lodging resistance in foxtail millet germplasm (Tian et al. 2010; Dwivedi et al. 2012). Path analysis studies in foxtail millet genotypes revealed that the most important component influencing the lodging resistance is the breaking strength of stem, which is correlated with larger culm diameter and culm wall thickness (Tian et al. 2015). Mechanical strength of the second internode from ground, fresh weight per plant and height of the centre of gravity were important traits to evaluate lodging resistance (Du et al. 2019). China has developed two cultivars (Longgu 28 and Nenxian 13) that are resistant to lodging (Dwivedi et al. 2012).

4.4.4 Biotic Stress Resistance

Climate-resilient millets experience fewer negative effects due to biotic stress than other crops. According to Dwivedi et al. (2012), millet crops are mostly affected by fungal diseases, whereas bacterial and viral diseases are minimum. Foxtail millet is mostly affected by blast, downy mildew, rust and other fungal diseases. Blast fungi (*Magnaporthe grisea*) can infect all parts of foxtail millet plant in different growing stages, resulting in seedling blast, leaf blast, stem blast and panicle blast, but panicle blast is the most harmful for grain yield (Liang et al. 1959; Yan et al. 1985, 1988). In China, foxtail millet blast is one of the most severe diseases and usually causes yield losses above 30%, although losses can be more than 80% for the most sensitive cultivars in certain environments (Li et al. 2021). *Bipolaris australiensis*, a novel leaf and sheath brown spot fungus, has also been found to infect this crop recently (Mirzaee et al. 2010). The most significant bacterial disease infecting foxtail millet is bacterial blight (*Xanthomonas* spp.). In addition, this crop harbours the virus itself as well as wheat curl mite (*Eriophyes tulipae* Keifer), which carries the wheat streak mosaic virus (Baltensperger 2002). These viral diseases, though do not impact the well-adapted foxtail millet cultivars, serve as an over-summering host and found to affect adjacent wheat fields in the USA. Rice black-streaked dwarf virus, rice stripe virus and barley virus G (BVG; in Korea) and barley yellow dwarf virus and barley yellow striate mosaic virus causing dwarf disease (in China) have been detected on foxtail millet (Oh et al. 2017; Shen et al. 2020). Suspected barley virus G (BVG) infection on foxtail millet was also reported by Baswaraj et al. (2023), which needs to be ascertained further.

In contrast to major cereal crops, foxtail millet has not received much research attention on disease management as severe incidence of disease or yield loss is not reported so far in India. In a study by Sharma et al. (2014), 155 germplasm accessions were screened against blast (Patancheru isolate Fx 57). Twenty-one accessions in field and 11 accessions in greenhouse were reported with neck and head blast resistance. Additional screening revealed that two accessions (ISe 1181 and ISe 1547) showed resistance to leaf, neck and sheath blasts, when tested against four isolates (Fx 57, Fx 58, Fx 60 and Fx 62). Accessions ISe 1067 and ISe 1575 exhibited strong blast resistance. Munirathnam et al. (2015) screened the foxtail millet genotypes for 2 years against leaf blast and rust. The study recommended a moderate resistance genotype DHFtMV 2–5 for the areas that are prone to both blast and rust and other genotypes SiA 3221, SiA 3222 (Garuda) and SiA 3223 (Renadu) for the rust-prone areas. After assessing 15 genotypes of foxtail millet against rust disease, Saralamma et al. (2019) found that 7 entries (DHFt 5-6, DHFt 77-3, PKS 22, SiA 3212, SiA 3219, DHFt 35-3, DHFt 2-5-3) were vulnerable to rust disease and none of the entry had a resistant reaction. In another study, two entries SiA 3205 and SiA 3164 exhibited highly resistant reaction against blast and rust disease (Rajesh et al. 2019). In a study by Ramana et al. (2022), seven genotypes, viz. SiA 1135, SiA 3292, SiA 3318, SiA 3399, SiA 3595, SiA 4193 and Suryanandi, were found resistant to blast. Similarly, nine germplasm accessions, viz. GS9, GS15, GS19 ISe1026, ISe1136, Se1227, ISe1638, ISe1647 and ISe1808, were found highly resistant to blast (Anand et al. 2023). Makwana et al. (2023) reported 23 foxtail millet genotypes as highly resistant to blast.

In China, field estimation of blast resistance in 1637 accessions of foxtail millet showed that 15% of the tested accessions were resistant to blast (Wu 1985). A similar study was reported by Liu et al. (1990). As there is no control over field conditions, measurement of blast resistance was carried out under artificial greenhouse conditions by Xie et al. (1984) and Wang et al. (1985), and many foxtail millet landraces with blast resistance in the seedling stage were identified. The blast resistance capacity of foxtail millet landraces in the late growing stage was positively correlated with that in the seedling stage. After inoculating a core collection of foxtail millet comprising 888 accessions with the blast strain HN-1, Li et al. (2021) reported less than 1.60% of the samples as highly resistant, 35.25% as moderately resistant, 57.09% as moderately susceptible and 6.08% as susceptible. Analysis of the blast-resistant genotypes revealed that a high proportion of resistant accessions were located in lower latitude regions where high rainfall occurs during the foxtail millet growing season. Genome-wide association studies (GWAS) using 720000 SNP markers showed that two genomic regions from chromosomes 2 and 9 were strongly linked with blast disease resistance in foxtail millet, and by utilizing transcriptome data associated to the rice blast, eight putative candidate genes were identified (Li et al. 2021).

Reports on screening accessions for other diseases are very limited. Two accessions, SiA 2863 and ISC 74A, were found resistant to banded blight disease in foxtail millet (Patro et al. 2018). For sheath blight disease, none of the genotypes out of 16 screened were highly resistant, and TNSi 417 and TNSi 419 were recorded to have susceptible reaction (Patro et al. 2020). Much higher than the potential disease damage, the bird damage is now considered a potential threat to foxtail millet, depending on growing conditions. The crop loss may go up to 100% in isolation and unprotected conditions irrespective of varietal differences or panicle morphology.

4.4.4.1 Nutritional Traits

In one of the earliest studies, Seetharam et al. (1983) reported that the protein content in foxtail millet ranged from 7.16% to 15.73% and identified lines with high protein content for both direct exploitation and use in breeding. Protein content in the three selected varieties of foxtail millet ranged from 9.17% to 11.2% (Kumar and Parmeswaran 1998), while a higher average protein content of 15.9% in foxtail millet was reported by Ravindran (2003). Yang et al. (2013) assessed seed protein, fat, starch and amino acid content in 259 foxtail millet cultivars from 6 provinces in China using Fourier transform near-infrared spectroscopy, and the results showed a wide range of variation for protein (11.85–20.58%), fat (2.82–4.47%), starch (65.59–74.12%) and amino acids.

In one of the early studies, Phillip and Maloo (1996) had evaluated 40 foxtail millet genotypes for their iron (Fe) content using calorimetric method and grouped the varieties into high (0.09–0.11% Fe), medium (0.07–0.08% Fe) and low (0.05–0.06% Fe) based on grain Fe content. Further, 21 diverse accessions promising for different grain nutrients were identified in the core collection of foxtail millet at ICRISAT (Table 4.4) (Upadhyaya et al. 2011). The grain nutrient content assessment showed significant variability. Grain calcium (Ca) content varied from 90.3 to 288.7 mg/kg, Fe from 24.1 to 68.0 mg/kg, zinc (Zn) from 33.6 to 74.2 mg/kg and protein from 10.7 to 18.5% (Upadhyaya et al. 2011). Grain protein (%) and minerals (Zn, Fe, Ca) were studied in 25 genotypes of foxtail millet chosen from 149 genotypes based on their diversity and superiority for various traits. Good range was recorded for grain nutrients, viz. protein (6.2–10.89%), Zn (22.9–114 mg/kg), Fe (3.3–162.6 mg/kg) and Ca (19.9–226.9 mg/kg) content, among the selected genotypes (Thippeswamy et al. 2017). Accession GS 95 had the highest protein content (10.89%), higher than the check cultivars PS 4, HMT 100-1 and SiA 326. Two agronomically superior lines, GS 78 and GS 71, were also found to be nutritionally superior.

Foxtail millet accessions with high vitamin E contents among 400 lines had been identified using high-performance liquid chromatography (Li et al. 2009). Shao et al. (2014) evaluated folic acid (vitamin B9) content of 245 foxtail millet traditional varieties originating from different regions in Shanxi, China, and observed wide genetic variations (0.37–2.37 μg/g with a mean of 1.53 μg/g), and the folic acid content of Jingu 21, a major leading cultivar, was recorded at 2 μg/g among a total of 24 varieties with higher folic acid content.

4.5 Pre-breeding Efforts

The wild relative of foxtail millet, *S. viridis*, belonging to the primary gene pool could be an interesting source for improvement of cultivated species. Naciri et al. (1992) performed an interspecific cross between *S. italica* and *S. viridis*, to introduce triazine resistance into the cultivated foxtail millet. Only two backcross generations along with selection were sufficient to eliminate weedy characters in the segregants. Darmency and Pernes (1987) demonstrated the potential of interspecific hybridization and studied the inheritance of eight morphological characters in the F_2 progeny of an interspecific cross between *S. viridis* and *S. italica*. Darmency et al. (1987) also reported polyploidization following interspecific hybridization between cultivated and green foxtail millet, to improve the traits of cultivated foxtail millet. The tetraploidization resulted in a shift in characteristics towards the crop species, especially, a twofold increase in seed weight. Nonadditive gene effects were found for most characters, except for the seed shedding, which was under the control of at least four loci (Darmency et al. 1987).

Interspecific hybridizations between foxtail millet cultivars and a green foxtail accession resistant to the herbicides for transfer of resistance into cultivated foxtail millet were successful (Darmency and Pernes 1985; Wang and Darmency 1997; Darmency et al. 1987). QTL that contributes to germination and early seedling drought tolerance was identified through interspecific hybridization between *S. italica* × *S. viridis* (Qie et al. 2014). Both *S. viridis* and *S. italica* contributed favourable alleles for drought tolerance indicating that wild progenitor *S. viridis* populations may provide novel stress tolerance alleles for foxtail millet improvement. Successful crossing between genotypes of cultivated foxtail millet and a wild accession, SIE 1598, (*S. pumila*) could be achieved at ICAR-Indian Institute of Millets Research with an objective of combining higher grain yield with disease resistance. In the early segregating generations (F_2–F_3), the wild characteristics like profuse tillering and narrow panicle size were prominent. However, phenotypic selection for improved morpho-agronomic traits has resulted in the advanced breeding lines with higher grain yield and reduced tillering (unpublished). Some of the promising breeding lines are in different stages of evaluation for grain yield and related traits (Hariprasanna, unpublished data).

4.6 Genomic Resources

Identification and utilization of genetic variability is the foundation of crop breeding. The tedious, time-consuming and resource-intensive nature of phenotype-based selection demands for the DNA marker-based precise and rapid selection of superior genotypes in the genetic improvement of crops. Advancements in sequencing technologies and computational methods over the last decade resulted in the rapid generation of genomic resources in foxtail millet, which are being stored in the public databases for its further utilization in various genetics and breeding applications such as genetic diversity analysis, construction of linkage maps, trait mapping, comparative genomics and marker-assisted breeding. This section focusses on the whole genome sequences, genetic maps and DNA markers and their utility in the genetic improvement of foxtail millet.

4.6.1 Whole Genome Sequences

Whole genome sequence is considered as the basic framework of any organism that possesses DNA sequences, both coding and non-coding. DNA markers are vital for various genetics and breeding applications such as assessment of genetic diversity among germplasm accessions, understanding evolutionary relationships, construction of linkage maps, identification of QTL or marker-trait associations for important target traits, genomic selection and marker-assisted breeding. The rapid accumulation of genome sequence data due to the advances in NGS technologies and bioinformatics tools during the past decade accelerates the development of genome-wide DNA markers such as SSR, In-Del, ILP and SNP markers for the above-mentioned applications.

Among the small millets, foxtail millet was the first genome to be sequenced (Bennetzen et al. 2012; Zhang et al. 2012b). Draft genome of foxtail millet was published simultaneously by two independent research groups (Bennetzen et al. 2012; Zhang et al. 2012b). The whole genome sequence in foxtail millet started with whole genome shotgun sequencing involving Sanger sequencing for the cultivar Yugu1 and with the rapid advancements in the DNA sequencing technologies. Further draft genomes of foxtail millet were sequenced on Illumina, Roche 454 and

PacBio sequencing platforms. All these draft genomes have gaps in them, which need to be filled. Towards this, a high-density bin map of Zhanggu was constructed based on the recombination data generated from recombinant inbred lines (RILs) of the cross between Zhanggu and A2 (Ni et al. 2017), which was used to update the reference genome of Zhanggu with 416 Mb after the addition of unanchored scaffolds (16 Mb) and to correct the assembly error in the reference genome of Yugu1 by filling 3158 gaps. Very recently, a pan-genome of *Setaria* was developed by He et al. (2023) using the protein-coding genes predicted from de novo assembly of 110 *Setaria* accessions representing wild, landrace and modern cultivated accessions, which were selected based on their phylogenetic relationships, geographical distribution, breeding utility and subgroup distribution representing the genetic diversity within foxtail millet and green foxtail. The number of gene families increased in the pan-genome (73,528) as additional genomes were included in the analysis. Among these gene families, the core genes, soft core genes, dispensable genes and private genes accounted for 23.8%, 42.9%, 29.4% and 3.9%, respectively. Interestingly, an additional 14,283 gene families were identified in the pan-genome that are not present in the Yugu1 reference genome. The details of the draft genome sequences published in foxtail millet are given in Table 4.5. Due to the smaller size of genome among the millets, all the four foxtail millet draft genomes sequenced accounted for >80% of the estimated genome size.

The whole genome sequence information is useful for the development of DNA markers and functional annotation genes, which will help in understanding evolutionary relationships among cereals, functional characterization of candidate genes associated with important target traits, marker-assisted breeding and identification of gene targets for genome editing.

4.6.2 DNA Markers

Prior to the NGS era, DNA markers such as SSR, InDel, ILP and SNP markers were developed through conventional approaches involving the development of genomic libraries followed by sequencing through Sanger method. The pace of marker development was slow due to the tedious and time-consuming methods involved in the process. Later, there was continuous accumulation of DNA sequence data such as nucleotides, ESTs/cDNAs, genome survey sequence (GSS) and gene sequence, which were targeted for DNA marker development. By using ESTs and employing computational strategy, about 26 SSR markers were developed from 1213 ESTs of foxtail millet (Jia et al. 2007). Further, 100 polymorphic SSR markers were developed from $(GA)_n$ and $(CA)_n$ repeat motif-enriched genomic libraries using the Sanger sequencing of genomic library strategy (Jia et al. 2009). In another study, RAPD-enriched library was used to develop a set of 45 polymorphic SSR markers (Heng et al. 2011), which were used for genetic diversity assessment and cross-species transferability.

Table 4.5 Details of whole genome sequencing in foxtail millet

Particulars	Yugu1	Zhanggu	TT8	Huagu11
Sequencing platforms used	Sanger sequencing on ABI 3730xl capillary sequencer	Illumina second-generation sequencer	Illumina paired-end reads, mate-pair reads, synthetic long reads and Roche 454 long distance paired-end reads	PacBio, Illumina and Hi-C
Assembly type	Whole genome shotgun sequencing assembly	Whole genome shotgun-next-generation sequencing assembly	Contig assembly	Hybrid de novo assembly
Total assembly size	396.7 Mb in 9 chromosomes and an additional 4.2 Mb in 327 scaffolds	~423 Mb	477 Mb	408.37 Mb
Genome coverage	~80% of the estimated size (~510 Mb)	~86% of the estimated size (~490 Mb)	97% of the estimated size (~490 Mb)	89% of the estimated size (456 Mb)
Number of contigs N90/scaffolds N90	3251/8	15,986/420	4315/9	86/9
Length of contig/scaffold	0.034 Mb/47.25 Mb	0.006 Mb/0.27 Mb	0.017 Mb/39.55 Mb	1.41 Mb/36.33 Mb
GC content	46.14%	45.89%	46.18%	46.16%
Number of protein-coding genes predicted	24,000–29,000	38,801	40,599	42,932
Chromosome length	401.29 Mb	400.13 Mb	458.46 Mb	401.57 Mb
Research group	Bennetzen et al. (2012)	Zhang et al. (2012b)	Tsai et al. (2016)	Wang et al. (2021)

Hi-C high-throughput chromosome conformation capture

With the help of $(CA)_n$-, $(AAC)_n$- and $(ATG)_n$-enriched genomic libraries of drought- and salt-tolerant foxtail millet cultivar (Prasad), a new set of 172 genomic SSR markers was developed (Gupta et al. 2012) followed by 78 SSR markers using $(GA/CT)_n$ microsatellite-enriched genomic library (Gupta et al. 2013). Similarly, using 66,027 ESTs of foxtail millet available in the public database, a set of 447 EST-SSR markers were developed and validated in green foxtail and other millets and cereals (Kumari et al. 2013).

Rapid advancements in the sequencing technologies, development of computational algorithms, reduction in cost of sequencing and ease of assembling of large genomes led to the whole genome sequencing of many of the millet species. This resulted in the generation of genomic resources such as genome-wide SSR, InDel, ILP and SNP markers, which are useful for various genetic and breeding applications such as genetic diversity and population structure assessments, fingerprinting, linkage map construction, identification of QTL or marker-trait associations, genomic selection and marker-assisted selection that can lead to the development of nutrient-rich climate-resilient foxtail millet cultivars. DNA markers developed by various researchers across the globe have been listed in Table 4.6.

Table 4.6 DNA markers developed in foxtail millet

Type of DNA marker	No. of markers	References
Genomic SSR	193	Jia et al. (2009)
	45	Heng et al. (2011)
	172	Gupta et al. (2012)
	64	Gupta et al. (2013)
	733	Zhang et al. (2014)
	74	Liu et al. (2022)
EST-SSR	26	Jia et al. (2007)
	447	Kumari et al. (2013)
Genome-wide SSR	28,342	Pandey et al. (2013)
	733	Zhang et al. (2014)
Genome-wide SNP	2,90,317	Bennetzen et al. (2012)
	5,42,322	Zhang et al. (2012b)
	7,62,082	Bai et al. (2013)
	25,84,083	Jia et al. (2013)
	4,83,414	Ni et al. (2017)
	10,367	Jaiswal et al. (2019)
	9,69,596	Wang et al. (2021)
Genome-wide InDels	33,587	Zhang et al. (2012b)
	26,802	Bai et al. (2013)
	1,56,282	Wang et al. (2021)
Genome-wide ILPs	5123	Muthamilarasan et al. (2014)
Genome-wide miRNA-based markers	176	Yadav et al. (2014a)
Genome-wide TE-based markers	20,278	Yadav et al. (2014b)

SSR simple sequence repeats, *ESTs* expressed sequence tags, *SNP* single nucleotide polymorphism, *InDel* insertion-deletion, *ILP* intron length polymorphism, *miRNA* microRNA, *TE* transposable element

4.6.3 Genetic Maps

Genetic linkage maps are considered as the basic framework for the mapping of major genes and QTL for agro-morphological traits, nutritional traits and biotic and abiotic stress tolerance; in addition, these genetic maps are also used for comparative genome mapping with other millets and related cereal crops. The first linkage map comprising nine linkage groups was constructed in foxtail millet by Wang et al. (1998) using an intervarietal cross (Longgu 25 × Pagoda Flower Green) and 160 RFLP marker loci. A greater conservation of marker order and genetic distances was observed between the loci when the above-mentioned linkage map was compared with another linkage map constructed using the interspecific cross, *S. italica* × *S. viridis*. A linkage map was constructed by Jia et al. (2009) using an F_2 population derived from the interspecific cross, B100 (*S. italica*) × A10 (*S. viridis*), based on 81 new SSR markers with 20 RFLP anchored markers accounting for the length of 1654 cM spanning 9 chromosomes. Later, a genetic linkage map (2082.5 cM) with 9 linkage groups was constructed comprising 192 different kinds of markers and 33 new markers from foxtail millet and pearl millet using an F_2 mapping population derived from the cross between Gao 146A and K 103 (Wang et al. 2014b).

The publication of reference genomes of foxtail millet during the previous decade (Zhang et al. 2012b; Bennetzen et al. 2012) helped in the detection of genome-wide SNPs and other DNA sequence variations leading to the construction of high-resolution genetic linkage maps. A genetic linkage map was constructed using 751 markers comprising of SNPs and SVs by genotyping the F_2 population generated from the cross, Zhang gu × A2 (Zhang et al. 2012b). During the same period, another genetic linkage map was constructed by genotyping the RILs from the cross B100 (*S. italica*) × A10 (*S. viridis*) with 992 SNPs across the 9 linkage groups (Bennetzen et al. 2012). Recently, a physical map was constructed using the 733 polymorphic SSR markers developed based on the comparison of foxtail millet and green foxtail genome (Zhang et al. 2014). A high-density genetic map was constructed using the F_2 population derived from the cross, Yugu1 × Longgu7, with 1035 SSR loci spanning 1318.8 cM (Fang et al. 2016). In another study, a high-density genetic map spanning 1648.8 cM was constructed by Wang et al. (2017b) using 9968 SNP markers by genotyping F_2 population derived from the cross between Hongmiaozhangu × Changnong35 through RAD-seq approach.

4.7 Future Prospects

Considerable gap exists between availability of genetic and genomic resources and actual utilization in foxtail millet improvement, first because of difficulty in making crosses between accessions and secondly because of limited access to many researchers from non-traditional areas or limited information about the characterized germplasm as well as lack of data on traits of interest to the breeder. Most breeders regularly use only small working collections available with them, and less diversity in the working collection leads to narrow genetic base even when they are able to make successful crosses. Hence, more efforts on characterization and evaluation of germplasm collection and identification of trait-specific accessions are necessary to enhance the use of germplasm in foxtail millet crop improvement.

The future traits to look for in foxtail millet collection are resistance to blast, leaf spot, rust and downy mildew diseases, tolerance to shoot fly, drought and high temperature, stay-green character, high biomass, higher content of protein and micronutrients, bigger seed size, non-shattering habit, exserted stigma and protruded anthers for higher success in recombination breeding. Another trait of interest is male sterility, which, if available, can be exploited in recombination as well as heterosis breeding. Rich germplasm diversity is available in the north-eastern region of the country, though there is no commercial cultivation of the crop in these regions. More collection efforts are necessary to collect diverse germplasm from unexplored areas. With the use of more diverse germplasm in breeding, crop improvement efforts will be more rewarding in future. Another important aspect with respect to foxtail millet is the availability of genomic sequence and large quantum of genomic resources. Successful deployment of genomic resources in basic studies related to water- and nutrient-use efficiency, nutritional traits, bioavailability, etc. in the germplasm will help in better understanding of the basic biology of millets and cereals as a whole. Deployment of molecular tools along with genomic resources shall help in realizing higher genetic gains. Standardization of novel genome editing tools such as zinc-finger nucleases (ZFNs), transcription activator-like effector nucleases (TALENs) and CRISPR/Cas in small millets may revolutionize the crop improvement approaches in future.

References

Akter S, Mannan MA, Mamun MAA, Islam MS (2019) Physiological basis of salinity tolerance in foxtail millet. Bangladesh Agron J 22(2):11–24

Albacete AA, Martínez-Andújar C, Pérez-Alfocea F (2014) Hormonal and metabolic regulation of source–sink relations under salinity and drought: From plant survival to crop yield stability. Biotechnol Adv 32:12–30

Amadou I, Amza T, Yong-Hui S, Guo-Wei L (2011) Chemical analysis and antioxidant properties of foxtail millet bran extracts. Songklanakarin J Sci Technol 33:509–515

Amarnath K, Durga Prasad AVS, Reddy CVCM (2019) DUS characterization of germplasm accessions in foxtail millet (*Setaria italica* (L.) Beauv.). Int J Curr Microbiol App Sci 8(2):2410–2422

Anand MM, Venkataravana P, Palanna KB, Devaraja (2023) Screening of foxtail millet germplasm against blast caused by *Pyricularia setariae*. Pharma Innov J 12(11):1934–1936

Anitha S, Kane-Potaka J, Tsusaka TW, Botha R, Rajendran A, Givens DI, Parasannanavar DJ, Subramaniam K, Prasad KDV, Vetriventhan M, Bhandari RK (2021) A systematic review and meta-analysis of the potential of millets for managing and reducing the risk of developing diabetes mellitus. Front Nutr 8:687428. https://doi.org/10.3389/fnut.2021.687428

Anju T, Sarita S (2010) Suitability of foxtail millet (*Setaria italica*) and barnyard millet (*Echinochloa frumentacea*) for development of low Glycemic Index biscuits. Malays J Nutr 16:361–368

Ardie SW, Khumaidaa N, Nurb A, Fauziah N (2015) Early identification of salt tolerant foxtail millet (*Setaria italica* L. Beauv). Procedia Food Sci 3:303–312

Austin DF (2006) Foxtail millets (*Setaria: Poaceae*) – abandoned food in two hemispheres. Econ Bot 60:143–158

Ayesha MD, Ratna Babu D, Babu JDP, Srinivasa Rao V (2019) Genetic parameters for grain yield and nutritional quality traits in foxtail millet [*Setaria italica* (L.) Beauv.]. Int J Curr Microbiol App Sci 8(2):4–9

Bai H, Cao Y, Quan J et al (2013) Identifying the genome-wide sequence variations and developing new molecular markers for genetics research by re-sequencing a landrace cultivar of foxtail millet. PLoS One 8(9):e73514

Baltensperger DD (1996) Foxtail and proso millet. In: Janick J (ed) Progress in new crops. ASHS Press, pp 182–190

Baltensperger DD (2002) Progress with proso, pearl, and other millets. In: Janick J, Whipley A (eds) Trends in new crops and new uses. ASHS Press, pp 100–103

Baswaraj R, Das IK, Manjunatha N, Thorat U, Patroti P, Basavaraj YB, Avinash S, Stanley J, Tara Satyavati C (2023) Investigating viral infection in foxtail millet by symptomatology, electron microscopy and molecular detection. In: Abstracts, VIROCON-2023, advancements in global virus research towards one health, 1–3 December 2023. ICAR-NRCB, Tiruchirappalli

Begum F, Sultana R, Nessa A (2013) Screening of drought tolerant foxtail millet (*Setaria italica Beauv.*) germplasm. Bangladesh J Scient Ind Res 48(10):3329

Benabdelmouna A, Shi Y, Abirached-Darmency M, Darmency H (2001) Genomic in situ hybridization (GISH) discriminates between the A and the B genomes in diploid and tetraploid *Setaria* species. Genome 44:685–690

Bennetzen JL, Schmutz J, Wang H et al (2012) Reference genome sequence of the model plant *Setaria*. Nat Biotechnol 30:555–561

Boudry P, Mörchen M, Saumitou-Laprade P, Vernet P, Van Dijk H (1993) The origin and evolution of weed beets: Consequences for the breeding and release of herbicide-resistant transgenic sugar beets. Theor Appl Genet 87:471–478

Bramel P, Giovannini P, Eshan Dulloo M (2022) Global strategy for the conservation and use of genetic resources of selected millets. In: Global Crop Diversity Trust, Bonn, p 85. https://doi.org/10.5281/zenodo.7798294

Bray F (1981) Millet cultivation in China: a historical survey. J D'agriculture Tradit de Bot Appl 28(3–4):291–307

Brink M (2006) *Setaria italica* (L.) P. Beauv. In: Brink M, Belay G (eds) PROTA (Plant Resources of Tropical Africa/Ressources végétales de l'Afrique tropicale). Wageningen. Retrieved from http://www.prota4u.org

Brunda SM, Kamatkar MY, Hundekar R, Naveenkumar KL (2014) Studies on genetic variability, heritability and genetic advance in foxtail millet in both rainy and post rainy season. J Agric Vet Sci 7(2):34–37

Cattivelli L, Rizza F, Badeck FW, Mazzucotelli E, Mastrangelo AM, Francia E, Marè C, Tondelli A, Stanca AM (2008) Drought tolerance improvement in crop plants: An integrated view from breeding to genomics. Field Crop Res 105:1–14

Ceasar SA, Ramakrishnan M, Vinod KK, Roch GV, Upadhyaya HD et al (2020) Phenotypic responses of foxtail millet (*Setaria italica*) genotypes to phosphate supply under greenhouse and natural field conditions. PLoS One 15(6):e0233896. https://doi.org/10.1371/journal.pone.0233896

Chang KC (ed) (1977) Food in Chinese culture: Anthropological and historical perspectives. Yale University Press

Chen J, Qi Y (1993) Recent developments in foxtail millet cultivation and research in China. In: Riley KW, Gupta SC, Seetharam A, Mushonga JN (eds) Advances in small millets. IBH Publishing Co, Oxford, pp 101–107

Cheng RH, Dong ZP (2010) Breeding and production of foxtail millet in China. In: He Z, Bonjean APA (eds) Cereals in China. CIMMYT, pp 87–95

Chhavi A, Sarita S (2012) Evaluation of composite millet breads for sensory and nutritional qualities and glycemic response. Malays J Nutr 18:89–101

Choi Y, Osada K, Ito Y, Nagasawa T, Choi M, Nishizawa N (2005) Effects of dietary protein of Korean foxtail millet on plasma adiponectin, HDL-cholesterol, and insulin levels in genetically type 2 diabetic mice. Biosci Biotechnol Biochem 69:31–37

Darmency H, Pernes J (1985) Use of wild *Setaria viridis (L.) Beauv.* to improve triazine resistance in cultivated S. italica (L.) by hybridization. Weed Res 25:175–179

Darmency H, Pernes J (1987) An inheritance study of domestication in foxtail millet using an interspecific cross. Plant Breed 99:30–33. https://doi.org/10.1111/j.1439-0523.1987.tb01146.x

Darmency H, Ouin C, Pernes J (1987) Breeding foxtail millet (*Setaria italica*) for quantitative traits after interspecific hybridization and polyploidization. Genome 29:453–456

de Wet JMJ, Harlan JR (1975) Weeds and domesticates: evolution in the man-made habitat. Econ Bot 29:99–107

de Wet JMJ, Oestry-Stidd LL, Cubero JI (1979) Origins and evolution of foxtail millet (*Setaria italica*). J d'Agric Trad et de Bota 26:53–64

Dharnishkumar M, Suthamathi P, Gnanamalar RP, Renuka R, Sathiya K, Vaithiyalingan M, Hemalatha G (2023) Genetic variability studies in foxtail millet (*Setaria italica* L.). Biol Forum Int J 15(9):212–216

Diao X (2017) Production and genetic improvement of minor cereals in China. Crop J 5:103–114

Doust AN, Kellogg EA (2002) Inflorescence diversification in the panicoid "bristle grass" clade (Paniceae, Poaceae): evidence from molecular phylogenies and developmental morphology. Am J Bot 89:1203–1222

Doust AN, Devos KM, Gadberry MD, Gale MD, Kellogg EA (2004) Genetic control of branching in foxtail millet. Proc Natl Acad Sci USA 101:9045–9050

Doust AN, Penly AM, Jacobs SW, Kellogg EA (2007) Congruence, conflict, and polyploidization shown by nuclear and chloroplast markers in the monophyletic "Bristle clade" (Paniceae, Panicoideae, Poaceae). Syst Bot 32:531–544

Doust AN, Kellogg EA, Devos KM, Bennetzen JL (2009) Foxtail millet: a sequence-driven grass model system. Plant Physiol 149:137–141

Du Y, Zhao J, Wang G, Li Y, Zhao G, Yan X (2019) Study of lodging resistance of spring-sowing foxtail millet in maturity stages. Crops 35(1):141–145

Dwivedi SL, Upadhyaya HD, Senthilvel S, Hash CT, Fukunaga K, Diao X, Santra D, Baltensperger D, Prasad M (2012) Millets: Genetic and genomic resources. In: Janick J (ed) Plant breeding reviews, vol 35. Wiley-Blackwell, pp 247–375

Ecoport (2011) Ecoport database. Retrieved from http://www.ecoport.org. Accessed 1 Jan 2024

Fang X, Dong K, Wang X, Liu T, He J, Ren R, Zhang L, Liu R, Liu X, Li M (2016) A high-density genetic map and QTL for agronomic and yield traits in foxtail millet [*Setaria italica* (L.) P. Beauv.]. BMC Genomics 17(1):336

Geervani P, Eggum BO (1989) Nutrient composition and protein quality of minor millets. Plant Foods Hum Nutr 39:201–208

Ghatak A, Chaturvedi P, Nagler M, Roustan V, Lyon D, Bachmann G, Postl W, Schrofl A, Desai N, Varshney R, Wwckwerth W (2016) Comprehensive tissue-specific proteome analysis of drought stress responses in *Pennisetum glaucum* (L.) R. Br. (Pearl millet). J Proteome 143:122–135

Giussani LM, Cota-Sanchez JH, Zuloaga FO, Kellogg EA (2001) A molecular phylogeny of the grass subfamily Panicoideae (Poaceae) shows multiple origins of C4 photosynthesis. Am J Bot 88:1993–2012

Gupta S, Kumari K, Sahu PP et al (2012) Sequence based novel genomic microsatellite markers for robust genotyping purposes in foxtail millet [*Setaria italica* (L.) P. Beauv.]. Plant Cell Rep 31:323–337

Gupta S, Kumari K, Muthamilarasan M et al (2013) Development and utilization of novel SSRs in foxtail millet [*Setaria italica* (L.) P. Beauv.]. Plant Breed 132:367–374

Hariprasanna K (2017) Foxtail millet, *Setaria italica* (L.) P. Beauv. In: Patil JV (ed) Millets and sorghum: biology and genetic improvement, 1st edn. Wiley, pp 112–149

Harlan JR (1992) Crops and Man, 2nd edn. American Society of Agronomy

Harlan JR, de Wet JMJ (1971) Towards a rational taxonomy of cultivated plants. Taxon 20(4):509–517

He H, Hui F (2015) The history of millet cultivation in ancient China. China Agricultural Science and Technology Press

He Q, Tang S, Zhi H et al (2023) A graph-based genome and pan-genome variation of the model plant *Setaria*. Nat Genet 55:1232–1242

Hendawy SF, El-Sherbeny SE, Hussein MS, Youssef AA (2012) Evaluation of some cultivars of foxtail plants under salinity conditions. J Appl Sci Res 8:620–627

Heng L, Chih C, Song C et al (2011) Development of simple sequence repeats (SSR) markers in *Setaria italica* (Poaceae) and cross-amplification in related species. Int J Mol Sci 12:7835–7845

Heuzé V, Sauvant D, Tran G, Bastianelli D, Lebas F (2015) Foxtail millet (*Setaria italica*), grain. Feedipedia.org. Retrieved from http://www.feedipedia.org/node/725. Last updated on March 26, 2015

Hussain RA, Ahmad R, Waraich EA, Nawaz F (2015) Nutrient uptake, water relations and yield performance of different wheat cultivars (*Triticum aestivum* L.) under salinity stress. J Plant Nutr 8:2139–2149

ICRISAT & FAO (1996) The world sorghum and millet economies: facts, trends and outlooks. Retrieved from http://www.fao.org/es/esc/common/ecg/63/en/Sor_Mil.pdf

Iftekhar Ahmed M, Molla MR, Rohman MM, Hossain MA (2017) Morphological characterization of foxtail millet germplasm. Int J Appl Res 3(2):1–9

Itagi S, Naik R, Bharati P, Sharma P (2012) Readymade foxtail millet mix for diabetics. Int J Sci Nat 3:131–134

Jaiswal V, Gupta S, Gahlaut V et al (2019) Genome-wide association study of major agronomic traits in foxtail millet (*Setaria italica* L.) using ddRAD sequencing. Sci Rep 9:Article 5020

Jali M, Kamatar M, Jali S, Hiremath M, Naik R (2012) Efficacy of value-added foxtail millet therapeutic food in the management of diabetes and dyslipidemia in type 2 diabetic patients. Recent Res Sci Technol 4:7

Jia XP, Shi YS, Song YC et al (2007) Development of EST-SSR in foxtail millet (*Setaria italica*). Genet Resour Crop Evol 54:233–236

Jia XP, Zhang ZB, Liu YH et al (2009) Development and genetic mapping of SSR markers in foxtail millet [*Setaria italica* (L.) Beauv.]. Theor Appl Genet 118(4):821–829

Jia G, Huang X, Zhi H et al (2013) A haplotype map of genomic variations and genome-wide association studies of agronomic traits in foxtail millet (*Setaria italica*). Nat Genet 45:957–961

Jiang ZK, Liu JR, Wang SY, Lu ZG, Yan HS, Liu HP, Liu RF, Li DH (2008) Analysis on genetic bases and pedigree evolution of foxtail millet variety in Henan province. Anhui Agric 5:124–127

Jusuf M, Pernes J (1985) Genetic variability of foxtail millet (*Setaria italica* (L.) P. Beauv): Electrophoretic study of five isoenzyme systems. Theor Appl Genet 71:385–391

Karvar SH, Vaidya ER, Ghorade RB, Gomashe SS, Kohakade SN, Rajane AR (2021) Extent of genetic variability for morpho-nutritional traits in foxtail millet [*Setaria italica* (L.) P. Beauv.]. Pharma Innov J 10(1):576–579

Kawase M, Fukunaga K, Kato K (2005) Diverse origins of waxy foxtail millet crops in East and Southeast Asia mediated by multiple transposable element insertions. Mol Gen Genomics 274:131–140

Krishnakumari S, Thayumanvan B (1997) Comparative study of resistant starch from minor millets on intestinal response of blood glucose, serum cholesterol and triglycerides in rats. J Sci Food Agric 73:296–302

Krishnamurthy L, Serraj R, Rai KN, Hash CT, Dakheel AJ (2007) Identification of pearl millet [*Pennisetum glaucum* (L.) R. Br.] lines tolerant to soil salinity. Euphytica 158:179–188

Krishnamurthy L, Upadhyaya HD, Gowda CLL, Kashiwagi J, Purushothaman R, Singh S, Vadez V (2014) Large variation for salinity tolerance in the core collection of foxtail millet (*Setaria italica* (L.) P. Beauv.) germplasm. Crop Pasture Sci 65:353–361

Krishnamurthy L, Upadhyaya HD, Kashiwagi J, Purushothaman R, Dwivedi SL, Vadez V (2016) Variation in drought-tolerance components and their interrelationships in the core collection of foxtail millet (*Setaria italica*) germplasm. Crop Pasture Sci 67:834–846

Kumar KK, Parmeswaran KP (1998) Characterization of storage protein from selected varieties of foxtail millet [*Setaria italica* (L.) Beauv]. J Sci Food Agric 77(4):535–542

Kumari K, Muthamilarasan M, Misra G et al (2013) Development of eSSR-markers in *Setaria italica* and their applicability in studying genetic diversity, cross-transferability and comparative mapping in millet and non-millet species. PLoS One 8(6):e67742

Lata C, Gupta S, Prasad M (2013) Foxtail millet: a model crop for genetic and genomic studies in bioenergy grasses. Crit Rev Biotechnol 33:328–343

Le Thierry d'Ennequin M, Toupance B, Robert T, Godelle B, Gouyon PH (1999) Plant domestication: a model for studying the selection of linkage. J Evol Biol 12:1138–1147

Li YM (1997) Breeding for foxtail millet drought tolerant cultivars (in Chinese). In: Li Y (ed) Foxtail millet breeding. Chinese Agricultural Press, Beijing, pp 421–446

Li Y, Wu SZ (1996) Traditional maintenance and multiplication of foxtail millet [*Setaria italica* (L.) Beauv.] landraces in China. Euphytica 87:33–38

Li Y, Wu S, Cao Y (1995) Cluster analysis of an international collection of foxtail millet [*Setaria italica* (L.) P. Beauv]. Euphytica 83:79-85

Li HW, Meng CJ, Liu TN (1935) Problems in the breeding of millet (*Setaria italica* (L.) Beauv.). J Am Soc Agron 27:963–970

Li CH, Pao WK, Li HW (1942) Interspecific crosses in *Setaria*. II. Cytological studies of interspecific hybrids involving 1, S. faberii and S. italica, and 2, a three-way cross, F2 of S. italica, S. viridis, and S. faberii. J Hered 33:351–355

Li HW, Li CH, Pao WK (1945) Cytological and genetical studies of the interspecific cross of the cultivated foxtail millet, *Setaria italica* (L.) Beauv., and the green foxtail millet, S. viridis L. J Am Soc Agron 37:32–54

Li Y, Jia JZ, Wang YR, Wu SZ (1998) Intraspecific and interspecific variation in *Setaria* revealed by RAPD analysis. Genet Resour Crop Evol 45:279–285

Li GY, Fan ZY, Liu F, Zhang P, Li WX, Liu QS, Zhu ZH (2009) Studies on vitamin E in foxtail millet determined by HPLC method. J Agric Sci Technol 11:129–133. (in Chinese)

Li W, Zhi H, Wang Y, Li H, Diao X (2012) Assessment of genetic relationship of foxtail millet with its wild ancestor and close relatives by ISSR markers. J Integr Agric 11:556–566

Li C, Yue J, Wu X, Xu C, Yu J (2014) An ABA-responsive DRE-binding protein gene from Setaria italica, SiARDP, the target gene of SiAREB, plays a critical role under drought stress. J Exp Bot 65(18):5415–5427

Li Z, Jia G, Li X, Li Y, Zhi H, Tang S, Ma J, Zhang S, Li Y, Shang Z, Diao X (2021) Identification of blast-resistance loci through genome-wide association analysis in foxtail millet (*Setaria italica* (L.) Beauv.). Journal of Integrative. Agriculture 20(8):2056–2064

Liang P, Li Y, Shen L (1959) Studies on millet blast caused by Pyricularia setariae Nishikado. Acta Phytopathol Sin 5:89–99. (in Chinese)

Liu H, Lin R, Xu A, Yang S, Li Z (1990) Identification of millet blast resistance in foxtail millet germplasm resources. Shanxi Agric Sci 3:24–25. (in Chinese)

Liu T, He J, Dong K et al (2022) Genome-wide identification of quantitative trait loci for morpho-agronomic and yield-related traits in foxtail millet (*Setaria italica*) across multi-environments. Mol Gen Genomics 297:873–888

Loni F, Ismaili A, Nakhoda B, Ramandi HD, Shobbar Z (2023) The genomic regions and candidate genes associated with drought tolerance and yield-related traits in foxtail millet: an integrative meta-analysis approach. Plant Growth Regul 101:169–185

Lu HY, Zhang JP, Liu KB, Wu NQ, Li YM, Zhou KS, Ye ML, Zhang TY, Zhang HJ, Yang XY, Shen LC, Xu DK, Li Q (2009) Earliest domestication of common millet (*Panicum miliaceum*) in East Asia extended to 10,000 years ago. Proc Natl Acad Sci USA 106(18):7367–7372

Maas EV (1985) Crop tolerance to saline sprinkling water. Plant Soil 89:273–284

Makwana K, Tiwari S, Tripathi MK, Patel V (2023) Selection of blast resistant lines from diverse germplasm set of foxtail millet. Biol Forum Int J 15(1):1–6

Mirzaee MR, Zare R, Nasrabad AA (2010) A new leaf and sheath brown spot of foxtail millet caused by Bipolaris australiensis. Aust Plant Dis Notes 5:19–20

Movsisyants AP (1950) Foxtail millet—a valuable forage crop. Sotsialist Zhivotn 12:60–63

Munirathnam P, Venkatramanamma K, Anusha A (2015) Evaluation of foxtail millet genotypes for blast and rust diseases under field conditions. Curr Biotica 9:263–268

Murugan R, Nirmala Kumari A (2006) Genetic divergence in Italian millet (*Setaria italica* (L.) Beauv). Indian. J Genet 66(4):339–340

Muthamilarasan M, Venkata Suresh B, Pandey G, Kumari K, Parida SK et al (2014) Development of 5123 intron-length polymorphic markers for large-scale genotyping applications in foxtail millet. DNA Res 21:41–52

Naciri Y, Darmency H, Belliard J, Dessaint F, Pernès J (1992) Breeding strategy in foxtail millet, *Setaria italica* (L.) P. Beauv., following interspecific hybridization. Euphytica 60(2):97–104

Nandini C, Bhat S, Tippeswamy V, Prabhakar (2018) Characterization of foxtail millet (*Setaria italica* (L.) Beauv.) germplasm for qualitative and quantitative traits to enhance its utilization. Academia. J Agric Res 6(5):121–129

Neethu Mohan, Ravishankar CR, Lohithaswa HC, Vijayakumar L, Raveendra HR (2019) Characterisation of new germplasm accessions for yield and yield related traits in foxtail millet [*Setaria italica* (L.) P. Beauv]. J Pharmacogn Phytochem SP1:649–652

Ni X, Xia Q, Zhang H et al (2017) Updated foxtail millet genome assembly and gene mapping of nine key agronomic traits by resequencing a RIL population. GigaScience 6:1–8

Nirmalakumari A, Vetriventhan M (2010) Characterization of foxtail millet germplasm collections for yield contributing traits. Electron J Plant Breed 1(2):140–147

Oduori C (1993) Small millets production and research in Kenya. In: Riley KW, Gupta SC, Seetharam A, Mushonga JN (eds) Advances in small millets. Oxford & IBH Publ Co Pvt Ltd, New Delhi, pp 67–73

Oelke EA, Oplinger ES, Putnam DH, Durgan BR, Doll JD, Undersander DJ (1990) Millets. In: Alternative field crops manual. University of Wisconsin, Cooperative Extension and University of Minnesota, Center for Alternative Plants & Animal Products and Minnesota Extension Service. [https://www.hort.purdue.edu/newcrop/afcm/millet.html]

Oh J, Park CY, Min HG, Lee HK, Yeom YA, Yoon Y, Lee SH (2017) First report of barley virus G in foxtail millet (*Setaria italica*) in Korea. Plant Dis 101(6):1061

Padulosi S, Mal B, Ravi S, Gowda J, Gowda KTK, Shanthakumar G, Yenagi N, Dutta M (2009) Food security and climate change: role of plant genetic resources of minor millets. Indian J Plant Genet Resour 22:1–16

Panaud O (2006) Foxtail millet. In: Kole C (ed) Genome mapping and molecular breeding in plants, Cereals and Millets, vol 1. Springer-Verlag, Berlin/Heidelberg, pp 325–332

Panchal A, Singh RK, Prasad M (2023) Recent advancements and future perspectives of foxtail millet genomics. Plant Growth Regul 99:11–23

Pandey G, Misra G, Kumari K et al (2013) Genome-wide development and use of microsatellite markers for large-scale genotyping applications in foxtail millet [*Setaria italica* (L.)]. DNA Res 20:197–207

Paroda RS (1989) Research advances in crop sciences. In: 40 Years of Agricultural Research and Extension in India. Indian Council of Agricultural Research, New Delhi, pp 35–81

Parvathi MS, Nataraja KN, Reddy YAN, Naika MBN, Gowda MVC (2019) Transcriptome analysis of finger millet (*Eleusine coracana* (L.) Gaertn.) reveals unique drought-responsive genes. J Genet 98(2):46

Patro TSSK, Madhuri J (2014) Identification of resistant sources for sheath blight in foxtail millet incited by Rhizoctonia solani Kuhn. Indian J Plant Sci 3(2):159–162

Patro TSSK, Meena A, Divya M, Anuradha N (2018) Evaluation of donor screening nursery (DSN) of foxtail millet against Rhizoctonia solani, the cause of sheath blight. Int J Chem Stud 6(3):2189–2191

Patro TSSK, Georgia KE, Raj Kumar S, Anuradha N, Sandhya Rani Y, Triveni U, Jogarao P (2020) Identification of foxtail millet varieties against banded blight disease incited by Rhizoctonia solani Kuhn. J Pharmacogn Phytochem 9(5):1265–1267

Pavan Kumar GK, Durga Prasad AVS, Reddy CVCM (2019) Association analysis in elite genetic resources of foxtail millet (*Setaria italica* (L.) Beauv) for yield and nutritional traits. J Pharmacogn Phytochem 8(5):414–419

Pawar VS, Pawar VD (1997) Malting characteristics and biochemical changes of foxtail millet. J Food Sci Technol 34:416–418

Pernes J (1983) La Génétique de la Domestication des Céréales. La Recherche 146:910–919. (in French)
Phillip J, Maloo SR (1996) An evaluation of *Setaria italica* for seed iron content. Int Sorghum Millets Newsl 82:37
Prasada Rao KE, de Wet JMJ, Brink DE, Mengesha MH (1987) Intraspecific variation and systematics of cultivated *Setaria italica*, foxtail millet (Poaceae). Econ Bot 41:108–116
Puranik S, Jha S, Srivastava PS, Sreenivasulu N, Prasad M (2011) Comparative transcriptome analysis of contrasting foxtail millet cultivars in response to short-term salinity stress. J Plant Physiol 168:280–287
Qie L, Jia G, Zhang W, Schnable J, Shang Z, Li W, Liu B, Li M, Chai Y, Zhi H, Diao X (2014) Mapping of quantitative trait locus (QTLs) that contribute to germination and early seedling drought tolerance in the interspecific cross *Setaria italica* x *Setaria viridis*. PLoS One 9(7):e101868
Rajesh M, Sudha A, Nirmalakumari A, Parasuraman P (2019) Identification of resistant sources for blast and rust in foxtail millet incited by Pyricularia setariae and Uromyces *setariae-italica*. Int J Curr Microbiol App Sci 8(3):1796–1800
Ramana DV, Reddy CVCM, Haritha T, Ahammed KS (2022) Blast screening and genetic variability studies for yield and yield contributing characters in foxtail millet (*Setaria italica* (L.)). Biol Forum Int J 14(4):1032–1039
Rao SVR, Reddy MR, Prarharaj NK, Shyam Sunder G (2000) Laying performance of broiler breeder chickens fed various millets or broken rice as a source of energy at a constant nutrient intake. Trop Anim Health Prod 32:329–338
Ravi SB (2004) Neglected millets save the poor from starvation. In: LEISA India, Valuing crop diversity, vol 1. AME Foundation, Bangalore, pp 34–36
Ravindran G (2003) Studies on millets: Proximate composition, mineral composition, phytate and oxalate contents. Food Chem 39(1):99–107
Reddy VG, Upadhyaya HD, Gowda CLL (2006) Characterization of world's foxtail millet germplasm collections for morphological traits. SAT eJournal 2(1):1–3
Rominger JM (1962) Taxonomy of *Setaria* (Gramineae) in North America, Illinois biological monographs no. 29. University of Illinois Press, Urbana
Santhakumar G (1999) Correlation and path analysis in foxtail millet. J Maharashtra Agric Univ 24(3):300–301
Saralamma S, Jyostna Kiranmai M, Reddy CVCM (2019) Screening of genotypes of *Setaria italica* (L.) against rust disease under field conditions during rabi season. J Res ANGRAU 47(4):84–87
Seetharam A (1997) Genetic resources in small millets and their utilization. In: Seethararn A, Rangaswamy M, Jayararnan N (eds) Extended summaries, national seminar on small millets – current research trends and future priorities as food, feed and processing for value addition. Tamil Nadu Agricultural University, Coimbatore, pp 42–44
Seetharam A (1998) Small millets research: achievements during 1947–97. Indian J Agric Sci 68(8):431–438. (Special issue)
Seetharam A, Prasada Rao KE (1988) Use of minor (small) millets germplasm and its impact on crop improvement in India. In: Proceedings of a Workshop on Germplasm Exploration and Evaluation in India. International Crops Research Institute for the Semi-Arid Tropics, Patancheru, pp 97–100
Seetharam A, Aradhya KM, Laxminarayana MR (1983) Variation for seed oil content in a world collection of foxtail millet germplasm. SABRAO J 15:99
Sema A, Sarita S (2002) Suitability of millet-based food products for diabetics. J Food Sci Technol 39:423–426
Shao LH, Wang L, Bai WW, Liu YJ (2014) Evaluation and analysis of folic acid content in millet from different ecological regions in Shanxi province. Sci Agric Sin 47(7):1265–1272. https://doi.org/10.3864/j.issn.0578-1752.2014.07.003

Sharma R, Girish AG, Upadhyaya HD, Humayun P, Babu TK, Rao VP, Thakur RP (2014) Identification of blast resistance in a core collection of foxtail millet germplasm. Plant Dis 98(4):519–524

Sheahan CM (2014) Plant guide for foxtail millet (*Setaria italica*). USDA-Natural Resources Conservation Service, Cape May Plant Materials Center, Cape May

Shen HR, Dong ZP, Wang YF, Quan JZ, Bai H, Li ZY (2020) First report of dwarf disease in foxtail millet (*Setaria italica*) caused by barley yellow striate mosaic virus in China. Plant Dis 104(4):1262–1262

Siles MM, Ken Russell W, Baltensperger DD, Nelson LA, Johnson B, Dale Van Vleck L, Jensen SG, Hein G (2004) Heterosis for grain yield and other agronomic traits in foxtail millet. Crop Sci 44:1960–1965

Singh D, Lawrence K, Marker S, Bhattacharjee I, Lawrence R, Choudhary R, Ercisli S, Karunakaran R (2023) Rainfed assessment of foxtail millet (*Setaria italica* L. Beauv) germplasms through genotyping and principal component analysis. Front. Plant Sci 14:1017652. https://doi.org/10.3389/fpls.2023.1017652

Thathola A, Srivastava S, Singh G (2010) Effect of foxtail millet (*Setaria italica*) supplementation on serum glucose, serum lipids and glycosylated hemoglobin in type 2 diabetics. Diabetol Croat 40:23–28

Thimmaiah SK, Viswanath DP, Vyakarnahal BS, Hunshal CS (1989) Effect of salinity on yield, seed quality and biochemical characteristics in *Setaria italica* L. Cereal Chem 66:525–527

Thippeswamy V, Sajjanar GM, Nandini C, Bhat S, Pushpa D (2017) Characterization of genotypes for nutritional traits in foxtail millet [*Setaria italica* (L.) Beauv.]. Int J Curr Microbiol App Sci 6(12):97–101

Tian BH, Wang J, Zhang L, Li Y, Wang S, Li H (2010) Assessment of resistance to lodging of landrace and improved cultivars in foxtail millet. Euphytica 172:295–302

Tian BH, Liu Y, Zhang LX, Song SX (2015) Characterization of culm morphology, anatomy and chemical composition of foxtail millet cultivars differing in lodging resistance. J Agric Sci 153:1437–1448. https://doi.org/10.1017/S0021859614001105

Tsai KJ, Lu MJ, Yang K et al (2016) Assembling the *Setaria italica* L. Beauv. genome into nine chromosomes and insights into regions affecting growth and drought tolerance. Sci Rep 6:Article number 35076

Umesh Chandra, Thomas TA, Sapra RL, Paroda RS (1987) Catalogue on foxtail millet (*Setaria italica* L.) germplasm. NBPGR Publication, New Delhi, p 80

Upadhyaya HD, Pundir RPS, Gowda CLL, Reddy VG, Singh S (2009) Establishing a core collection of foxtail millet to enhance the utilization of germplasm of an underutilized crop. Plant Genet Resour 7:177–184. https://doi.org/10.1017/S1479262108178042

Upadhyaya HD, Ravishankar CR, Narasimhudu Y, Sarma NDRK, Singh SK, Varshney SK, Reddy VG, Singh S, Parzies HK, Dwivedi SL, Nadaf HL, Sahrawat KL, Gowda CLL (2011) Identification of trait-specific germplasm and developing a mini core collection for efficient use of foxtail millet genetic resources in crop improvement. Field Crop Res 124:459–467

Van Delden S, Vos J, Ennos A, Stomph T (2010) Analysing lodging of the panicle bearing cereal teff (*Eragrostis tef*). New Phytol 186:696–707. https://doi.org/10.1111/j.1469-8137.2010.03224.x

Van K, Onoda S, Kim MY, Kim KD, Lee SH (2008) Allelic variation of the Waxy gene in foxtail millet (*Setaria italica* (L.) P. Beauv.) by single nucleotide polymorphisms. Mol Gen Genomics 279:255–266

Vavilov NI (1926) Studies on the origin of cultivated plants. Bull Appl Bot 16:1–248

Vetriventhan M, Upadhyaya HD, Dwivedi SL, Pattanashetti SK, Singh SK (2015) Finger and foxtail millets. In: Singh M, Upadhyaya HD (eds) Genetic and genomic resources for grain cereals improvement, vol 1. Academic, Oxford, pp 291–319. https://doi.org/10.1016/B978-0-12-802000-5.00007-1

Vinall HN (1924) Foxtail millet: its culture and utilization in the United States. USDA. Farmers Bull 793

Wang T, Darmency H (1997) Inheritance of sethoxydim resistance in foxtail millet, *Setaria italica* (L.) Beauv. Euphytica 94:69–73. https://doi.org/10.1023/A:1002989725995

Wang Y, Chu J, Song Y, Xie S, Yan W, Jin L, Liu H, Xie S (1985) Screening tests of millet varieties for resistance to blast pathogen. Acta Phytophylacica Sinica 12:175–180. (in Chinese)

Wang ZM, Devos KM, Liu CJ, Wang RQ, Gale MD (1998) Construction of RFLP-based maps of foxtail millet, *Setaria italica* (L.) P. Beauv). Theor Appl Genet 96:31–36

Wang C, Jia G, Zhi H et al (2012) Genetic diversity and population structure of Chinese foxtail millet [*Setaria italica* (L.) Beauv.] landraces. G3 (Bethesda) 2:769–777

Wang M, Li P, Li C, Pan Y, Jiang X, Zhu D, Zhao Q, Yu J (2014a) SiLEA14, a novel atypical LEA protein, confers abiotic stress resistance in foxtail millet. BMC Plant Biol 14:290

Wang ZL, Wang J, Yuan F et al (2014b) Construction of genetic map of foxtail millet [*Setaria italica* (L.) P. Beauv.] using PCR-based molecular markers. Sci Agric Sinica 47(17):3492–3500

Wang C, Zhou LB, Zhang GB, Xu Y, Zhang LY, Gao X, Gao J, Jiang N, Shao MB (2017a) Drought resistance identification and drought resistance indices screening of liquor-making waxy sorghum resources at adult plant stage. Sci Agric Sin 50:1388–1402. (in Chinese)

Wang J, Wang Z, Du X, Yang H, Han F, Han Y et al (2017b) A high-density genetic map and QTL analysis of agronomic traits in foxtail millet [*Setaria italica* (L.) P. Beauv.] using RAD-seq. PLoS One 12(6):e0179717. https://doi.org/10.1371/journal.pone.0179717

Wang J, Li S, Lan L et al (2021) De novo genome assembly of a foxtail millet cultivar Huagu11 uncovered the genetic difference to the cultivar Yugu1, and the genetic mechanism of imazethapyr tolerance. BMC Plant Biol 21:Article number 271

Wei S (1994) The origin, domestication and spreading of Chinese foxtail millet. Anc Curr Agric 2:6–18. (in Chinese)

Wu X (1985) A preliminary study on the identification and screening of millet blast. Heilongjiang Agric Sci:46–50. (in Chinese)

Wu C, Zhang M, Liang Y, Zhang L, Diao X (2023) Salt stress responses in foxtail millet: Physiological and molecular regulation. Crop J 11:1011–1021. https://doi.org/10.1016/j.cj.2023.06.001

Xiao J, Sun ZX, Chen GQ, Liu Z, Xin ZX, Kong FX (2021) Evaluation of drought tolerance in different genotypes of foxtail millet during the entire growth period. Agron J 114:340–355

Xiao JB, Liu ZX, Zong X, Li JZ, Chen GQ, Zhu XD, Sun ZX (2022) Identification of drought tolerance in foxtail millet during its entire growth period based on principal component analysis and membership function. Agric Res Arid Areas 40:34–44. (in Chinese)

Xie S, Yan W, Jin X, Liu H, Hu J (1984) A preliminary study on the varieties from Jilin Province resistance to millet blast. Jilin Agric Sci 2:74–77. (in Chinese)

Yadav CB, Muthamilarasan M, Pandey G et al (2014a) Development of novel microRNA-based genetic markers in foxtail millet for genotyping applications in related grass species. Mol Breed 34:2219–2224

Yadav CB, Bonthala VS, Muthamilarasan M et al (2014b) Genome-wide development of transposable elements-based markers in foxtail millet and construction of an integrated database. DNA Res 22:1–12

Yan W, Xie S, Jin X, Liu H, Hu J (1985) A preliminary study on the physiological races of millet blast (*Pyricularia setariae Nishikado*). Sci Agric Sin 18:57–62. (in Chinese)

Yan W, Xie S, Liu H, Li N (1988) Evaluation for the specialized resistance of millet germplasm resources and elite Chinese varieties to millet blast in Jilin Province. Jilin Agric Sci 1:6–10. (in Chinese)

Yang Z (1997) Food storage in pre-Qin days. Agricultural Archaeology 3:2-8 (in Chinese)

Yang XY, Wan ZW, Perry L, Lu HY, Wang Q, Zhao CH, Li J, Xie F, Yu JC, Cui TX, Wang T, Li MQ, Ge QS (2012) Early millet use in northern China. Proc Natl Acad Sci USA 109:3726–3730

Yang XS, Wang LL, Zhou XR, Shuang SM, Zhu ZH, Li N et al (2013) Determination of protein, fat, starch, and amino acids in foxtail millet [*Setaria italica* (L.) Beauv.] by Fourier transform near-infrared reflectance spectroscopy. Food Sci Biotechnol 22(6):1495–1500

Zhang JP, Liu TS, Zheng J, Jin Z, Zhu Y, Guo JF, Wang GY (2007) Cloning and characterization of a putative 12-oxophytodienoic acid reductase cDNA induced by osmotic stress in roots of foxtail millet. DNA Seq 18:138–144

Zhang WY, Liu BH, Xie JX, Li JM, Gu GQ, Wang YF, Li HQ, Li YQ, Diao XM (2012a) Screening of indices for drought tolerance test at booting stage in foxtail millet. J Plant Genetic Res 13:765–772. (in Chinese)

Zhang G, Liu X, Quan Z, Cheng S, Xu X, Pan S, Xie M, Zeng P, Yue Z, Wang W (2012b) Genome sequence of foxtail millet (*Setaria italica*) provides insights into grass evolution and biofuel potential. Nat Biotechnol 30:549–554

Zhang S, Tang C, Zhao Q et al (2014) Development of highly polymorphic simple sequence repeat markers using genome-wide microsatellite variant analysis in foxtail millet [*Setaria italica* (L.) P. Beauv.]. BMC Genomics 15:Article number 78

Zhang X, Shabala S, Koutoulis A, Shabala L, Zhou M (2017) Meta-analysis of major QTL for abiotic stress tolerance in barley and implications for barley breeding. Planta 245(2):283–295

Zhao ZJ (2014) The process of origin of agriculture in China: archaeological evidence from flotation. Quat Sci 34:73–84. (in Chinese with English abstract)

Zhao X, Ma K, Li Z, Li W, Zhang X, Liu S, Meng R, Lu B, Li X, Ren J, Zhang L, Yuan X (2023) Transcriptome analysis reveals brassinolide signaling pathway control of foxtail millet seedling starch and sucrose metabolism under freezing stress, with implications for growth and development. Int J Mol Sci 24:11590. https://doi.org/10.3390/ijms241411590

Little Millet

5

Lalit Arya, Monika Singh, Sunil Shriram Gomashe,
Chithra Devi Pandey, Dinesh Prasad Semwal,
Manjusha Verma, Mamta Singh, Badal Singh,
K. N. Ganapathy, Sushil Pandey, and Akansha Bajpai

Abstract

Little millet (*Panicum sumatrense* Roth. ex Roem. & Schult.; Family: Poaceae; tetraploid; self-pollinated; common names: Indian millet or *sama* or *Kutki*) is a C4 grass and have small-seeded grains. It is nutritionally rich, can be grown under marginal soil conditions, and is known for resistance to pests and tolerance to drought and salt stress. These qualities make it a crop to work on and can play a vital role in achieving sustainable development goals and in combating malnutrition and hunger under changing climatic conditions. The studies have been carried out to phenotype and genotype little millet, but the comprehensive characterization and evaluation including high-throughput genotyping and phenotyping and large-scale biochemical profiling of little millet accessions is lacking. Such studies need to be carried out in view of best utilization and management of little millet genetic resources for future food and nutritional security and agrarian sustainability. This chapter provides information on little millet importance; taxonomy, origin, and history of cultivation; national and global germplasm hold-

L. Arya (✉) · M. Singh · C. D. Pandey · D. P. Semwal · M. Verma · M. Singh · B. Singh
ICAR-National Bureau of Plant Genetic Resources, New Delhi, Delhi, India
e-mail: lalit.arya@icar.gov.in

S. Pandey
Division of Germplasm Conservation, ICAR-National Bureau of Plant Genetic Resources, New Delhi, Delhi, India

S. S. Gomashe
ICAR-National Bureau of Plant Genetic Resources, Regional Station, Akola, Maharashtra, India

K. N. Ganapathy
ICAR-Indian Institute of Millets Research, Hyderabad, Telangana, India

A. Bajpai
ICAR-National Bureau of Plant Genetic Resources, New Delhi, India

K. Tripathi et al. (eds.), *Plant Genebank Utilization for Trait Discovery in Millets*, https://doi.org/10.1007/978-981-96-4043-0_5

ings; germplasm collection, mapping, and gap analysis; agro-morphological and molecular characterization for identification of trait-specific germplasm, core set, and diverse accessions; genomic resources (SSR/SNP markers, chloroplast genome sequence, draft whole genome sequence, transcriptomic resources, etc.); pre-breeding approaches, production, cultivation, and consumption constraints. And future prospective of little millet research like application of GWAS (genome-wide association studies), transcriptome wide association studies (TWAS), pan-genomes, nested association mapping (NAM), Genomic Selection (GS), and genome editing is also discussed.

Keywords

Agro-morphological · Genomic resources · Genome-wide SNP markers · Germplasm holdings · Little millet · Management · SSRs · Utilization

5.1 Introduction

Little millet (*Panicum sumatrense* Roth. ex Roem. & Schult.), also called Indian millet or *sama* or Kutki, is a tetraploid ($2n = 4x = 36$) (Wanous 1990). It belongs to the family Poaceae, subfamily Panicoideae. It is a C4 grass and have small-seeded grains. It is cultivated in India, Pakistan, Nepal, Sri Lanka, Myanmar, and other Southeast Asian Countries. In India, it is largely cultivated in Central India including Odisha, Maharashtra, Gujarat, Madhya Pradesh, Karnataka, and Andhra Pradesh. It can grow up to 2000 m above sea level. The crop requires very less water and can be grown on marginal lands and is adaptable to grow under different climatic conditions. The millet is also known for its resistance to pests, drought, and salt (Sivakumar et al. 2006; Ajithkumar and Panneerselvam 2014) and is considered under the category of climate-resilient crops. Most of little millet cultivars gets matured in about 90 days (Hiremath et al. 1990) and are environment friendly and sustainable.

Little millet can be used as a substitute for rice, and all the dishes or eatables that can be made using rice can also be prepared with little millet. Its flour is also used in making baked goods. Little millet grains are a rich source of carbohydrates (78%), protein (7%), fat (4.26%) (Mallikarjun et al. 2013), dietary fiber, and phytochemicals (phytates, tannins, flavonoids, and phenolic acids). They are known for many health benefits due to the presence of gamma-aminobutyric acid (GABA) (84.6 nm/100 mg), phenolic compounds (521 mg GAE/100 g), tocopherols (1.3 mg/100 g), and carotenoids (104 μg/100 g) (Pradeep and Guha 2011; Guha et al. 2015). Furthermore, its gluten-free nature and high dietary fiber content make it popular in combating lifestyle diseases.

A considerable number of little millet germplasm resources are available in national and international genebanks/organizations and have been characterized by some to form core sets (Gowda et al. 2008; Upadhyaya et al. 2014) and to identify trait-specific germplasm. These core sets and trait-specific germplasm need to be evaluated across multiple locations and over different seasons for proper utilization in crop improvement. Since little millet can be grown under adverse and diverse environmental conditions and is nutritionally superior to other cereals, its management (exploration, collection, characterization, and evaluation) and utilization are crucial in the context of changing climatic conditions and future food and nutritional security.

5.2 The Crop

5.2.1 History, Origin, Taxonomy, Domestication

Little millet (*Panicum sumatrense*) is native to India. It is particularly a crop of hill tribes and the Eastern Ghats of Andhra Pradesh and adjacent Odisha (de Wet et al. 1983). The Eastern Ghats of India are considered the center of domestication for little millet (Bramel et al. 2022). Little millet is widely distributed across most agricultural zones of India, suggesting its ancient domestication (~5000 years ago in India). Harappa and Farmana are two Indus Valley Civilization sites where little millet dominated the millet collection. Around 2600 BC, little millet cultivation peaked at Harappa, with over 10,000 grains recovered from this site (Weber 2013). During the Mature Harappan period, 2500–2000 BC, browntop millet, native small *Setaria* species, and little millet dominated the Saurashtra region. In the Shavanger district of Gujarat, excavations at Oriyo Timbo dating to 2000–1500 BC revealed that 77% of seeds recovered were millets, including little millet. By 1500 BC in eastern India, Neolithic people also focused on summer crops, including *Vigna* and little millet (Bhat et al. 2018).

Little millet (*Panicum sumatrense*) originated from *Panicum psilopodium* through selection and further cultivation. A single reciprocal translocation might be responsible for the divergence between these two species. The chromosomes of hybrids between *Panicum sumatrense* and *Panicum psilopodium* pair almost perfectly, with only a single quadrivalent, producing vigorous and fertile plants with non-shattering spikelets (de Wet et al. 1983; Hiremath et al. 1990).

5.2.1.1 Taxonomy

Panicum sumatrense Roth. ex Roem. & Schult., Syst. Veg. 2: 434. 1817. Type: Sumatra, *Heyne*, E. Coromandelia quoque specimen possidemus, photograph (K). *Panicum miliare* Lamk., Tab. Encycl. Meth. Bot. 1: 175. 1791, Nomen illegitimum; *P. attenuatum* Wild., Enum. Hort Pl. Berol. 1033. 1809, non *P. attenuatum* Moench; *P. simplex* Rottl. Ex Trin., Gram. Panic. 216. 1826; *P. psilopodium* Trin., Gram. Panic. 216. 1826. (de Wet et al. 1983).

The correct name of little millet is *Panicum sumatrense* and species name has been derived from the specimen collected from Sumatra (Indonesia) and might have been introduced to Indonesia by Indian immigrants. Little millet (*Panicum sumatrense*) has two subspecies, namely, *Panicum sumatrense* Roth. ex Roem. & Schult. subsp. *psilopodium* (Trin.) Wet. and *Panicum sumatrense* Roth. ex Roem. & Schult. subsp. *sumatrense*. Nana and robusta are the two races of *Panicum sumatrense*

Roth. ex Roem. & Schult. subsp. *sumatrense*, which have been categorized based on its panicle morphology; further, nana matures early as compared to robusta (high biomass) (de Wet et al. 1983; Saha et al. 2016). Extensive variability exists with respect to growth habit and inflorescence morphology. Plants are decumbent (60–170 cm tall, produce up to 46 flowering culms, open inflorescences up to 50 cm long) to erect (120–190 cm tall, produce up to 46 flowering culms, open to compact inflorescences up to 46 cm long).

5.3 Production, Cultivation, and Consumption Constraints

The area, production, and yield of little millet was 2.34 lakh hectare, 1.27 lakh tons, and 544 kg/ha, respectively, in the year 2015–2016 in India (Hariprasanna 2023). But if we see the overall scenario for millets, the area under millet cultivation is decreasing considerably over the last six decades. From 1951 to 2021, there is 89.7% decline in cultivated area for small millets and 44% decline in total production of small millets. There were fluctuating levels of small millet production other than finger millet and production levels fell down to more than 80%, but there was a marginal rise of productivity from 380 kg/ha in 1950–1951 to 800 kg/ha in 2019–2020 at all India level.

There are several constraints associated with millet cultivation and consumption. There has been a decline in the area under millet cultivation due to low yields, low productivity, low profitability, labor-intensive and time-consuming processing of millets, disliking the taste of millets, lack of awareness about their health benefits, high cost compared to wheat and rice, and limited availability in local markets. A lack of government support was also a constraint in the cultivation and consumption of millets.

However, the government is now taking initiatives to promote millet cultivation and consumption. The year 2018 was declared the *"National Year of Millets,"* and millets were notified as *Nutri-cereals* by the Government of India. Furthermore, the year 2023 was observed as the *"International Year of Millets"* by the United Nations at the request of the Government of India to generate awareness about the health benefits of millets and to promote their production and consumption. Other initiatives include the introduction of millets in the Public Distribution System (PDS), encouragement to develop value-added products, the National Millet Mission (NMM), the Price Support Scheme (PSS), Minimum Support Price (MSP), and their inclusion in the midday meal program. Additional factors that can enhance millet production and consumption include the engagement of the private sector in popularizing millets, Farmer Producer Organization (FPO)-led farming, organic farming of millets, and strengthening research to improve little millet varieties in terms of yield, nutrition, and other traits of economic importance. Efforts should also focus on developing varieties with improved tolerance to biotic and abiotic stresses.

5.4 Status of Germplasm Collection

5.4.1 Little Millet Germplasm Holdings at Global, National Genebanks, and Institutions/Organizations

Diversity in any crop is important for sustainable agriculture. Plant Genetic Resources (PGR) are the basic raw material and provides the required variability for crop improvement programs and are key in ensuring future food and nutritional security. Little millet is an indigenous crop of India and efforts were made to collect the available variability from different parts of the country and are conserved in the National Genebank, ICAR-National Bureau of Plant Genetic Resources (NBPGR), New Delhi. The total number of little millet accessions globally conserved in various genebanks/organizations are listed in Table 5.1, and source-wise distribution of little millet accessions at National Genebank, ICAR-National Bureau of Plant Genetic Resources (NBPGR), New Delhi is given in Fig. 5.1. Some portion of germplasm may be duplicates within and/or between the genebanks.

ICRISAT Genebank conserves 473 accessions of little millet. Traditional cultivars/landraces collected from different parts of India (466) and a few from Cameroon (03), Myanmar (02), Syria (01), and Sri Lanka (01).

5.4.2 Contribution of Indian Accessions to the Global Core Collections

The Indian millet germplasm accessions contributed significantly to the global germplasm diversity which can indirectly be measured by assessing the proportion of Indian accessions in the global millet core collections (Upadhyaya et al. 2014; Elangovan et al. 2022). The total number of core collection in little millet is 56 accessions, and collected and exhibited maximum variability from India are such as Andhra Pradesh (15), Telangana (05), Bihar (03), Karnataka (02), Madhya Pradesh (01), Chhattisgarh (01), Maharashtra (15), Odisha (09), Tamil Nadu (03), West

Table 5.1 Status of little millet genetic resources conserved in genebanks globally

Name of the country	Number of accessions	Organization
India	2208	ICAR-National Bureau of Plant Genetic Resources, New Delhi
	1003	University of Agricultural Sciences, Bangalore
	694	ICAR-Indian Institute of Millets Research, Hyderabad
	473	International Crop Research Institute for the Semi-Arid Tropics, Patancheru, Hyderabad
United States	226	USDA Agricultural Research Service, Washington

Sources: http://pgrportal.nbpgr.ernet.in; https://demo.uasbangalore.edu.in/en/small-millets-zars-gkvk-bengaluru/; https://www.genesys-pgr.org/partners; http://genebank.icrisat.org/IND/Core?Crop=Little%20millet; Vetriventhan et al. (2020)

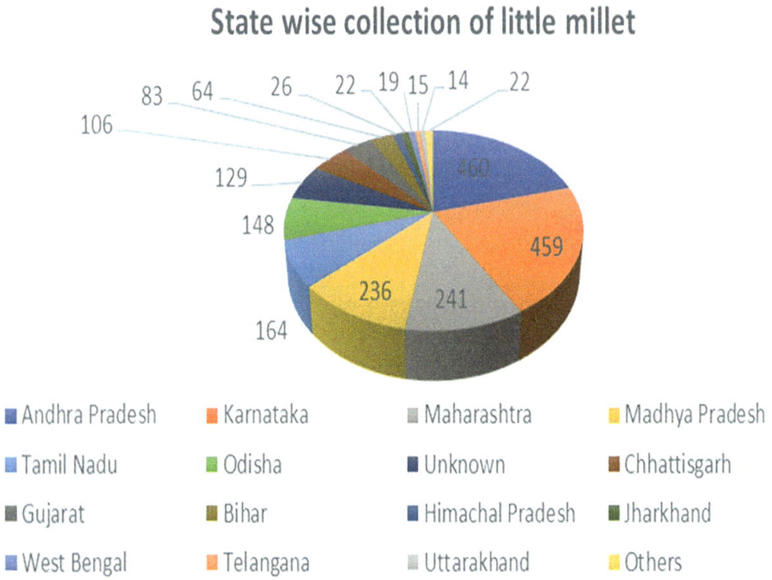

Fig. 5.1 State-wise collections of *Panicum* germplasm conserved in the National Genebank, ICAR-NBPGR, New Delhi, India

Bengal (01), and Myanmar (01). The percentage contribution by accessions of Indian origin is 98.21%.

5.4.3 Germplasm Collection, Mapping, and Gap Analysis

In little millet, a total of 2772 accessions have so far been collected from various parts of the country belonging to more than 1500 collection sites. Among these, 2567 accessions had passport data with locality details of collection sites (state, district and village) and these were further screened for availability in the National Genebank. Analyses of passport data shows that majority of little millet germplasm accessions have been collected from Andhra Pradesh (597), followed by Karnataka (475), Maharashtra (271), Madhya Pradesh (258), Tamil Nadu (238), Odisha (188), Chhattisgarh (143), Gujarat (91), Bihar (88), Himachal Pradesh (78), and Jharkhand (51), while minimum accessions were collected from Rajasthan (6), Meghalaya (1), Arunachal Pradesh (1), and Jammu and Kashmir (1). In order to know the diversity distribution pattern, DIVA-GIS software version 7.5 (www.diva-gis.org) was used to generate the potential distribution map. Geo-referenced maps were prepared using WGS84 datum and Everest projection systems (Hijmans et al. 2001; Semwal et al. 2018). Geo-referencing and mapping of collected diversity of little millet germplasm is shown in Fig. 5.2.

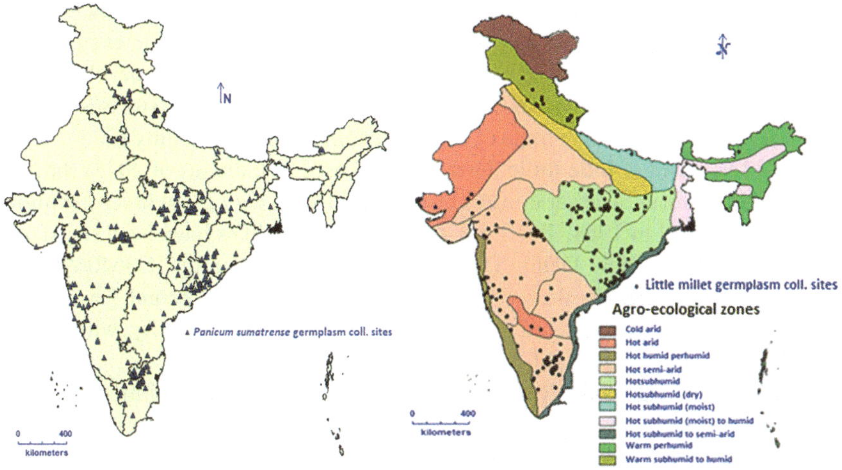

Fig. 5.2 Geo-referenced map of little millet germplasm collected from different states of the country

5.4.3.1 Gaps Identified for Future Collection from Different States of the Country

Based on gap analysis, the proposed area for future collection included Chhattisgarh (Bastar, Bijapur, Narayanpur, and Kondagaon), Gujarat (Tapi and The Dangs), Odisha (Nabarangpur, Naupada, Kalahandi, and Balangir), and Karnataka (Dharwad, Belgaum, and Chikkamagaluru).

5.5 Germplasm Use in Crop Improvement

5.5.1 Characterization and Evaluation of Little Millet and Development of Representative Subsets for Enhanced Use

Characterization and evaluation of germplasm are prerequisites for effective utilization in crop improvement programs. Morphological characterization involves recording highly heritable characters that can be easily visualized and are expressed under variable environmental conditions. Evaluation, characterization, and identification of useful gene sources are key factors for the effective utilization of germplasm in crop improvement. Multilocation evaluation of potential accessions across different agroclimatic conditions helps identify economically important accessions for release as varieties or for use in little millet improvement. Crop improvement in millets gained momentum with the establishment of the *All India Coordinated Small Millets Improvement Project* (AICSMIP) in 1986, headquartered at the University of Agricultural Sciences, Bangalore.

Little millet germplasm has been characterized and evaluated by various researchers, and a core set has also been developed to increase the efficiency of little millet germplasm utilization. The aim of crop improvement is to develop high-yielding varieties with resistance to various biotic and abiotic stresses and to enhance morpho-agronomic traits of economic importance and quality traits. So far, around 32 varieties of little millet have been released in India (as of 2023) (https://www.seednet.gov.in/SeedVarieties/CentralVariety.aspx). *Co 1* is the first little millet variety, developed in 1954. Two little millet germplasm lines have been registered by ICAR-NBPGR, New Delhi. The first, with national identity IC0640696, has been registered for early flowering and early maturity. The second germplasm, IC0483093, has been registered for early flowering and early maturity, with advantages in grain and fodder yield.

Channappagoudar et al. (2007) reported significant differences for morphophysiological traits, including plant height, number of green leaves, number of tillers per plant, leaf area index, leaf area duration, and photosynthetic rate among 14 genotypes grown at Main Research Station, University of Agricultural Sciences, Dharwad during *kharif* season under rainfed condition. Further, a set of 109 accessions of little millet germplasm was characterized during *rabi*, 2008–2009 at Department of Millets, Centre for Plant Breeding and Genetics, Tamil Nadu Agricultural University (Nirmalakumari et al. 2010). High level of variability along with high heritability and genetic advance were observed for total number of tillers per plant, number of productive tillers per plant, flag leaf width and sheath length, and grain yield.

Katara et al. (2019) observed significant variability among 30 genotypes of little millet at Junagadh, Gujarat, during *Kharif* 2017, as assessed using characters, namely, plant height, days to 50% flowering, days to maturity, number of productive tillers per plant, panicle length, grain weight per main panicle, grain yield per plant, biological yield per plant, harvest index, 1000 seed weight, chlorophyll content, and specific leaf weight along with seven nonmetric characters, namely, plant growth habit, inflorescence shape, panicle compactness, grain shape and color, lodging, and plant pigmentation. Natesan et al. (2020) studied 21 morphological descriptors in BL6, ATL 1, TNPsu 176, Co (Samai) 4, Paiyur 2 genotypes and more than 16 descriptors could differentiate ATL1, a high-yielding variety of little millet from other studied genotypes.

Vetriventhan et al. (2021) characterized 200 landraces of little millet and highly significant variability was observed for the studied agronomic and grain nutrient traits. Accessions of race *robusta* were found high yielding and those of *nana* were found rich in grain nutrients. Promising trait-specific accessions were identified for seed weight (10 accessions), high grain yield (15), high biomass yield (15), and consistently high grain nutrients over 2 years (30). A total of 1635 accessions of little millet conserved in National Genebank at ICAR-NBPGR, New Delhi, were characterized at two locations in 2018–2019, and significant variability was observed with respect to days to 50% flowering (DFF), plant height, panicle length, panicle compactness, tiller number, flag leaf blade length and width, and thousand grain weight (TGW) (Arya et al. 2021a). Further from this study, a total of 82 accessions having 39 to 49 days of days to 50% flowering, 34 accessions with high

Fig. 5.3 Little millet accessions showing high thousand grain weight

thousand grain weight and bold seeds (Fig. 5.3), and 118 NIRS (Near Infrared Spectroscopy) spectral data-based diverse accessions were identified for evaluation at multilocations to select promising accessions for use in little millet breeding programs (Arya et al. 2021b).

Amaravel et al. (2023) evaluated 323 genotypes of little millet for 10 quantitative traits, based on which they were grouped into 13 distinct clusters. The largest cluster comprised 243 genotypes whereas 5 smallest clusters had only 1 genotype each. Sellakumar et al. (2023) characterized 17 landraces and varieties of little millet for 11 quantitative and 6 qualitative traits. Based on the quantitative traits, the genotypes were grouped into five clusters, with major contributors as 1000 seed weight and plant height. The qualitative traits, namely, grain color, panicle compactness, and pigmentation of leaf sheath, were found highly variable among the genotypes studied, which were grouped into six clusters. These highly variable qualitative traits could be employed for the identification of these landraces.

Plant genetic resources are the essential basic raw material for crop improvement programs to meet the demands of increasing population and for mitigating climate change. To study the diverse set of germplasm, development of subsets like core, mini core, and reference collections is the best entry point to search genetic variability for agronomic and nutritional traits, tolerance to biotic and abiotic stresses, and their effective use in breeding programs (Upadhyaya et al. 2014). Developing core collection (about 10% of the entire collection) or mini core (about 1% of the entire collection) has been suggested to represent the diverse germplasm with agronomically beneficial traits for enhancing their use in crop improvement. Upadhyaya et al. (2014) developed core set comprising of 56 accessions (12% of total) representing two races and four subraces, after comprehensive evaluation of entire germplasm collections of little millet (460 accessions) at ICRISAT for more than 20 morpho-agronomic traits including 6 qualitative and 14 quantitative traits. The statistical analysis indicated that the variation in the entire collection has been retained in the core collection. The diverse representative subsets can be further evaluated at multilocations to identify trait-specific diverse germplasm for future use.

5.6 Pre-breeding in Little Millet

Little millet is a predominantly self-pollinated crop. Little millet floral biology and duration of anthesis and pollination events makes it very difficult to execute any pre-breeding activity for developing trait-specific genetic stocks. Till date, there are no concrete or focused efforts made for pre-breeding in little millet. However, the methods used in other such highly self-pollinated crops can be useful for accelerating the pre-breeding or wide hybridization efforts. Artificial hybridization techniques in small millets are elaborated by Gupta et al. 2012, Nandini et al. 2019, Patil 2021, and Nagaraja et al. 2023. The study of floral biology will help in planning the most appropriate methods of emasculation and pollination.

5.6.1 Little Millet: Floral Biology

Little millet inflorescence is a panicle, much branched erect or nodding up to 15–45 cm long and 1–5 cm wide (Seetharam et al. 2003). Spikelets are solitary, sometimes two in number. Each spikelet comprised of two very small flowers. The lower side flower is sterile and the upper one is fertile or a bisexual. The upper lemma and its palea encloses the sterile flower and lower lemma and palea enclose the fertile of bisexual flower (Sundararaj and Thulasidas 1976). The spikelets are elliptical in shape and dorsally compressed. The spikelets contain three anthers about 1.5 mm in length. The glumes are lengthy and reach apex of florets and are thinner than fertile lemma. The lower glume is ovate in shape (0.7–1.2 mm long), membranous, and devoid of keels with 1–3 veins. The upper glume is also similar in shape but larger in size than the lower glume and has 11–15 veins (Nanda and Agarwal 2008) (Fig. 5.4).

5.6.2 Little Millet: Anthesis and Pollination

After 2–3 days of panicle emergence, the opening of the spikelet occurs from top to bottom of the panicle. Maximum bloom will progress usually on sixth or seventh day, and overall, it takes a fortnight for complete flowering (Sundararaj and Thulasidas 1976). During anthesis period, which occurs in between 9:30 am and 10:30 am (Jayaraman et al. 1997), glumes open for a very short time (2–5 min.) and it results in self-pollination (Seetharam et al. 2003).

Fig. 5.4 Little millet panicle, spikelets, and floret showing glumes, lemma, palea, and feathery stigma

Difficulties in Hybridization and Crossing Techniques in Little Millet

(a) Very small florets make it almost difficult for hand emasculation and pollination.
(b) The anthesis period is very short (2–5 min) which limits the success rate.
(c) Pollen grains are very small and light weight to handle it easily.
(d) Availability of genic markers is also very less so as to identify the true hybrids after effecting the crosses.

There is greater extent of variability readily available in National Genebank, ICAR-NBGPR, New Delhi, and it can be efficiently used only when there is efficient artificial hybridization technique developed and it should have a simple and effective method of emasculation and pollination.

Different emasculation and crossing methods used in millets are (1) contact method, (2) hand emasculation followed by pollination, (3) hot water treatment, (4) USSR method (induced opening of the flower), and (5) modified crossing method (SMUASB method). Out of these five methods, SMUASB was tested in little millet by Nandini et al. (2019) and Patil (2021), and it was found highly advantageous as compared to other methods. The modified method of crossing (SMUASB) was introduced in proso millet and little millet during 2014 at Project Coordinating Unit on Small Millets, University of Agricultural Sciences, GKVK, Bengaluru, hence named this method as SMUASB (Small millets, University of Agricultural Sciences, Bengaluru). In little millet, lemma and palea are very tight and any attempt to open the floret prior to normal anthesis results in damage to the flower and no seed set. In SMUASB method, cold water is used as a mechanical stimulator for opening of florets. Little millet flower opening starts from top to bottom of the panicle. Male and female parents are planted in the alternate rows for crossing. Best time for crossing is 8 AM to 9 AM (Nandini et al. 2019).

5.6.3 Pre-breeding Efforts in Little Millet

There are two predominant races in little millet, i.e., *nana* and *robusta*. These are majorly based on the panicle morphology where *nana* race matures very early with lower biomass yields, whereas *robusta* represents typically a vigorously growing late maturing types.

5.6.4 Interspecific Hybridization

Early domestications from weedy species *Panicum psilopodium* gave rise to little millet (de Wet et al. 1983). It is evident that the chromosome pairing among the hybrids (*Panicum sumatrense* and *P. psilopodium*) is almost perfect depicting only one quadrivalent and revealed that the divergence between the two species occurred initially through a single reciprocal translocation (Hiremath et al. 1990).

5.6.5 Intraspecific Hybridization

In recent past, efforts were made by few researchers to make the crosses in little millet which revealed good success in obtaining the true hybrids. Nandini et al. (2019) used the modified crossing method (SMUASB) in little millet for making crosses among two parents in little millet. JK-8 (green color) was used as female parent and Peddasame (purple pigmented plants) was used as male parent. Purple pigmentation in Peddasame was used as marker for identification of true hybrids. On an average, 56% success rate was observed using this SMUASB method. Patil (2021) used the SMUASB method in little millet. The cultivars GV-1 (green color) and MV-126 (purple plant stem pigmented) were crossed. This method was also used in other two cross combinations. On an average, 50% success rate was observed for getting true F1s.

5.7 Genetic and Genomic Resources

5.7.1 Molecular Markers for Characterization, Utilization, and Management of Little Millet Germplasm

5.7.1.1 Marker-Based Diversity/Polymorphism Studies

Tiwari et al. (2018) utilized 175 RAPD marker loci (26 primers) for genotyping of 32 genotypes of little millet and found polymorphism level of 88.58% with an average of 4.31 polymorphic bands per primer. The polymorphism information content (PIC) ranged from 0.102 to 0.517. Genotypes were clustered into two major and further into subgroups based on their geographical regions. The identified diverse genotypes can be used in little millet improvement.

Rajput et al. (2019) used 10 ISSR markers to characterize 40 landraces of little millet from Madhya Pradesh and observed 76.19% polymorphism and PIC value ranging from 0.37 to 0.53. Landraces belonging to Dindori region followed by Betul and Chhindwara region showed highest frequency of polymorphism. Utility of ISSR markers in genetic relationship studies in little millet was emphasized.

Natesan et al. (2020) characterized five genotypes of little millet using morphological- and cereal-based SSR markers. Out of 25 maize SSR markers used, two markers were polymorphic and out of 25 pearl millet SSR markers used, one was polymorphic and the identified markers can be used for varietal identification.

Subsequently, Desai et al. (2021) reported a total of 30,220 unigenes and 4443 genic-SSR motifs with an average of 1 SSR locus per 10 kb. Trinucleotide repeats were the most abundant at the rate of 66.67%. Functional annotation of microsatellite containing unigenes revealed their linkage to gene expression regulation, metabolism, and response to environmental stresses. Out of 50 primers validated, 48% showed polymorphism, with a transferability rate of 70% to other millets. Selected microsatellite markers can be utilized in the management and characterization of little millet genetic resources.

Genotyping of 96 accessions of little millet exhibiting variation for agronomic traits, including days to 50% flowering (39–118 days), thousand grain weight (1.12–3.487 g) was carried out with SSR markers. Phylogenetic analysis revealed close groupings of accessions based on thousand grain weight. Late flowering accessions were also closely grouped. Thus, markers selected can be utilized for large-scale characterization, DNA fingerprinting studies, and other molecular studies of little millet (ICAR-NBPGR 2023).

5.7.1.2 Genome-Wide SNP Markers

Johnson et al. (2019) presented first genome-wide analysis of 190 and finally of 165 post-filtered little millet accessions. GBS (genotype by sequencing) was used to generate raw 14,473 SNPs and finally 2245 filtered SNPs. These little millet accessions were grouped into eight putative subpopulations and weaker population structure was observed compared to proso millet and kodo millet used in their study. Population structure of little millet did not correlate with existing race designations. And the identified SNPs accurately captured the population structure of little millet, as indicated by the estimates of narrow-sense heritability and geographic correlations. The information generated can be used in selecting accessions for breeding programs and prioritizing collecting missions. This study further revealed that race is not a good determining factor of genetic relatedness.

5.7.2 Genome Sequence

Genome sequencing data provides information at DNA sequence level and facilitates the development of genome-wide markers (SSRs/SNPs, etc.) from coding and noncoding regions of the genome and help in understanding diversity level and population structure of the particular species. Further, the genome sequence data generated can be utilized for genome-wide association studies, markers, or genomics-assisted breeding for crop improvement. Not much information is available on genome sequencing projects in little millet. To name a few are as follows:

5.7.2.1 Chloroplast Genome Sequence

In little millet, complete chloroplast genome of Accession No. IT261894 was sequenced having a size of 139,384 bp and 125 genes including 4 rRNA genes, 20 duplicated gene in the inverted repeat region, 30 tRNA genes, and 91 protein-coding genes (Sebastin et al. 2018). Little millet chloroplast genome sequence could serve as a source for little millet crop improvement.

5.7.2.2 Whole Genome Sequence

Partial genome sequence of little millet cultivar Lit 171 with Bioproject ID: PRJDB15472; Biosample ID SAMD00587101 is available at NCBI. This is the only genome sequence information available in the public domain and a lot of research work is needed to generate genomic resources in little millet for its improvement.

5.7.2.3 RNA Sequencing/Transcriptomics and Databases

RNA sequencing has been carried out using tissues of different development stages to generate genic/EST-SSRs and other genomic resources, and for gene expression studies and transcriptome database development in little millet (Table 5.2).

5.8 Future Prospects

High-yielding, bold seeded, non-shattering, non-lodging, early maturing, nutritionally rich, biotic (shoot fly, rust) and abiotic stress-tolerant/stress-resistant varieties are the breeding objectives in little millet crop improvement. Efforts have been made for improving little millet by developing varieties, characterizing germplasm, and identifying core set and trait-specific accessions, but comprehensive studies involving high-throughput phenotyping (morphophysiological and biochemical traits) and genotyping studies are lacking. Genomics, transcriptomics, metabolomics, and comparative genomic studies are also in infancy in little millet. Reference genome sequence is yet not available. So whole genome sequencing for generating little millet reference genome followed by pan-genome sequencing is the need of hour to accelerate crop improvement programs. Genetic linkage maps are also lacking may be due to small-flowered nature of crop. GWAS (genome-wide association studies) with the use of high-throughput sequencing technologies can be applied for

Table 5.2 Genomic resources (https://www.ncbi.nlm.nih.gov/search/all/?term=little%20millet) and transcriptome databases in little millet

Little millet sample ID	Project data type	Development stage/tissue	Genomic platform	Reference
PRJNA885279	Transcriptome or gene expression	Vegetative/leaf and root	Complete Genomics	NCBI
PRJNA880733	Transcriptome or gene expression	Vegetative/leaf and root	Complete Genomics	NCBI
PRJNA729217	Transcriptome sequencing to develop EST-SSR markers	Vegetative stage, reproductive stage, and maturity stage	Ion Torrent S5	NCBI
PRJNA554415	Transcriptome sequencing subjected to drought and salt stress	Seedling/leaves	Illumina HiSeq 2500	NCBI
PRJNA267462	Transcriptome or gene expression to generate genomic resources	Leaves	Illumina HiSeq 2000	NCBI
LMTdb (https://igkv.ac.in/xenom/index.aspx)	Little millet transcriptome database	RLM-37 secondary leaf under drought conditions and secondary leaf, flag leaf, and panicle under control conditions	Illumina HiSeq Platform A Web interface to access information related to DEGs, microsatellite markers, gene functions, and plant metabolic pathway information	Shekhar et al. (2023)

NCBI, Accessed on 4 Dec 2023

the identification of novel QTLs/QTNs/mapping loci associated with the traits of interest. Transcriptome wide association studies (TWAS) can also be used to find gene trait associations. Nested association mapping (NAM) can also be applied to detect powerful QTLs. MutMap+, QTLseq approach could be applied to rapidly identify casual mutations and genes/the exact location and sequence of candidate genes involved in controlling agronomically important traits. Genomic Selection (GS) models can be developed to accelerate breeding programs. Extensive research pertaining to functional genomics and computational biology is also required for trait improvement in little millet. Further, genome editing tools based on site-directed nucleases (SDNs) and clustered regularly interspaced short palindromic repeats (CRISPR)/CRISPR-associated (Cas) could be employed to modify target genes for improving crop productivity, quality, and resistance/tolerance to various biotic and abiotic stresses. By combining high-throughput phenotyping, genotyping, and advanced computational and artificial intelligence (AI) approaches, it would be easy to identify molecular mechanisms controlling complex traits and expedite the management and utilization of little millet genetic resources for future food and nutritional security and agricultural sustainability.

References

Ajithkumar IP, Panneerselvam R (2014) ROS scavenging system, osmotic maintenance, pigment and growth status of *Panicum sumatrense Roth.* under drought stress. Cell Biochem Biophys 68:587–595. https://doi.org/10.1007/s12013-013-9746-x

Amaravel M, Nirmalakumari A, Geetha S, Sathiya K, Renuka R (2023) Understanding the genetic basis of yield-related traits in little millet (*Panicum sumatrense* Roth. ex. Roem. and Schult.) germplasm through association and diversity analysis. Agric Sci Digest 43(6):733–740. https://doi.org/10.18805/ag.D-5766

Arya L, Singh M, Pandey S, Verma M, Singh M, Gomashe S, Pandey CD, Bhardwaj R (2021a) Little millet, a nutri-cereal: glimpses of variability in the National Genebank. ICAR-NBPGR, New Delhi. http://www.nbpgr.ernet.in/Divisions_and_Units/Downloadfile.aspx?EntryId=9128

Arya L, Singh M, Gomashe S, et al (2021b) Identification of trait specific accessions of little millet (*Panicum sumatrense*) for its enhanced utilization. Presented at the 2nd International Agrobiodiversity Congress (virtual mode), 15–18 November 2021

Bramel P, Giovannini P, Eshan Dulloo M (2022) Global strategy for the conservation and use of genetic resources of selected millets. Global Crop Diversity Trust, Bonn, Germany. https://doi.org/10.5281/zenodo.7798294

Channappagoudar BB, Hiremath SM, Biradar NR, Koti RV, Bharamagoudar TD (2007) Morpho-physiological traits influencing the grain yield potential in little millet. Karnataka J Agric Sci 20(3):473–476

de Wet JMJ, Prasada Rao KE, Brink DE (1983) Systematics and domestication of *Panicum sumatrense* (Graminae). J Agric Tradit Bot Appl 30:159–168

Desai H, Hamid R, Ghorbanzadeh Z et al (2021) Genic microsatellite marker characterization and development in little millet (*Panicum sumatrense*) using transcriptome sequencing. Sci Rep 11:20620. https://doi.org/10.1038/s41598-021-00100-4

Elangovan M, Venkatesh K, Pandey S, Pandey CD (2022) International year of millets 2023: opportunity for enhancing the use of Indian millets germplasm. Indian J Plant Genet Res 35(3):90–94

Gowda J, Rekha D, Somu G, Bharathi S, Krishnappa M (2008) Development of core set in little millet (*Panicum sumatrense* Roth. ex Roem. & Schult.) germplasm using data on twenty-one morpho-agronomic traits. Environ Ecol 26:1055–1060

Guha M, Sreerama YN, Malleshi NG (2015) Influence of processing on nutraceuticals of little millet (*Panicum sumatrense*). In: Processing and impact on active components in food. Academic Press, pp 353–360

Gupta A, Sood S, Agrawal PK, Bhatt JC (2012) Floral biology and pollination system in small millets. Eur J Plant Sci Biotechnol 6(Special Issue 2):80–86

Hariprasanna K (2023) Small millets in India: current scenario and way forward. Indian Farming 73(01):38–41

Hijmans RJ, Guarino L, Cruz M, Rojas E (2001) Computer tools for spatial analysis of plant genetic resources data: 1. DIVA-GIS. Plant Genet Resour Newslett 127:15–19

Hiremath SC, Patil GNV, Salimath SS (1990) Genome homology and origin of *Panicum sumatrense* (Gramineae). Cytologia (Tokyo) 55:315–319. https://doi.org/10.1508/cytologia.55.315

ICAR NBPGR (2023) Annual report 2022. ICAR-National Bureau of Plant Genetic Resources, New Delhi, India, pp 86–87. http://www.nbpgr.ernet.in/Downloadfile.aspx?EntryId=9329

Jayaraman N, Suresh S, Nirmala A, Ganeshan NM (1997) Genetic enhancement and breeding strategies in small millets. In: National seminar on small millets, pp 19–21. 23–24 April 1997, Coimbatore, India

Johnson M, Deshpande S, Vetriventhan M, Upadhyaya HD, Wallace JG (2019) Genome-wide population structure analyses of three minor millets: kodo millet, little millet, and proso millet. Plant Genome 12:190021. https://doi.org/10.3835/plantgenome2019.03.0021

Katara D, Kumar R, Rajan D, Chaudhari SB, Zapadiya VJ (2019) Genetic-morphological analysis in little millet (*Panicum sumatrense* Roth. ex Roem. & Schult.) under different sown conditions. Int J Curr Microbiol App Sci 8(5):177–189

Mallikarjun YK, Hemalatha S, Meghana DR, Sharanappa T, Rama KN (2013) Evaluation of little millet (*Panicum sumatrense*) landraces for cooking and nutritional composition. Curr Res Biol Pharm Sci 2(1):7–11

Nagaraja TE, Nandini C, Bhat S, Parveen SG (2023) Artificial hybridization techniques in small millets—a review. Front Plant Sci 14:1112117. https://doi.org/10.3389/fpls.2023.1112117

Nanda JS, Agarwal PK (2008) Botany of field crops. Kalyani Publishers, I

Nandini C, Bhat S, Srinathareddy J, Prabhakar. (2019) Modified crossing (SMUASB) method for artificial hybridization in proso millet (*Panicum miliaceum* L.) and little millet (*Panicum sumatrense*). Electron J Plant Breed 10(3):1161–1170

Natesan S, Venugopalan G, Selvamani SB, Krishnamoorthy M, Sarankumar Chandran S, Angamuthu N (2020) Characterization of little millet (*Panicum sumatrense*) varieties using morphological descriptors and SSR based DNA fingerprinting. J Phytology 12:29–34. https://doi.org/10.25081/jp.2020.v12.6317

Nirmalakumari A, Salini K, Veerabadhiran P (2010) Morphological characterization and evaluation of little millet (*Panicum sumatrense* Roth. ex. Roem. and Schult.) germplasm. Electron J Plant Breed 1(2):148–155

Patil R (2021) Reproductive biology, breeding behaviour, emasculation, pollination techniques and hybridization in little millet (*Panicum sumatrense* L.). Pharma Innova 10(11):1153–1159

Pradeep SR, Guha M (2011) Effect of processing methods on the nutraceutical and antioxidant properties of little millet (*Panicum sumatrense*) extracts. Food Chem 126:1643–1647. https://doi.org/10.1016/j.foodchem.2010.12.047

Rajput LPS, Tantwai K, Pooniya SK, Tsuji K (2019) Assessment of genetic variability among the landraces of little millet *Panicum sumatrense* from different districts of Madhya Pradesh. Int J Curr Microbiol App Sci 8(4):2686–2693. https://doi.org/10.20546/ijcmas.2019.804.312

Saha D, Gowda MVC, Arya L, Verma M, Bansal KC (2016) Genetic and genomic resources of small millets. Crit Rev Plant Sci 35:56–79. https://doi.org/10.1080/07352689.2016.1147907

Sebastin R, Lee G, Lee KJ, Shin M, Lee J, Ma K et al (2018) The complete chloroplast genome sequences of little millet (*Panicum sumatrense* Roth. ex Roem. & Schult.) (Poaceae). Mitochondrial DNA B Resour 3:719–720

Seetharam A, Gowda J, Halaswamy JH (2003) Small millets. In: Chowdhury SK, Lal SK (eds) Nucleus and breeder seed production manual. Indian Agricultural Research Institute, New Delhi, India, pp 54–67

Sellakumar S, Sneha A, Mano N, Lydia PJ, Petchiammal KI, Wilson D, Kumar PD, Francis N, Sheriff SK (2023) Characterization of little millet (*Panicum sumatrense* Roth. ex. Roem. & Schult.) landraces and varieties for genetic diversity and association of traits. Electron J Plant Breed 14(2):561–571

Semwal DP, Ahlawat SP, Pradheep K (2018) Pigeon pea (*Cajanus cajan* L.) and its wild species germplasm collection status, diversity distribution, and trait-specific germplasm mapping using GIS tools in India. Legum Res 41(5):656–662

Shekhar S, Prasad AS, Banjare K et al (2023) LMTdb: a comprehensive transcriptome database for climate-resilient, nutritionally rich little millet (*Panicum sumatrense*). Front Plant Sci 14:1106104. https://doi.org/10.3389/fpls.2023.1106104

Sivakumar S, Mohan M, Franco OL, Thayumanavan B (2006) Inhibition of insect pest α-amylases by little and finger millet inhibitors. Pestic Biochem Physiol 85:155–160. https://doi.org/10.1016/j.pestbp.2005.11.008

Sundararaj DP, Thulasidas G (1976) Botany of field crops. Macmillan

Tiwari N, Tiwari S, Tripathi N (2018) Genetic characterization of Indian little millet (*Panicum sumatrense*) genotypes using random amplified polymorphic DNA markers. Agric Nat Resour 52(4):347–353

Upadhyaya HD, Dwivedi SL, Singh SK, Singh S, Vetriventhan M, Sharma S (2014) Forming core collections in barnyard, kodo, and little millets using morphoagronomic descriptors. Crop Sci 54:2673–2682

Bhat BV, Rao BD, Tonapi VA (2018) The story of millets (Ed.). Karnataka State Department of Agriculture, Bengaluru, and ICAR-Indian Institute of Millets Research, Hyderabad, India, p. 110

Vetriventhan M, Azevedo VCR, Upadhyaya HD et al (2020) Genetic and genomic resources, and breeding for accelerating improvement of small millets: current status and future interventions. Nucleus 63:217–239. https://doi.org/10.1007/s13237-020-00322-3

Vetriventhan M, Upadhyaya HD, Azevedo VC, Allan V, Anitha S (2021) Variability and trait-specific accessions for grain yield and nutritional traits in germplasm of little millet (*Panicum sumatrense* Roth ex. Roem. & Schult.). Crop Sci 61:2658–2679. https://doi.org/10.1002/csc2.20527

Wanous MK (1990) Origin, taxonomy, and ploidy of the millets and minor cereals. Plant Variet Seeds 3:99–112

Weber S (2013) The vanishing millets of the Indus civilization. Archaeol Anthropol Sci 3(1):9–15. https://doi.org/10.1007/s12520-013-0143-6

Barnyard Millet

6

Badal Singh, Mithilesh Kumar, Lalit Arya,
Amit Kumar Singh, Neelam Shekhawat,
Pavan Kumar Malav, Chithra Devi Pandey, Kirti Rani,
and Sushil Pandey

Abstract

Barnyard millet, a vital crop in warm and temperate regions like Asia, offers food security to underprivileged populations due to its nutritional richness and adaptability to challenging environments. Cultivated for both human consumption and animal feed, barnyard millet plays a significant role in regions unsuitable for rice production, with India being a major producer. The genus *Echinochloa* encompasses over 250 species, with Indian and Japanese barnyard millets (*E. frumentacea* and *E. esculenta*) as key varieties. Both species, domesticated thousands of years ago, exhibit traits typical of domestication syndrome, showing parallel evolution with their wild ancestors and offering insights into the domestication process. The diversity of barnyard millet is declining due to reduced cultivation and socioeconomic shifts in India. Conservation efforts primarily focus on ex situ collections in Japan and India, neglecting on-farm conservation. Addressing

B. Singh (✉)
Division of Germplasm Evaluation, ICAR-NBPGR, New Delhi, India
e-mail: badal.singh@icar.gov.in

M. Kumar
Agricultural Research Station, Mandor, Agriculture University, Jodhpur, Rajasthan, India

L. Arya · A. K. Singh
Division of Genomic Resources, ICAR-NBPGR, New Delhi, India

N. Shekhawat · K. Rani
ICAR-NBPGR, Regional Station, Jodhpur, Rajasthan, India

P. K. Malav
Division of Plant Exploration and Germplasm Collection, ICAR-NBPGR, New Delhi, India

C. D. Pandey · S. Pandey
Division of Germplasm Conservation, ICAR-National Bureau of Plant Genetic Resources, New Delhi, India

K. Tripathi et al. (eds.), *Plant Genebank Utilization for Trait Discovery in Millets*, https://doi.org/10.1007/978-981-96-4043-0_6

this gap is crucial for preserving the genetic diversity of barnyard millet in its natural environment. Expanding the genetic diversity of barnyard millet is essential, particularly for traits like disease resistance and seed size. Utilizing wild relatives' gene pools and employing markers for trait incorporation are warranted. Interspecific hybridization has potential, but fertility barriers must be addressed with modern biotechnological tools. Systematic genetic analysis is vital for identifying agriculturally significant traits. Despite facing challenges like genetic erosion and limited access, collaborative efforts in conservation and innovation are essential to fully harness the potential of barnyard millet genetic resources for sustainable agriculture and food security.

Keywords

Barnyard millet · Plant genetic resources · Germplasm · Breeding · Utilization · Pre-breeding · Trait-specific sources

6.1 Introduction

Millets are a diverse collection of small-seeded grasses produced worldwide as cereal crops or grains for human use and livestock feed. They are especially useful in arid and semiarid locations because of their drought tolerance and low soil fertility. Millets have long been a traditional food in nations like India, highlighting their significance both nutrient-wise and culturally. Particularly in rural and tribal communities, millets are frequently utilized in traditional recipes. The cultural significance of millets is highlighted by the frequent inclusion of meals produced from them in many Indian festivals, such as Sankranti in South India. Millets fall into two categories: major and minor. Pearl millet (*Pennisetum glaucum*) and sorghum (*Sorghum bicolor*) are major millets, while other millets are in the minor millet class. Significant minor millets include foxtail millet (*Setaria italica*), finger millet (*Eleusine coracana*), proso millet (*Panicum miliaceum*), kodo millet (*Paspalum scrobiculatum*), small millet (*Panicum sumatrense*), and barnyard millet (*Echinochloa esculenta*) (Ajeesh Krishna et al. 2022).

Barnyard millet (*Echinochloa* species) is an ancient millet crop produced in warm and temperate regions of the world, with a strong presence in Asia, particularly India, China, Japan, and Korea. It is the fourth most produced minor millet, giving many underprivileged people worldwide food security. Barnyard millet is a short-duration crop that can tolerate a variety of biotic and abiotic challenges and grow in unfavorable environmental circumstances with very little input. The grains are prized for their high nutritional content and reduced cost as compared to major cereals like rice, wheat, and maize. Protein, carbs, fiber, and, most significantly, micronutrients like iron (Fe) and zinc (Zn) are abundant in it (Singh et al. 2010; Saleh et al. 2013; Chandel et al. 2014). These nutrients have been linked to a host of health advantages (Saleh et al. 2013; Ugare et al. 2014). Because of all these benefits, barnyard millet is a great option to grow as a supplementary crop for subsistence farmers or as a fallback crop when the monsoon fails in places that grow main crops like rice (Gupta et al. 2009).

Barnyard millet is typically grown in regions with edaphic and climatic conditions unsuited for rice production (Yabuno 1987). Barnyard millet, with production and productivity of 87,000 tons and 857 kg/ha, respectively, is the second most important small millet in India, after finger millet (Padulosi et al. 2009). It is mostly grown in two distinct agroecologies in India: the Deccan plateau in Tamil Nadu's southern region and the mid-hills of Uttarakhand's Himalayan region in the north. In many Indian states, wild barnyard millet, or *Echinochloa colona*, is often seen growing as a weed in rice fields and is used as food during drought years (Padulosi et al. 2009). Barnyard millet can yield up to eight harvests annually and has garnered some interest as fodder in the United States and Japan owing to its quick growth and early maturity (http://www.fao.org/docrep/T0818E/T0818E01.htm). Because of its higher protein and calcium content, crop straw is regarded as better fodder than rice, oat, or Timothy straw (Obara 1938). In the United States, ducks and other birds primarily eat its grains (Mitchell and Tomlinson Jr 1989).

Managing and utilizing barnyard millet genetic resources faces challenges like genetic erosion, limited access, lack of research, inadequate infrastructure, conservation issues, and climate change impacts. Collaborative efforts are needed to address these challenges and unlock the full potential for enhancing agricultural sustainability, food security, and nutritional diversity. This chapter initiates an exploration of the diverse domain concerning barnyard millet genetic resources, emphasizing a dedicated adherence to conservation, innovation, and equitable access. Through an investigation of strategies encompassing conservation, utilization, and enhancement, our objective is to facilitate the development of a more robust and sustainable agricultural landscape.

6.2 The Crop

6.2.1 History, Origin, Domestication, and Evolution

More than 250 annual and perennial species of the genus *Echinochloa* can be found throughout the planet's warmer and more temperate regions (Bajwa et al. 2015). Morphological, cytological, and molecular marker research have been used to identify 35 species based on their taxonomic and evolutionary relationships (Yamaguchi et al. 2005). *E. frumentacea* Link; syn. *E. colona* var. *frumentacea* (Link) Ridl. (Indian barnyard millet) and *E. esculenta* (A. Braun) H. Scholz; syn. *E. Utilis* Ohwi et Yabuno (Japanese barnyard millet) are grown as minor cereals in Japan, Korea, the northeastern regions of China and India, Pakistan, and Nepal, respectively (Yabuno 1987; Yamaguchi et al. 2005). The Indian barnyard millet (*Echinochloa frumentacea*) was grown in India, whereas the Japanese barnyard millet (*Echinochloa esculenta*) originated in eastern Asia. *Echinochloa esculenta* has four races: *crus-galli*, *macrocarpa*, *utilis*, and *intermedia* (Fig. 6.1).

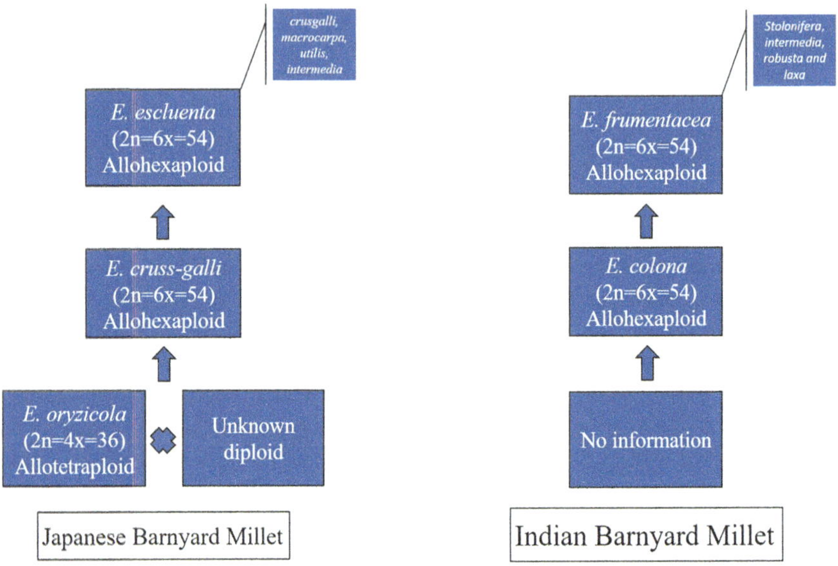

Fig. 6.1 Evolution of Indian and Japanese barnyard millet. (Infographics by Badal Singh)

Echinochloa frumentacea, cultivated annually in India, Central African Republic, Tanzania, and Malawi, shares a parallel evolutionary lineage in both India and Africa (Doggett 1989). Its wild progenitor, *E. colona* (L.) Link, known as Jungle rice, serves as its precursor, though the precise domestication period remains uncertain. Conversely, the annual *Echinochloa esculenta* grows best in temperate climates such as those in Germany, Japan, Korea, China, Russia, and so on (De Wet et al. 1983). Japanese barnyard millet was domesticated straight from *E. crus-galli* (L.) Beauv., or barnyard grass, about 4000 years ago in Japan (Doggett 1989).

Archaeological findings suggest barnyard millet's cultivation in Japan as early as the Yayoi period, around 4–5 millennia ago (Watanabe 1970). Alternatively, records from the Jomon period, dating back to 2000 BC, also document its domestication (Nesbitt 2005). Nozawa et al. (2004) utilized 13 SSR markers to analyze genetic diversity among 170 *Echinochloa* accessions, revealing two classes for *E. esculenta* and 11 for its ancestral species, *E. crus-galli*.

The domestication syndrome, comprising traits acquired during the transition from wild to cultivated forms due to selection pressure (Hammer 1984, 2003), remains insufficiently explored in barnyard millet. However, significant differences from their wild ancestors, including reduced vegetative branching, compact growth, larger inflorescences, reduced shattering, and larger seeds, are evident in both Indian barnyard and Japanese barnyard (Yabuno 1975). These characteristics, akin to the domestication syndrome observed in foxtail millet and pearl millet (Defelice 2002; Doust et al. 2005; Poncet et al. 1998, 2000), suggest a similar pattern for barnyard millet. Archaeological evidence from Northern Japan indicates a gradual increase in seed size over millennia, supporting the notion of human selection (Crawford 1983, 2011; Takase 2009).

The cross-compatibility between domesticated barnyard millet and their wild counterparts, alongside naturally occurring intergrades, offers insights into the domestication process and the genetic underpinnings of domestication traits in this crop.

6.2.1.1 Phylogeny

The intricate relationships among *Echinochloa* species are revealed through crossing experiments and meiotic chromosomal behavior. Traditional taxonomic methods relying on morphology, reproduction, and ecology may yield ambiguous results due to domestication effects. Molecular marker analysis confirms three cross-compatible groups: *E. oryzicola*, *E. crus-galli* complex, and *E. colona-frumentacea*. These groups exhibit distinct phylogenetic lineages, as demonstrated by nucleotide sequencing of cpDNA regions, highlighting species divergence. Phylogenetic analysis also supports the hypothesis that *E. crus-galli* originated from hybridization between *E. oryzicola* and an unidentified diploid species (Fig. 6.1). Additionally, the sh4 gene analysis in *Oryza* suggests a genomic relationship among Asian *Echinochloa* species, indicating that the allohexaploid *E. crus-galli* shares two genomes with its parental donor, *E. oryzicola*. Furthermore, the tetraploid perennial species *E. stagnina* shares one genome with *E. oryzicola* and possesses an unknown genome, *E. crus-pavonis*, showing close affinity with *E. crus-galli* and *E. oryzicola*. Conversely, *E. colona* exhibits distant affinities in all homoeologous copies. These findings elucidate the evolutionary processes and genetic relationships within the *Echinochloa* genus (Yabuno 1966, 1984, 2001; Yasuda et al. 2002; Yamaguchi et al. 2005; Aoki and Yamaguchi 2008, 2009).

6.2.2 Taxonomy, Cultivation, and Uses

6.2.2.1 Botanical Description

Barnyard millet, known for its wide adaptation, thrives in diverse environments, including altitudes up to 2000 m above sea level during the summer season (Gupta et al. 2009). This tall, robust annual crop reaches heights of up to 220 cm and boasts a remarkably short generation time, completing its life cycle in just 45–60 days, though this may vary (Denton 1987; Padulosi et al. 2009). In certain northern hill ecosystems, the maturation process may extend beyond this timeframe (Hulse et al. 1980). Its distinctive features include flat, wide leaf blades without ligules and terminal panicles ranging from cylindrical to elliptic, densely packed with spikelets of varying colors (Napper 1965). Spikelets are typically two-flowered, awnless or awned, with sterile lower florets and fertile upper florets (Gupta et al. 2010; Mitich 1990). The grain, enclosed within shiny, hardened lemma and palea, measures 2–3 mm in length and 1–2 mm in width (Sundararaj and Thulasidas 1976).

6.2.2.2 Reproductive Biology and Flowering Mechanism

The shape of the inflorescence, the timing of flowering, and other morphological characteristics vary in barnyard millet (Obara 1938). Flowering progresses from the top to the bottom of the inflorescence over 10–15 days, peaking between 6 and 7 am (Sundararaj and Thulasidas 1976; Jayaraman et al. 1997). Flowers are hermaphrodite, primarily self-pollinating but with some wind-mediated outcrossing (Maun and Barrett 1986). Indian and Japanese barnyard millet can be differentiated by panicle morphology and spikelet characteristics (Yabuno 1971; De Wet et al. 1983). *E. frumentacea* is characterized by smaller, awnless spikelets with membranous glumes compared to *E. esculenta's* larger, usually awned spikelets with chartaceous glumes and lemma (Yabuno 1987).

6.2.2.3 Cultivation and Uses

Barnyard millet, a fast-growing annual summer crop, is prized for its rapid growth, drought tolerance, and adaptability to marginal environments, making it vital for famine-prone areas (De Wet et al. 1983; Kono et al. 1987; Majid and Hamid 1989). Germination occurs swiftly, typically within 24 h of water imbibition, with adventitious roots appearing after 5 days and the first tillers emerging 10 days post-germination (Rahn et al. 1968; Holm et al. 1991). Temperature significantly influences growth, with optimal ranges of 27–33 °C during the day and 15–22 °C at night for *E. frumentacea* (Muldoon et al. 1982). In Uttarakhand's high hills, *E. frumentacea* and *E. esculenta* exhibit significant variation in seedling emergence and growth characteristics due to differences in cold tolerance (Muldoon et al. 1982; Bandyopadhyay 1999). Vegetative growth transitions to inflorescence development approximately 40 days after emergence, with both cultivated species displaying quantitative short-day plant characteristics but exhibiting adaptability to various photoperiods (Maun and Barrett 1986; Mitich 1990). Late planting shortens the vegetative phase, impacting yield adversely. Seeds mature within 20–40 days post-reproductive phase initiation, with optimal harvesting time at 30–35 days after heading for Japanese barnyard millet (Kumagai et al. 2011). In *Echinochloa crus-galli*, freshly harvested seeds display different degrees of intrinsic dormancy that lasts from 4 to 48 months (Sung et al. 1987; Manidool et al. 1992). Possessing a C4 photosynthetic pathway, *Echinochloa* species enjoy competitive advantages when grown alongside C3 crops and outperform rice under water stress conditions due to higher water-use efficiency and a 60% greater carbon efficiency ratio (Bouhache and Bayer 1993; Migo et al. 1991). In the Indian Himalayan region, barnyard millet serves as a traditional rice substitute, where dehulled grains are cooked and consumed akin to rice. Popular dishes like "*madira ki kheer*" are made from barnyard millet porridge in Uttarakhand. Southern India utilizes barnyard millet in traditional dishes such as *idli*, *dosa*, and *chakli*. Additionally, various processed products like biscuits, sweets, noodles, and ready-to-use mixes have been developed from barnyard millet (Arora and Srivastava 2002; Poongodi et al. 2003; Veena et al. 2004; Ugare 2008). However, industrial involvement is crucial for large-scale production and global commercialization. Rich in iron, barnyard millet holds potential for commercial foods targeting diabetics, infants, and pregnant women. Nevertheless, the lack of readily available processed products hampers its utilization despite its nutritional superiority. Developing millet-based functional foods and ready-to-use products is essential to meet modern consumer demands and enhance acceptability. Value addition not only provides consumers with diverse, convenient, and quality foods but also contributes to the revival of barnyard millet cultivation. Moreover, barnyard millet serves as a significant fodder source in the Himalayan region due to its rapid growth and voluminous fodder production from broad leaves. Highly palatable, barnyard fodder can be utilized for hay or silage making, further emphasizing its agricultural significance (Arora and Srivastava 2002; Poongodi et al. 2003; Veena et al. 2004; Ugare 2008).

6.3 Germplasm Conservation

6.3.1 Status of Barnyard Millet Germplasm Holdings at Global, National, and Institutional Genebanks

The diversity of barnyard millet is rapidly diminishing as a result of a significant decrease in the amount of land dedicated to its cultivation and the changing social, cultural, and economic aspects of the agricultural community in India (Maikhuri et al. 2001). Significant endeavors have been undertaken to safeguard the variety of crops outside their natural habitat; however, there is a lack of information addressing the conservation of crops on farms, in their natural environment (Padulosi et al. 2009; Vetriventhan et al. 2016).

Around 8920 accessions of barnyard millets germplasm are globally conserved at 34 genebanks in different countries. The major collections of barnyard millet germplasm accessions are conserved in India and Japan (Vetriventhan et al. 2016). ICAR-National Bureau of Plant Genetic Resources, New Delhi, India; University of Agricultural Sciences, Bangalore, India; Vivekananda Parvatiya Krishi Anusandhan Sansthan (VPKAS), India; Indian Institute of Millets Research (IIMR), India; International Crop Research Institute for the Semi-Arid Tropics (ICRISAT), India; and Department of Genetic Resources, National Institute of Agrobiological Sciences (NIAS), Japan, conserve nearly 40% of the global accession (Renganathan et al. 2020; Vetriventhan et al. 2020; and Bramel et al. 2022). The status of *Echinochloa* species genetic resources conserved in the different genebanks are represented in Table 6.1.

ICAR-NBPGR is actively working on germplasm collection, characterization, and evaluation for various agronomic, biotic, and abiotic stresses, grain, and nutritional traits in barnyard millet. Series of exploration missions were undertaken to collect the barnyard millet germplasm from different parts of the country and conserved in the National Genebank, ICAR-National Bureau of Plant Genetic Resources (NBPGR), New Delhi. Species-wise status of *Echinochloa* germplasm conserved in NGB are given in Table 6.2.

Table 6.1 Status of *Echinochloa* species conserved at global, national, and institutional genebanks

Name of the country	Number of accessions	Organization accessions
India	2001	National Bureau of Plant Genetic Resources, New Delhi
	965	University of Agricultural Sciences, Bangalore
	525	Vivekananda Parvatiya Krishi Anusandhan Sansthan, Almora
	1159	Indian Institute of Millets Research, Hyderabad
	749	International Crop Research Institute for the Semi-Arid Tropics, Patancheru, Hyderabad
Japan	3671	Department of Genetic Resources I National Institute of Agrobiological Sciences, Kannondai
	65	Plant Germplasm Institute, Kyoto University
United States	292	USDA Agricultural Research Service, Washington
	306	National Centre for Genetic Resources Conservation, Collins
	304	North Central Regional Plant Introduction Station, Ames
China	717	Institute of Crop Science, Chinese Academy of Agricultural Sciences, Beijing
Kenya	208	National Genebank of Kenya, Crop Plant Genetic Resources Centre, Muguga
Ethiopia	92	International Livestock Research Institute, Addis Ababa
Australia	67	Australian Plant Genetic Resource Information Service, Queensland
Pakistan	56	Plant Genetic Resources Program, Islamabad
Norway	44	Svalbard Gene Bank, Spitsbergen
United Kingdom	44	Millennium Seed Bank Project, Seed Conservation Department, Royal Botanic Gardens, London
Germany	36	Leibniz Institute of Plant Genetics and Crop Plant Research, Gatersleben

Source: Data for the accessions: http://pgrportal.nbpgr.ernet.in; https://demo.uasbangalore.edu.in/en/small-millets-zars-gkvk-bengaluru/; https://www.genesys-pgr.org/partners; Renganathan et al. (2020)

Table 6.2 Status of *Echinochloa* germplasm conserved at the National Genebank, ICAR-NBPGR, New Delhi, India

S. no.	Common name	Botanical name	Total accessions
1	Jungle rice	*Echinochloa colonum* (L.) Link	110
3	Barnyard grass	*Echinochloa crus-galli* (L.) P. Beauv	42
4	Japanese barnyard millet	*Echinochloa esculenta* (A. Braun) H. Scholz	6
5	Indian barnyard millet	*Echinochloa frumentacea* Link	1825
6	Early barnyard grass	*Echinochloa oryzoides* (Ard.) Fritsch	6
7	Barnyard grass	*Echinochloa* sp.	12
		Grand total	**2001**

6.3.2 Germplasm Collection, Mapping, and Gap Analysis in National Genebank, India

A total of 2588 accessions have so far been collected from various parts of the country. Among these, 1078 accessions had passport data with locality details of collection sites (state, district, and village) and these were further screened for availability in the national genebank. Geo-referencing and mapping of collected diversity of barnyard millet germplasm was done. Analyses of passport data shows that majority of barnyard millet germplasm accessions have been collected from Uttarakhand (353) followed by Uttar Pradesh (144), Maharashtra (137), Madhya Pradesh (116), Andhra Pradesh (117), Tamil Nadu (105), Chhattisgarh (69), Karnataka (64), and Telangana (52) while minimum accessions were collected from Odisha (12), Jammu and Kashmir (5), Delhi (5), Andaman and Nicobar Islands (2), Kerala (2), Meghalaya (2), Punjab (2), and West Bengal (1).

6.3.2.1 Gap Analysis

A total of 1985 accessions including locality information of 810 accessions were conserved in national genebank. Maximum accessions were conserved from Uttarakhand (218) followed by Andra Pradesh (102), Maharashtra (96), Madhya Pradesh (96), Uttar Pradesh (93), Tamil Nadu (64), and Chhattisgarh (43). However, there is none conserved from West Bengal, Punjab, and Delhi; hence, there is also vast gap between collected and conserved germplasm (Fig. 6.2) particularly from Meghalaya, Jammu and Kashmir, Kerala, Karnataka, Himachal Pradesh, and Odisha.

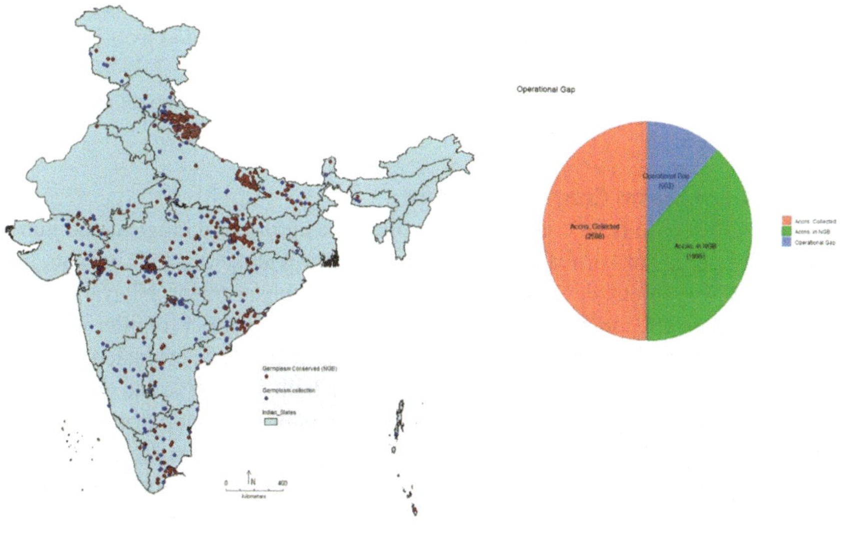

Fig. 6.2 Geo-referenced map of barnyard millet germplasm collected v/s conserved from different states/UT of the country and gap analysis. (Infographics by Pavan Kumar Malav and Badal Singh)

6.4 Germplasm Use in Crop Improvement

Effective utilization of barnyard millet (*Echinochloa* spp.) germplasm is essential for improving the crop's adaptability to agricultural demands and environmental stressors. The underutilization of germplasm in barnyard millet presents a significant obstacle to crop improvement efforts. Several factors contribute to this challenge, including limited awareness of available germplasm resources, inadequate infrastructure for germplasm conservation and evaluation, and insufficient integration of modern breeding techniques. With a limited genetic base, accessing diverse genetic resources, including wild relatives, landraces, and modern cultivars, offers a promising approach. These resources contain valuable traits like disease resistance and high-yield potential, crucial for breeding superior varieties. Methodical germplasm utilization involves collecting, conserving, and evaluating diverse accessions, followed by controlled crosses and selection. Modern biotechnological tools, such as molecular markers, expedite trait identification and introgression. Leveraging barnyard millet germplasm strategically can significantly enhance crop resilience, productivity, and nutritional quality, ensuring agricultural sustainability and food security.

6.4.1 Characterization and Evaluation

Germplasm characterization involves documenting heritable traits that are distinctly identifiable, distinguishing it from preliminary evaluation, which focuses on recording a select few agronomic traits vital for crop improvement. Standardized descriptors for plant genetic resource characterization have been provided by a number of international organizations, including Bioversity International, UPOV, PPVFR (India), and NBPGR (India). Germplasm diversity in barnyard millet can be characterized and accessible using descriptors (Figs. 6.3, 6.4, 6.5 and 6.6).

6.4.1.1 Germplasm Characterization

Several studies have been conducted to explore the agro-morphological and nutraceutical properties of barnyard millet germplasm, providing valuable insights into its potential for varietal development and nutritional enhancement. Nilavarasi et al. (2020) conducted a comprehensive study to characterize barnyard millet varieties, cultivars, and mutants based on morphological descriptors. They evaluated 11 genotypes using 12 morphological characters, including both qualitative and quantitative traits. The study identified key diagnostic characters of the genotypes, paving the way for precise varietal identification and breeding efforts. In a similar vein, Joshi et al. (2015) assessed 49 landraces of barnyard millet for agro-morphological traits and grain smut resistance. Significant variation was observed in various traits, highlighting the genetic diversity present within the germplasm. The study identified potential donor lines for varietal development and emphasized the importance of disease resistance in breeding programs. Sood et al. (2015a, b) evaluated 95 germplasm lines of barnyard millet from global collections and analyzed their qualitative

Fig. 6.3 Inflorescence shape and inflorescence compactness variation

Fig. 6.4 Inflorescence color variation

and quantitative traits. The study revealed high variation in days to maturity, grain weight, and yield components, with distinct clusters formed based on trait morphology. This underscores the importance of utilizing diverse germplasm for breeding

Fig. 6.5 Lower racemes character variation

Fig. 6.6 Seed color variation

initiatives and selecting donor lines based on specific traits. Furthermore, Panwar et al. (2016) compared the nutraceutical and antinutritional properties of barnyard millet and finger millet varieties. Barnyard millet was found to be superior in terms of crude fiber, total carotenoids, and polyphenol content, highlighting its potential as a nutritious food crop. This study underscores the importance of evaluating germplasm for nutritional traits to enhance the value of barnyard millet as a functional food. In a recent study by Nandini et al. (2020), 494 barnyard millet germplasm accessions were characterized for quantitative traits using multivariate analysis. Significant variation was observed, particularly in days to flowering, plant height, and days to maturity. The study identified key traits contributing to genetic diversity and highlighted the potential for creating variability in breeding programs.

6.4.2 Developing Representative Subsets for Enhanced Use

Core collections comprise about 10% of the total collection which represent the entire germplasm collection with maximum genetic diversity and minimum reiteration of a crop species and its relatives. It raises the effective use and protection of germplasm and provide an opportunity to explore complete variability for various traits at one and the same time. A mini core collection (~1% of total collections) gives a quick approach to the researchers for comprehensive evaluation and genetic improvement in crops due to its manageable size. Further, the *composite collection* represents a wide spectrum of genetic diversity captured from the entire collection of germplasm preserved in national/global genebanks. A composite collection of crop species is a set of accessions, selected primarily to represent the entire collection, ecologically, taxonomically, and phenotypically. The composite collection consists of accessions from the mini core collection, added representative accessions of core collection, control cultivars, wild species, released cultivars, and accessions with distinct agro-morphological traits (Upadhyaya et al. 2008b).

6.4.2.1 Contribution of Indian Accessions to the Global Core Collections

Based on phenotypic and genotypic characterization, the barnyard millet core collection was created by ICRISAT Genebank to facilitate the effective use of genetic resources and to improve breeders' access to them (Upadhyaya et al. 2014). The core comprises of 89 accessions, out of which 49 are collected from different parts of India and exhibited maximum variability for different traits. The percentage contribution by Indian origin accessions is 55.1%, followed by Japan (20) and few from Cameroon (01), Egypt (01), Malawi (01), Pakistan (01), Sudan (01), Syrian Arab Republic (01), and Union of Soviet Socialist Republics (03). In barnyard millet, the largest collection is maintained by CGIAR at different genebanks with a total of 2365 accessions (Sood et al. 2015a). In India, the largest barnyard millet collection is maintained at NBPGR with 1718 accessions followed by All India Coordinated Small Millet Improvement Project at Bengaluru with 985 accessions (Gomashe 2017).

Core collections consisting 50 (Gowda et al. 2009) and 89 (Upadhyaya et al. 2014) genotypes have been developed in barnyard millet. A core set constituting 50 accessions was formed using 729 accessions of barnyard millet (*Echinochloa frumentacea*) germplasm in Karnataka, India, using data on 24 morpho-agronomic traits (Gowda et al. 2009). Other core subset in barnyard millet comprising of 89 accessions was structured from 766 germplasm accessions based on 21 agro-morphological traits, depicting 3 subspecies and 8 races. This core subset described 12% of the 766 germplasm collection of barnyard millet at genebank of ICRISAT, Hyderabad, India (Upadhyaya et al. 2014). Core subset of 89 accessions in barnyard millet belongs to 11 countries of diverse geographical origin (India-51, Japan-18, Russia-3, Egypt-1, Pakistan-1, Malawi-1, Cameroon-1, Syrian Arab Republic-1, and origin unknown-12). These core collections can be evaluated in different locations to identify location-specific desirable genotypes and to isolate other trait-specific stable genotypes which can be of great value for the overall genetic improvement of this crop. Sood et al. (2015a, b) evaluated global core collection of barnyard millet in the submontane Himalayan region of India using multivariate analysis, and they found that the accessions with more plant height, broad and long leaves, long inflorescence, and higher numbers of racemes need to be given preference as donors/parents in varietal improvement programs of barnyard millet.

6.4.3 Trait Discovery Using Core and Mini Core Collection

Each significantly unique trait can be precisely determined through the evaluation and characterization of germplasm. The barnyard millet core collection was developed by Gowda et al. (2009) utilizing 729 accessions for 24 traits, of which 50 were in a subset. Similarly, Upadhyaya et al. (2014) used 736 barnyard accessions for 21 traits, with 89 accessions in the subset, to create a core collection. These core collections can be evaluated in several locations to delineate other trait-specific stable genotypes that can be highly useful for the crop's overall genetic improvement and to identify desirable genotypes particular to a certain area (Table 6.3). Halaswamy et al. (2001) conducted a study to analyze the national collection and identified 7 promising accessions with larger plant height, 9 accessions with a higher number of

Table 6.3 Trait-specific barnyard millet germplasm identified

S. no.	Genotypes	Traits
1.	HCBM 1016, HCBM 1018, HCBM 1019, HCBM 1020	Grain smut, head smut, sheath blight
2.	TNAU 25, TNAU 63, VL 10, PRJ 1, PRB 903, DHBMV 56-6, DHBMV93-3, RBM 82, RBM 45, RBM 78, RBM 85, RBM 57, RBM 77, RBM 81	Grain smut
3.	VL 207, VL 172, TNAU 151, TNAU 155, KOPBM 46	Shoot fly
4.	DHBM 996 and TNEF 204	Shoot fly and stem borer
5.	IEC 336, IEC 343, IEC 344, IEC 330, IEC 217, JEC 340 and JEC 338, IC257799	Flowering
6.	IEC I57, IEC 159, IEC 71, IEC 78, IEC 257, IEC 271, IEC 158, IEC 156, IEC 282, IC426595, IC436926, IC426594, IC257799	Maturity duration
7.	IC257799, IC24848, VL 29, IC52701, IC338896	Basal tillers
8.	IC426595, IC97034, VL 29, IC97031	Inflorescence length
9.	IC426594	Inflorescence thickness
10.	IC326725, KL I	1000 grain weight
11.	IC548641, IC538089, IC548697	100 seed weight
12.	IC97034, VL 29 IC338960, IC97031	Yield per plant
13.	IC548696, IC548635, IC548697	Ear head length
14.	IC548681, IC548696, IC548658	Plant height
15.	CO(Kv)2, TNEF204, MDU 1	Salinity

Source: Gomashe (2017)

basal tillers, 10 accessions with longer inflorescence, and 27 accessions with early flowering. Additional lines, specifically IEC 566 and IEC 566/2, have been discovered. These lines differ from cultivated varieties in various ways. They produce a large amount of pollen grains and have flowers that remain open for an extended period of time. The stigma of these flowers protrudes enough to make emasculation and pollination easier (Nirmalakumari and Vetriventhan 2009). Furthermore, a simple dehulling variant known as B29 was discovered in accessions originating from the Uttarakhand hills. The dehulled grain recovery of B29 was 40% greater than the check variety VL 172 and 140% higher than the check variety PRJ 1. The accession has been registered with NBPGR under the reference number INGR09023 (Gupta et al. 2015).

In Himalayan region of Uttarakhand, India, Mehta et al. (2005) identified diverse genotypes for the morpho-physiological yield and yield component traits in barnyard millet. Gupta et al. (2010) identified 5 genotypes for shorter plant height (<120 cm), 4 genotypes for taller plant height (>200 cm), 3 genotypes for large number of productive tillers, 3 genotypes for longer inflorescence, 3 genotypes for more raceme number, 6 genotypes for longer raceme, and 5 genotypes for higher grain yield in barnyard millet. They also screened barnyard millet germplasm accessions against grain smut disease. Trivedi et al. (2017) evaluated 178 accessions for morpho-physiological and biochemical traits like flowering, maturity, 1000 grain weight, yield potential, and biochemical traits such as lipid peroxidation, total glutathione, and total ascorbate content in barnyard millet. Manimekalai et al. (2018) identified single genotype as primary sources for 10 morpho-nutritional traits, i.e., for earliness, plant height, high basal tillers, longer ear length, high ear weight, fodder type, bold seeded type, high grain Fe and Zn content, and high grain yield in barnyard millet. According to morpho-agronomic features, 89 accessions' worth of germplasm from core collections were divided into three groups (Sood et al. 2015a, b). Considerable variation was found for yield-related traits in the barnyard millet germplasm accessions. Through the characterization of barnyard millet landraces, resistant lines for low levels of phytic acid and tannin (NDS6, NDS3), high levels of protein (NDS6, NDS2), high lysine (NDS6, NDS5), and grain smut (RBM 57, RBM77, and RBM58) were reported.

The germplasm collected in the hills of Uttarakhand had an easy dehulling accession (B29) according to Gupta et al. (2010). This accession has recovered by around 140% over PRJ 1, a variety selected from *E. esculenta* accessions, and by 40% over the national check VL 172. A small number of unique lines that produce an abundance of pollen and have a projecting stigma that makes them easily accessible for pollination and emasculation have been identified. When compared to cultivated varieties, these lines (IEC 566 and IEC 566/2) show variations for a number of traits (Nirmalakumari and Vetriventhan 2009). In a characterization experiment using a national collection, Halaswamy et al. (2001) found that certain accessions had superior agronomic features, such as larger stem length, early blooming, basal tillers, plant height, and long inflorescence. The material that is currently adapted in India's main growing regions is from *E. frumentacea*. PRJ 1 outperformed the current ruling checks varieties in terms of yield (45.4%) (Upadhyaya et al. 2008a). Mehta et al.

(2007) discovered that there was a strong direct impact on grain output from heat units at maturity (HUM), photothermal units at 50% flowering (PTUF), plant height, days to 50% flowering, and panicle weight. The evaluated germplasm showed significant variation in terms of days to maturity (73–90 days), number of productive tillers (2–15), number of leaves on the main tiller (2–18), ear length (9–21 cm), finger length (21–55 mm), number of grains/finger (26–117), grain yield/ear (0.13–18 g), and grain yield/plant (0.35–26 g) according to Joshi (2013). In an experiment comprising 4 locations and involving 65 barnyard germplasm samples of both indigenous and exotic origin, Prakash and Vanniarajan (2013) found a wide range of variability in both genotypic and phenotypic traits, with little environmental influence on the expression of most agronomic traits. For ear head width, grain weight per ear head, single ear head weight, ear head length, and grain yield per plant, strong heritability in conjunction with high genetic progress was noted.

6.4.4 Use of Barnyard Millets Germplasm in Crop Improvement

India grows barnyard millet in two distinct agroclimatic zones: the Deccan plateau in South India, particularly Tamil Nadu, and the Himalayan region in the country's north. Barnyard millet is a staple crop in the North Western Himalayan region and hilly areas with steep slopes and undulating topography that have little space for crop diversification (Sood et al. 2015a, b). It is also very easy to cultivate and produces a lot of fodder and an assured crop harvest. The northeastern provinces of Arunachal Pradesh and Sikkim are noted for cultivating this crop across the entire region (Anonymous 2010). Tamil Nadu, Madhya Pradesh, Maharashtra, and Uttarakhand all grow it as well. Barnyard millet is an important crop for food and nutritional security, but not much work has been done to improve it. The majority of the developed and released varieties (Table 6.4) are selections from local *E. frumentacea* germplasm and have low yield levels and are prone to a number of diseases. As was previously indicated, new opportunities for raising production and degrees of resistance to various diseases and pests have been made possible by the exotic introduction of distinct barnyard variety (PRJ-1) (Yadav et al. 2010). The high-yielding variety PRJ-1 was developed from the Japanese accession IEc542 in 2003 in barnyard millet (Table 6.3). It will need more of these initiatives to improve its ability to adapt to changing climate conditions.

The genetic base of barnyard millet is narrow, especially in terms of features like disease resistance and seed size. The genetic diversity of the crop can be expanded by utilizing the gene pool of wild relatives, especially the ancestors. The incorporation of characteristics from wild relatives can be expedited by creating markers that are closely associated with these traits. Interspecific hybridization between Indian and Japanese barnyard millet presents a favorable opportunity for reciprocal genetic enhancement of both species. An analysis of the fertility hurdles between the two species is necessary, and the utilization of modern biotechnological methods is required to overcome these barriers. It is necessary to methodically analyze the existing genetic diversity in order to identify the origins of agriculturally significant

Table 6.4 List of barnyard varieties developed from local collections

S. no.	Variety name	Pedigree	Institute developed	Year of release	Characteristics features
1.	K-1	A pure-line selection from Tenkasi local	TNAU	1970	Panicles loose and spear shaped Grains dull white in color
2.	Kudiraivali K2 (Kovilpatty-2)	A pure-line selection from a local cultivated type	TNAU	1978	–
3.	VL Madira-8	Isolated from local collection	VPKAS, Almora	1978	Grains light gray, smooth with white endosperm
4.	Anurag	Selection from genetic collection of Gorakhpur	ARS	1983	Resistant to pink borer and shoot fly
5.	Gujarat banti-1	F-2 is a selection from the segregating material of Banti	AICSMIP, GAU, Surat	1984	Grain color is yellowish white
6.	RAU-3	Pure-line selection	IDRO-minor millets improvement project, RAU, Dholi, Muzaffarpur (Bihar)	1985	Moderately resistant to *Helminthosporium frumentace*
7.	VL Madira-21	Local collection from Garhwal region of Uttar Pradesh	VPKAS, Almora	1988	Very low incidence of smut and leaf spot diseases
8.	VL Madira-29	Location collection (vhc 5360-1) from the Kumaon region of Uttar Pradesh	VPKAS, Almora	1988	Very low incidence of smut disease
9.	Chandan	Selection from genetic collections of Ballia district	–	1989	Plants remain erect with thick and strong stem synchrony in tillering spike
10.	Kanchan	Selection from germplasm	CSAUAT, Kanpur	1993	–
11.	VL Madira 172	EF 2 × VHC 5205	VPKAS, Almora	2000	Tolerant to grain smut

(continued)

Table 6.4 (continued)

S. no.	Variety name	Pedigree	Institute developed	Year of release	Characteristics features
12.	Sushrutha (RAU 11)	Introduction	RPCAU, Pusa and UAS, Dharwad	2000	Karnataka
13.	VL Madira 181	ECC 27 × VL 60	VPKAS, Almora	2001	Bihar, Karnataka, Madhya Pradesh, Tamil Nadu
14.	PRJ-1	Selection from germplasm accession IEc 542 obtained from ICRISAT, India	GBPUA&T, Uttarakhand	2003	Resistant to grain smut disease
15.	Pratap sanwa-1 (ER-64)	Selection from local germplasm of Agroclimatic zone Iva (Udaipur)	MPUA&T, Udaipur, Rajasthan	2004	Moderately resistant to the diseases like grain smut and brown spot
16.	VL Madira 207	Vl Madira 172 × GECH 506	VPKAS, Almora	2008	Tolerant to grain smut
17.	Co(KV)2	Pure-line selection from a genetic accession ef 79	TNAU, Coimbatore	2009	–
18.	Co 1	A selection from the germplasm collection from Palladam area in Coimbatore district	TNAU, Coimbatore	2009	–

Source: Gomashe (2017)

characteristics, such as increased yield, resistance to diseases, nutritional quality of the grain, and others.

6.5 Pre-breeding

Novel pathways for genetic improvement may be created via the introgression of desired traits from wild relatives and progenitors. The two species of Japanese and Indian barnyard millets may benefit genetically from interspecific hybridization. In order to find genetic resources for improved agronomic performance, disease resistance, and nutritional attributes, it is necessary to assess the present diversity (Gomashe 2017). Very limited information was available with regard to pre-breeding in barnyard millet. It has been found that the cultivated species of *Echinochloa* differ morphologically. They include *colona* and *frumentacea*'s high seed-shattering behavior and the other two species' low seed-shattering behavior (Seetharam et al. 2003). Variations in caryopsis coloration are also noted, with the straw brown coloration becoming darker in *crus-galli* and the yellow brown in *esculenta*. The maximum height of a plant is 250 cm, attained by the *crus-galli* and *frumentacea* species. The *colona* typically reaches 60 cm, while the esculenta species reaches intermediate heights (Joshi et al. 2015). Performing crossing and hybridization in barnyard millet presents challenges due to the small size of the inflorescence (Seetharam et al. 2003). The African and Indian germplasm have been used in recombination breeding which led to development of many varieties of small millets (Seetharam et al. 2003). Regarding registration of genetic stocks, fewer barnyard millet genetic resources are registered in India. The registered germplasm mostly consists of easy dehulling trait as in the case of genotype B29 (Gupta et al. 2010). The first variety released was K1 in India. Of the several varieties released in India, notable varieties are PRJ1, MDU1, Co2, VL207, VL-Madira172, and Gujarat Banti-1. In order to produce early maturing, high-yielding segregants, Mehta et al. (2005) highlighted an interspecific hybridization strategy including early maturing *E. frumentacea* and high-yielding *E. esculenta*. On the other hand, hybrids containing wild ancestors are fertile, while those involving *E. frumentacea* and *E. esculenta* are sterile in both respects. It was shown that two barnyard millet accessions from the species *E. esculenta*, PRB 9402 and PRB 9602, were resistant to the grain smut (AICSMIP 2001). It is important to breed genotypes that facilitate dehulling in order to lessen the tedious post-harvest processing. From the variety of germplasm available at ICAR-VPKAS, Almora, Uttarakhand, India, the easy dehulling genotype "B29" was chosen. It has been shown to have the benefit of cutting down on processing time overall (Gupta et al. 2010, 2015). Its extended lemma over the kernel, which applies pressure and splits readily under force when processed in the "Vivek Thresher 1" dehulling machine, is primarily responsible for this attribute. When compared to the processing of two other promising types, VL Madhira 172 and PRJ1, it considerably reported faster dehulling (42 and 146.4%). According to Mandelbaum et al. (1995), these wild ancestors can offer valuable genetic resources for the advancement of cultivated species. Significant differences in calcium and protein levels across accessions point to the possibility of selective breeding for nutritional qualities (Mandelbaum et al. 1995).

6.6 Genetic and Genomic Resources of Barnyard Millet

Barnyard millet is known for its nutritional importance and climate resilience. Limited genomic resources are available in barnyard millet and hence, there is a need to focus on developing more genomic resources in this millet. However, the progress made with respect to genome sequenced, markers generated, and molecular characterization carried out in barnyard millet is mentioned below:

6.6.1 RAPD (Random Amplified Polymorphic DNA) Markers and Their Utilization

Hilu (1994) used random amplified polymorphic DNA (RAPD) markers to assess the genetic diversity and phylogeny in domesticated and wild species of *Echinochloa* and shown the distinctness of *E. frumentacea* (Indian barnyard millet) and *E. utilis* (Japanese barnyard millet) and grouping with their respective wild progenitors. Utility of RAPD markers in barnyard millet cultivar fingerprinting and breeding programs was emphasized. Prabha et al. (2012) assessed the genetic diversity among 33 accessions comprising 15 accessions of *Echinochloa crus-galli* and 18 accessions of *E. frumentacea* based on morphological and RAPD data. Two groups were observed at a similarity level of 34%. And the level of within and between species variation was also mentioned with *E. crus-galli* accessions showing more diversity within the species. The data generated can be utilized in the management of barnyard millet resources. Moharil et al. (2016) analyzed 21 accessions of *Echinochloa* spp. complex (wild and cultivated species) belonging to adjoining regions of Vidarbha, Maharashtra, and Melghat using 20 RAPD primers. Two major groups comprising of *E. frumentacea* (15 accessions) and *E. colona* (6 accessions) were observed, thus showing clear-cut grouping between the two.

6.6.2 Simple Sequence Repeats (SSRs) Markers and Their Utilization

Kumari et al. (2013) reported the cross-genera transferability of 106 eSSRs from foxtail millet to 8 millets and 4 non-millet species and showed an average of ~88% transferability, while 90.6% transferability rate was observed with respect to barnyard millet. The markers transferred can be utilized in the characterization of barnyard millet germplasm. Arya et al. (2014) developed SSR markers in barnyard millet through cross-transferability from finger millet and pearl millet. In case of finger millet SSRs, 14.2% transferability rate was observed, while pearl millet SSRs showed 31.2% transferability. The markers developed can be used for characterization of larger set of barnyard millet germplasm. Babu and Chauhan (2017) downloaded 41 EST sequences from public domain and were used to develop 22 EST-SSR primer pairs. Tetranucleotide repeats were more frequent followed by the penta- and hexa-nucleotide repeats. One reliable and one potential SNP and two haplotypes were identified in these EST sequences. Comparative analysis showed homology to chromosomal regions 2, 5, 6, 8, 9, and 12 of rice and to *Zea mays Waxy* gene. The identified markers can be used in the molecular characterization of barnyard millet. Babu et al. (2018a) transferred and developed 39 SSRs out of 64 maize and finger millet genomic SSRs tested in barnyard millet and its wild species. SSR loci bnlg2323 and bnlg2123 were found highly polymorphic and PIC value ranged from 0.25 to 0.73 with an average of 0.35. Maize microsatellites were more polymorphic as compared to finger millet SSRs and were able to distinguish barnyard millet genotypes clearly. Population structure revealed two groups and two of the genotypes (IEC514 and IEC409) showed admixture. In another study, Babu et al. (2018b) observed 100% cross transferability rate of SSRs from finger millet and showed 91% polymorphism in 11 genotypes of barnyard millet with a mean of 4.3 alleles per locus. The range and average PIC value of 0.27 to 0.73 and 0.54, respectively, was observed with markers transferred from finger millet. And with respect to rice, 71% SSRs were cross transferable with 48% polymorphism and a mean of 3.38 alleles per locus. For rice SSRs, PIC value ranged from 0.15 to 0.67 with an average of 0.44.

Sixty-one barnyard millet genotypes belonging to southern India were evaluated by Manimekalai et al. (2018) based on 51 expressed sequence tag (EST)-simple sequence repeat (SSR) markers and 10 morpho-nutritional traits. Fourteen SSR markers were found polymorphic with PIC value ranging from 0.276 to 0.652 with an average of 0.43, and grain nutritional and morpho-agronomic traits showed large variability among 61 genotypes. Genetic relationships observed among these genotypes can be used in devising strategies for the management and utilization of barnyard millet from southern India. Jayakodi et al. (2019) assembled and characterized transcriptome (97,065 transcripts) of CO (KV) 2, a popular high-yielding *E. frumentacea* variety from southern India, and compared it with its wild relative *E. crus-galli* (93,725) and detected 3489 protein-coding transcripts unique to Indian barnyard millet. Key genes regulating a response in Zn and Fe accumulation were identified, and the role of photosynthesis in the drought adaptation was revealed. Thirty out of 10,881 SSR loci discovered (including di-49.02%; tri-47.74%; tetra-2.35%; penta-0.67%; and hexa- 0.21% repeat motifs) were screened in 30 accessions of *E. frumentacea* and 10 were found polymorphic. Genomic resources generated in this study will accelerate both genetic and functional genomic research in this crop. Renganathan et al. (2021) identified an EST-SSR marker, BMESSR 39 associated with the anthocyanin pigment trait after screening 50 EST-SSR markers in the pigmented (ACM 331 with purple inflorescence) and nonpigmented (ACM 333, ACM 335, and MA 10 with green inflorescence) bulks of the F2 population. The information generated would help in carrying out genetic studies in barnyard millet and breeding of anthocyanin pigmented genotypes. Recently, Meniya et al. (2023) identified a total of 46,157 SSRs and 15 SSR loci were validated in 30 accessions of barnyard millet with an average Polymorphism Information Content (PIC) of 0.28. Identified microsatellite markers can be used for genetic improvement of barnyard millet and related crops.

6.6.3 SNP Markers and Their Utilization

Wallace et al. (2015) reported the genotyping-by-sequencing (GBS)-based characterization of 95 barnyard millet including 89 of core collection available at ICRISAT Genebank and identified several 1000 SNPs to study population structure and phylogenetic relationships. Population clusters observed within the *E. colona* and within *E. crus-galli* match with the phylogenetic relationships. The data generated can be used for barnyard millet improvement.

6.6.4 Genome Sequence

6.6.4.1 Whole Genome Sequence

Chromosome level draft genome assembly of *E. colona* var. *frumentacea* (Link) Ridl. (synonym *E. frumentacea* Link) of genotype PI463783 was generated based on Illumina and PacBio Sequel II data. Genome size of 1.13 Gb was assembled (though K-mer analysis assessed 1.18 Gb genome size) along with 93,697 annotated genes. Further, 737 accessions of barnyard grasses were also re-sequenced and genomic information on dual roles of *Echinochloa* species as crops and as weeds as well as vital resources for studying adaptation, plant polyploidization, and precision weed control was generated (Wu et al. 2022). Another partial genome assembly of barnyard millet of cultivar CO(KV)2 was reported by Meniya et al. (2023). Ion S5 Next Generation Sequencer (NGS) was used to generate 11,39,481 contigs and draft genome was assembled.

6.6.4.2 Chloroplast Genome Sequence

Perumal et al. (2016) reported the complete chloroplast genome sequence (38.6% GC content) of Indian barnyard millet, having a genome size of 139,593 bp with a typical quadripartite structure and 112 genes (including 1 conserved open reading frame, 4 rRNA genes, 30 tRNA genes, 77 protein coding genes). Further, phylogenetic analysis with other grass species revealed its divergence around 1.9–2.7 million years with its close relatives *E. crus-galli* and *E. oryzicola*.

6.7 Future Prospects

Due to a paucity of genetic and genomic resources, breeding for barnyard millet is progressing very slow. Barnyard millet's core collection needs to be developed and utilized for developing breeding populations and selection of recombinants. Consequently, it is necessary to create and assess biparental and multiparent populations for diverse nutritional and agronomic features, as well as core and mini core collections that represent maximum diversity. These resources will be helpful for identifying potential genes as well as tracking the genomic regions linked to the desired attributes using genome-wide association studies (GWAS), genomic selection (GS), and linkage-based QTL mapping. The genetic diversity within cultivated barnyard millet is notably narrow, particularly concerning crucial traits like disease resistance and seed size. However, an avenue for broadening the genetic base lies in harnessing the wild relative gene pool, particularly the progenitors of barnyard millet. Additionally, interspecific hybridization between Indian and Japanese barnyard millet holds promise for mutual genetic enhancement of both species. Overcoming fertility barriers between the two species using modern biotechnological tools is imperative for successful hybridization. Furthermore, systematic analysis of available genetic diversity is crucial for identifying sources of agronomically important traits such as higher yield, disease resistance, and grain nutritional quality. Molecular breeding efforts in barnyard millet are currently lagging behind other small millets like foxtail millet, which has had its genome sequenced, and finger millet, which has had a genetic map developed. Developing a genetic map for barnyard millet would significantly accelerate breeding efforts toward targeted improvement.

However, on a global scale, barnyard millet, like other small millets, is facing a decline in importance as a food crop due to competition with major cereals such as wheat, rice, maize, and sorghum. This decline in production has resulted in reduced consumption, compounded by changing lifestyles and government policies. Further, the labor-intensive processing associated with the crop has also contributed to a decline in its cultivation area and production. Nonetheless, due to its short crop cycle and adaptability to a wide range of soil types, barnyard millet may continue to serve as a valuable crop in Asia, particularly on poor agricultural land in regions with low rainfall or a short growing season. Despite limited prospects in India, the crop may gain importance as a niche crop in dry regions at medium to high altitudes under changing climate conditions. To ensure the future viability of barnyard millet, collective actions are essential. These include raising public awareness about its nutritional value; intensifying research on crop production, processing, and utilization; implementing value addition strategies; and providing government support for marketing initiatives. By undertaking concerted research efforts and implementing supportive policies, barnyard millet can be preserved and its potential harnessed for future agricultural sustainability and food security.

References

AICSMIP (2001) Annual report 2000–2001. All India Coordinated Small Millet Improvement Project (ICAR), Bengaluru, p 6

Ajeesh Krishna TP, Maharajan T, Ceasar SA (2022) Improvement of millets in the post-genomic era. Physiol Mol Biol Plants 28(3):669–685

Anonymous (2010) Status paper on millets, Directorate of Millets Development, Department of Agriculture and Cooperation, Ministry of Agriculture, Government of India

Aoki D, Yamaguchi H (2008) Genetic relationship between *Echinochloa crus-galli* and *Echinochloa oryzicola* accessions inferred from internal transcribed spacer and chloroplast DNA sequences. Weed Biol Manag 8:233–242

Aoki D, Yamaguchi H (2009) *Oryza sh4* gene homologue represents homoeologous genomic copies in polyploid *Echinochloa*. Weed Biol Manag 9:225–233

Arora S, Srivastava S (2002) Suitability of millet-based food products for diabetics. J Food Sci Technol 39:423–428

Arya L, Chauhan D, Yadav Y, Verma M (2014) Transferability of simple sequence repeat (SSR) markers developed in finger millet and pearl millet to kodo millet and barnyard millet. In: Johri AK, Mishra GC (eds) Innovative approach in stem cell research, cancer biology and applied biotechnology. Excellent Publishing House, New Delhi, pp 60–64

Babu B, Chauhan R (2017) In-silico identification of EST-based microsatellite markers and SNPs, and comparative genomic analysis of ESTs in barnyard millet for their omics applications. Curr Agric Res J 5:279–287. https://doi.org/10.12944/CARJ.5.3.03

Babu B, Rashmi C, Sood S (2018a) Cross transferability of finger millet and maize genomic SSR markers for genetic diversity and population structure analysis of barnyard millet. Indian J Genet Plant Breed 78:364–372. https://doi.org/10.31742/IJGPB.78.3.5

Babu B, Sood S, Kumar D, Joshi A, Pattanayak A, Kant L et al (2018b) Cross-genera transferability of rice and finger millet genomic SSRs to barnyard millet (*Echinochloa spp.*). 3 Biotech 8:95. https://doi.org/10.1007/s13205-018-1118-1

Bajwa AA, Jabran K, Shahid M, Ali HH, Chauhan BS (2015) Eco-biology and management of *Echinochloa crus-galli*. Crop Prot 75:151–162

Bandyopadhyay BB (1999) Genotypic differences in relation to climatic adaptation of two cultivated barnyard millet species at Garhwal Hills. Indian J Genet Plant Breed 59(01):105–108

Bouhache M, Bayer DE (1993) Photosynthetic response of flooded rice (*Oryza sativa*) and three *Echinochloa* species to changes in environmental factors. Weed Sci 41(4):611–614

Bramel P, Giovannini P, Eshan Dulloo M (2022) Global strategy for the conservation and use of genetic resources of selected millets. Global Crop Diversity Trust, Bonn. https://doi.org/10.5281/zenodo.7798294

Chandel G, Meena R, Dubey M, Kumar M (2014) Nutritional properties of minor millets: Neglected cereals with potentials to combat malnutrition. Curr Sci 107:1109–1111

Crawford GW (1983) Paleoethnobotany of the Kameda Peninsula Jomon, vol 73. University of Michigan, Museum of Anthropological Archaeology

Crawford GW (2011) Advances in understanding early agriculture in Japan. Curr Anthropol 52(S4):S331–S345

De Wet J, Prasada Rao K, Mengesha M, Brink D (1983) Domestication of mawa millet (*Echinochloa colona*). Econ Bot 37:283–291. https://doi.org/10.1007/BF02858883

Defelice MS (2002) Green foxtail, *Setaria viridis* (L.) P. Beauv. Weed Technol 16(1):253–257

Denton DC (1987) Food crops for waterfowl. In: Wesley DE, Leitch WG (eds) Fireside waterfowler: fundamentals of duck and goose ecology. Stackpole Books, Harrisburg, p 352

Doggett H (1989) Small millets—a selective overview. In: Seetharam A, Riley KW, Harinarayana G (eds) Small millets in global agriculture. Oxford, pp 3–18

Doust AN, Devos KM, Gadberry MD, Gale MD, Kellogg EA (2005) The genetic basis for inflorescence variation between foxtail and green millet (Poaceae). Genetics 169(3):1659–1672

Gomashe SS (2017) Barnyard millet: present status and future thrust areas. In: Millets and Sorghum: biology and genetic improvement, pp 184–198

Gowda J, Bharathi S, Somu G et al (2009) Formation of core set in barnyard millet [*Echinochloa frumentacea* S. (Roxb.) Link] germplasm using data on twenty-four morpho-agronomic traits. Int J Plant Sci 4:1–5

Gupta A, Mahajan V, Kumar M, Gupta HS (2009) Biodiversity in the barnyard millet (*Echinochloa frumentacea* Link, Poaceae) germplasm in India. Genet Resour Crop Evol 56:883–889

Gupta A, Mahajan V, Gupta HS (2010) Genetic resources and varietal improvement of small millets for Indian Himalaya. In: Tewari LM, Pangtey YPS, Tewari G (eds) Biodiversity potentials of the Himalaya. Gyanodaya Prakashan, Nainital, pp 305–316

Gupta A, Sood S, Agrawal PK, Bhatt JC (2015) B29: An easy dehulling barnyard millet (*Echinochloa frumentacea* Link) genotype. Natl Acad Sci Lett 38(1):21–24

Halaswamy BH, Srinivas GV, Ramakrishna BM et al (2001) Characterization and preliminary evaluation of national collections of Barnyard millet (*Echinochloa* spp.) Germplasm. Indian J Plant Genet Resour 14:213–216

Hammer K (1984) The domestication syndrome. Kulturpflanze 32:11–34

Hammer K (2003) Evolution of cultivated plants and biodiversity. Nova Acta Leopoldina NF 87:133–146

Hilu KW (1994) Evidence from RAPD markers in the evolution of *Echinochloa* millets (Poaceae). Plant Syst Evol 189:247–257

Holm LG, Plunknett DL, Pancho JV, Herberger JP (1991) The World's worst weeds. Distribution and biology. Krieger Publishing Company, Malabar

Hulse JH, Laing EM, Peason OE (1980) Sorghum and millets: their composition and nutritive value. Academic, London

Jayakodi M, Madheswaran M, Adhimoolam K et al (2019) Transcriptomes of Indian barnyard millet and barnyard grass reveal putative genes involved in drought adaptation and micronutrient accumulation. Acta Physiol Plant 41:66. https://doi.org/10.1007/s11738-019-2855-4

Jayaraman N, Suresh S, Nirmala Ganeshan NM (1997) Genetic enhancement and breeding strategies in small millets. In: National seminar on small millets: current research trends and future priorities as food feed and in processing for value addition, 19–21. 23–24 April, 1997, Bangalore, India, (Extended summaries)

Joshi V (2013) Assessment of genetic variability and identification of genotypes for different traits in barnyard millet (*Echinochola* spp.). Int J Agric Food Sci Technol 4(2):65–67

Joshi RP, Jain AK, Chauhan SS, Singh G (2015) Characterization of barnyard millet (*Echinochloa frumentacea* (Roxb.) Link.) Landraces for agro-morphological traits and disease resistance. *Electronic Journal of.* Plant Breed 6(4):888–898

Kono Y, Yamaguchi A, Kawamura N, Tatsumi J, Nonoyama T, Inagaki N (1987) Interspecific differences of the capacities of waterlogging and drought tolerances among summer cereals. Jpn J Crop Sci 56(1):115–129

Kumagai S, Matusda H, Yoshida H, Abe A, Sagawa S, Hoshino T (2011) Optimum harvest time judged from maturation properties of Japanese barnyard millet in the late ripening stage. Jpn J Crop Sci 80:29

Kumari K, Muthamilarasan M, Misra G, Gupta S, Alagesan S, Parida S et al (2013) Development of eSSR-markers in *Setaria italica* and their applicability in studying genetic diversity, cross-transferability, and comparative mapping in millet and non-millet species. PLoS One 8:e67742. https://doi.org/10.1371/journal.pone.0067742

Maikhuri RK, Rao KS, Semwal RL (2001) Changing scenario of Himalayan agroecosystems: loss of agrobiodiversity, an indicator of environmental change in Central Himalaya, India. Environmentalist 21:23–39

Majid MA, Hamid MA (1989) Importance genetic resources and breeding of small millets in Bangladesh. In: Small millets in global agriculture. Oxford and IBH Publ. Co, New Delhi

Mandelbaum CI, Barbeau WE, Hilu KW (1995) Protein, calcium, and iron content of wild and cultivated species of *Echinochloa*. Plant Foods Hum Nutr 47:101–108

Manidool C, t'Mannetje L, Jones RM (1992) *Echinochloa crus-galli* (L.) P. Beauv. Plant Resour South-East Asia 4:126–127

Manimekalai M, Dhasarathan M, Karthikeyan A, Murukarthick J, Renganathan VG, Thangaraj K, Vellaikumar S, Vanniarajan C, Senthil N (2018) Genetic diversity in the barnyard millet (*Echinochloa frumentacea*) germplasms revealed by morphological traits and simple sequence repeat markers. Curr Plant Biol 14:71–78

Maun MA, Barrett SC (1986) The biology of Canadian weeds. 77. *Echinochloa crus-galli* (L.) Beauv. Can J Plant Sci 66:739–759

Mehta H, Tyagi PC, Mohapatra KP (2005) Genetic diversity in barnyard millet (*Echinochloa frumentacea* Roxb.). Indian J Genet Plant Breed 65(04):293–295

Mehta H, Tyagi PC, Mohapatra KP (2007) Genetic analysis of morpho-physiological and yield component traits in barnyard millet (I). Crop Improv 34(1):82–85

Meniya VH, Padhiyar SM, Desai H, Kheni JK, Tomar RS (2023) Development and validation of microsatellite markers for barnyard millet obtained by partial genome assembly. Ann Arid Zone 62(1):83–89. https://doi.org/10.59512/aaz.2023.62.1.9

Migo TR, Pamplona RR, Dingkuhn M, De Datta SK (1991) Interaction of water stress and nitrogen supply on the photosynthetic parameters of two upland rices and two upland weeds. Philippine J Weed Sci 18:pp. 69–89

Mitchell WA, Tomlinson WH Jr (1989) Browntop millet (*Panicum ramosum*): Section 7.1.5. USACE wildlife resources management manual. Technical report EL-89-12. US Army Engineer Waterways Experiment Station, Vicksburg, MS, USA

Mitich LW (1990) Intriguing world of weeds. Barnyard grass. Weed Technol 4:918–920

Moharil MP, Gawai D, Dikshit N, Dudhare MS, Jadhav PV (2016) Assessment of genetic diversity in Indian barnyard millet (*Echinochloa* spp. Complex) using morphological and molecular markers. J Appl Nat Sci 8(3):1643–1648

Muldoon DK, Pearson CJ, Wheeler JL (1982) The effect of temperature on growth and development of Echinochloa millets. Ann Bot 50(5):665–672

Nandini C, Bhat S, Saritha HS, Pandey CD, Sushil Pandey P, Bai L, Gowda J (2020) Characterization of barnyard millet (*Echinocloa frumentaceae* (Roxb.) Link) germplasm for quantitative traits to enhance its utilization. *Electronic Journal of*. Plant Breed 11(04):1066–1072

Napper DM (1965) Grasses of Tanganyaika: with keys for identification, Bulletin 18, Ministry of Agriculture Forest and Wildlife, Tanzania, 146 pp

Nesbitt M (2005) Grains. In: Prance G, Nesbitt M (eds) The cultural history of plants. Routledge Press, New York, pp 45–60

Nilavarasi R, Sujatha K, Geetha R, Vanniarajan C (2020) Morphological characterization and evaluation of barnyard millet varieties, cultivars and mutants for varietal identification. J Pharmacogn Phytochem 9(1):1600–1604

Nirmalakumari A, Vetriventhan M (2009) Phenotypic analysis of anther and pollen in diversified genotype of barnyard millet (*Echinochloa frumentacea*) floral characters. ICFAI Univ J Genet Evol 2:12–16

Nozawa S, Nakai H, Sato YI (2004) Characterization of microsatellite and ISSR polymorphisms among Echinochloa (L.) Beauv. spp. in Japan. Breed Res 6:187

Obara T (1938) Studies of barnyard millet. I. Classification of varieties by agricultural traits. Jpn J Crop Sci 9:471–518

Padulosi S, Bhag M, Bala Ravi S, Gowda J, Gowda KTK, Shanthakumar G, Yenagi N, Dutta M (2009) Food security and climate change: role of plant genetic resources of minor millets. Indian J Plant Genet Resour 22:1–16

Panwar P, Dubey A, Verma AK (2016) Evaluation of nutraceutical and antinutritional properties in barnyard and finger millet varieties grown in the Himalayan region. J Food Sci Technol 53:2779–2787

Perumal S, Jayakodi M, Kim DS, Yang TJ, Natesan S (2016) The complete chloroplast genome sequence of Indian barnyard millet, *Echinochloa frumentacea* (Poaceae). Mitochondrial DNA B: Resour 101(1):79–80. https://doi.org/10.1080/23802359.2015.1137832

Poncet V, Lamy F, Enjalbert J, Joly H, Sarr A, Robert T (1998) Genetic analysis of the domestication syndrome in pearl millet (*Pennisetum glaucum* L., Poaceae): Inheritance of the major characters. Heredity 81(6):648–658

Poncet V, Lamy F, Devos KM, Gale MD, Sarr A, Robert T (2000) Genetic control of domestication traits in pearl millet (*Pennisetum glaucum* L., Poaceae). Theor Appl Genet 100:147–159

Poongodi T, Vijayakumar JBM, Nazni P, Rajeshwari M (2003) Value addition for minor millets and its glycemic load among normal and type 2 diabetic subjects. In: 39th national nutritional conference of NSI, Hyderabad

Prabha D, Negi YK, Khanna VK (2012) Assessment of genetic diversity of barnyard millet accessions using molecular markers. Indian J Plant Genet Resour 25(2):174–179

Prakash R, Vanniarajan C (2013) Genetic variability for panicle characters in indigenous and exotic barnyard millet (*Echinochloa frumentacea* [Roxb.] Link) germplasm over environment. Vegetos 26(2):297–306

Rahn EM, Sweet RD, Vengris J, Dunn S (1968) Life history studies as related to weed control in the Northeast. 5. Barnyard grass. Agric Exp Stat Univ Delaware Bull 368:1–46

Renganathan VG, Vanniarajan C, Karthikeya A, Ramalinga J (2020) Barnyard millet for food and nutritional security: current status and future research direction. Front Genet 11:1–21. https://doi.org/10.3389/fgene.2020.00500

Renganathan VG, Vanniarajan C, Senthil N, Nirmalakumari A, Karthikeyan A, Veni K, Ramalingam J (2021) Genetics and molecular markers for anthocyanin pigmentation in barnyard millet (*Echinochloa frumentacea* [Roxb.] Link). Plant Breed 140:246–253. https://doi.org/10.1111/pbr.12892

Saleh SM, Zhang Q, Chen J, Shen Q (2013) Millet grains: nutritional quality, processing, and potential health benefits. Compr Rev Food Sci Food Saf 12:281–295

Seetharam A, Gowda J, Halaswamy JH (2003) Small millets. In: Chowdhury SK, Lal SK (eds) Nucleus and breeder seed production manual. Indian Agricultural Research Institute, New Delhi, pp 54–67

Singh KP, Mishra HN, Saha S (2010) Moisture-dependent properties of barnyard millet grain and kernel. J Food Eng 96:598–606. https://doi.org/10.1016/j.jfoodeng.2009.09.007

Sood S, Khulbe R, Gupta AK, Agrawal PK, Upadhyaya HD, Jagdish BC (2015a) Barnyard millet – a potential food and feed crop of the future. Plant Breed 134:135–147

Sood S, Khulbe RK, Kumar A, Agrawal PK, Upadhyaya HD (2015b) Barnyard millet global core collection evaluation in the submontane Himalayan region of India using multivariate analysis. Crop J 3(6):517–525

Sundararaj DP, Thulasidas G (1976) Botany of field crops. Macmillan Publisher, New Delhi, p 509

Sung SJS, Leather GR, Hale MG (1987) Development and germination of barnyard grass (*Echinochloa crus-galli*) seeds. Weed Sci 35(2):211–215

Takase K (2009) Prehistoric and Protohistoric plant use in the Japanese Archipelago. Meiji University Premodern Japan Research Exchange, 708

Trivedi AK, Arya L, Verma SK, Tyagi RK, Hemantaranjan A (2017) Evaluation of barnyard millet diversity in the central Himalayan region for environmental stress tolerance. J Agric Sci 155(10):1497–1507

Ugare R (2008) Health benefits, storage quality and value addition of barnyard millet (*Echinochloa frumentacaea* Link) (Doctoral dissertation, UAS, Dharwad)

Ugare R, Chimmad B, Naik R, Bharati P, Itagi S (2014) Glycemic index and significance of barnyard millet (*Echinochloa frumentacae*) in type II diabetics. J Food Sci Technol 51:392–395. https://doi.org/10.1007/s13197-011-0516-8

Upadhyaya HD, Dwivedi SL, Baum M, Varshney RK, Udupa SM, Gowda CLL, Hoisington D, Singh S (2008a) Genetic structure, diversity, and allelic richness in composite collection and reference set in chickpea (*Cicer arietinum* L.). BMC Plant Biol 8:106

Upadhyaya HD, Gowda CLL, Sastry DVSSR (2008b) Plant genetic resources management: Collection, characterization, conservation, and utilization. J SAT Agric Res 6:1–15

Upadhyaya D, Dwivedi SL, Singh SK, Singh S, Vetriventhan M, Sharma S (2014) Forming core collections in barnyard, kodo, and little millets using morpho-agronomic descriptors. Crop Sci 54:2673–2682

Veena B, Chimmad BV, Naik RK, Shantakumar G (2004) Development of barnyard millet-based traditional foods. Karnataka J Agric Sci 17:522–527

Vetriventhan M, Upadhyaya HD, Dwivedi SL, Pattanashetti SK, Singh SK (2016) Finger and foxtail millets. In: Genetic and genomic resources for grain cereals improvement. Academic, pp 291–319

Vetriventhan M, Azevedo VCR, Upadhyaya HD, Nirmalakumari A, Kane-Potaka J, Anitha S, Antony Ceasar S, Muthamilarasan M, Bhat BV, Hariprasanna K, Bellundagi A, Cheruku D, Backiyalakshmi C, Santra D, Vanniarajan C, Tonapi VA (2020) Genetic and genomic resources, and breeding for accelerating improvement of small millets: current status and future interventions. Nucleus 63(3):217–239

Wallace JG, Upadhyaya HD, Vetriventhan M, Buckler ES, Hash TC, Ramu P (2015) The genetic makeup of a global barnyard millet germplasm collection. Plant Genome 8(1):1–7. https://doi.org/10.3835/plantgenome2014.10.0067

Watanabe N (1970) A spodographic analysis of millet from prehistoric Japan. J Fac Sci Univ Tokyo 3:357–379

Wu D, Shen E, Jiang B et al (2022) Genomic insights into the evolution of *Echinochloa* species as weed and orphan crop. Nat Commun 13:689. https://doi.org/10.1038/s41467-022-28359-9

Yabuno T (1966) Biosystematic study of the genus *Echinochloa*. J Jap Bot 19:277–323

Yabuno T (1971) A note on barnyard millet. SABRAO Newslett 3:43–45

Yabuno T (1975) The classification and geographical distribution of the genus *Echinochloa*. Weed Res Jpn 20:97–104

Yabuno T (1984) A biosystematic study on *Echinochloa oryzoides* (Ard.) Fritsch. Cytologia 49:673–678

Yabuno T (1987) Japanese barnyard millet (*Echinochloa utilis*, Poaceae) in Japan. Econ Bot 41(4):484–493

Yabuno T (2001) Taxonomy and phylogeny of the genus *Echinochloa*. In: Yabuno T, Yamaguchi H (eds) Natural history of genus Echinochloa. Zennokyo Shuppan, Tokyo, pp 15–30. (in Japanese)

Yadav VK, Kumar B, Yadav R et al (2010) All India coordinated small millet improvement project, Ranichauri, GB Pant University of Agriculture and Technology, Pantnagar, Uttarakhand, India, p 79

Yamaguchi H, Utano AYA, Yasuda K, Yano A, Soejima A (2005) A molecular phylogeny of wild and cultivated *Echinochloa* in East Asia inferred from non-coding region sequences of trnt-L-F. Weed Biol Manag 5(4):210–218

Yasuda K, Yano A, Nakayama Y, Yamaguchi H (2002) Molecular identification of *Echinochloa oryzicola* and *E. crus-galli* using a polymerase chain reaction-restriction fragment length polymorphism technique. Weed Biol Manag 2:11–17

Proso Millet

7

Aviansh Singode, Rahul Sager,
Thirugnanasambandam Perumal Prathima,
Balakrishna Domathoti, Seva Nayak Dheeravathu,
Sushil Pandey, and C. Tara Satyavathi

Abstract

Proso millet (*Panicum miliaceum* L.) is an ancient crop cultivated by humankind in the Neolithic era. The crop was domesticated in China, spread to Central Asia, and later to Europe and North America. It is a short-duration C_4 crop with inherent biotic and abiotic stress tolerance and has vast potential for climate change and sustainable agriculture. Short-duration and shallow root systems make it suitable for use in any cropping system. Proso millet grains are rich in protein 13–18%, magnesium, and phosphorus. It is an important crop of the past that has found its place in genebanks worldwide. A large number of accessions collected from tropical, subtropical, and temperate regions are conserved in the genebank globally. These accessions are invaluable and harbor important genes for agronomic and adaptive traits. Collection, conservation, and characterization of germplasm are very important for cataloguing and further utilization for trait-specific breeding. Conventional breeding has provided many varieties of proso millet, utilizing the available genetic resources. However, the genetic gains were not significant in the yesteryears. Advances in genetics and plant breeding, including genomics and biotechnology, can complement the conventional breeding strategies. The use of

Dedicated to our beloved Dr. Maruthamuthu Elangovan

A. Singode (✉) · R. Sager · B. Domathoti · S. N. Dheeravathu · C. T. Satyavathi
ICAR- Indian Institute of Millets Research, Hyderabad, India
e-mail: avinash@millets.res.in

T. P. Prathima
ICAR- Sugarcane Breeding Institute, Coimbatore, India

S. Pandey
Division of Germplasm Conservation, ICAR-National Bureau of Plant Genetic Resources, New Delhi, Delhi, India

K. Tripathi et al. (eds.), *Plant Genebank Utilization for Trait Discovery in Millets*, https://doi.org/10.1007/978-981-96-4043-0_7

biotechnology, such as marker-assisted selection, genome-wide association studies, genomic selection, and genome editing, will enable a wider scope for trait-specific breeding in proso millet. The proso millet genome sequencing has opened avenues for "omics" studies for full utilization of genomic resources for plant breeding and development of climate-resilient and nutrient-rich cultivars.

Keywords
Proso millet · Plant Genetic Resources · Germplasm · Breeding · Utilization · Pre-breeding · Trait-specific germplasm

7.1 Introduction

Proso millet (*Panicum miliaceum* L.) is an allotetraploid species with a chromosome number of $2n = 4x = 36$ and $x = 9$. Paleobotanical studies have suggested that proso millet was cultivated as early as 10,000 BCE. The excavated proso millet samples from the Chishan riverbanks in China date back to 10,300–8700 years before present (BP) (Lu et al. 2009). Advanced molecular studies on crop remains from excavated sites revealed that proso millet underwent several discrete domestication events across the globe (Hunt et al. 2018). Proso millet is a global crop that is currently cultivated in nearly 30 countries. It is commonly known as *hog millet* or *common millet* in the United States, *broomcorn millet* in China, and *Hersey* in German, and in Korea, Japan, and other Asia Pacific nations, it is called common millet. In India, proso millet is known by numerous regional names: *Baragu* in Kannada, *Cheena* in Bengali, *Cheno* in Gujarati, *China bachari-bagmu* in Odia, *Panivaragu* in Tamil, *Vari* in Marathi, and *Variga* in Telugu. Proso millet is rich in proteins and minerals, such as magnesium and phosphorus, and is consumed as food globally, including India, China, Korea, Russia, and Germany, whereas it is consumed as a birdfeed in the United States (Das et al. 2019). Nutritionally, proso millet caryopsis has 68–75% carbohydrates and protein content up to 14% and contains vitamins and essential amino acids.

Morphologically, the proso millet is designed for lower water requirements; it has a shallow-rooted and short life cycle of 65–90 days. The crop shows remarkable adaptability to arid and warm climates (Rajasekaran and Francis 2021); however, it is also known to grow in temperate regions and high altitudes of up to 3500 m (Das et al. 2019) making it suitable for cultivation in marginal lands with less farm inputs (Dai et al. 2011). The plant exhibits notable drought resistance and is capable of tolerance in dry, semiarid regions that receive annual precipitation as low as 330–350 mm (Goron and Raizada 2015). The antiquity of proso millet in various parts of the world has led researchers to collect, conserve, and characterize the germplasm. Genebanks across the world conserve more than 28, 000 accessions of proso millet, including very few accessions of wild species. The *Panicum* genus has 600 species, as described by Hickey and King (1998); however, Clayton (1986) classified 460 species under *Panicum* genus. These species are annual or perennial and are conserved separately as grass species and not in the secondary and tertiary pools of proso millet, unlike wheat, rice, and maize. Proso millet and its races have a large number of accessions conserved in genebank. These collections were from diverse agroecological regions of the world depicting diverse climatic conditions, as mentioned above, and each landrace or population is endemic to region that have genes/allelic variants fixed to suit the local environmental conditions.

Proso millet is an important crop for sustainable agriculture and climate change scenarios. Breeding efforts must be strengthened to harness adaptability and health benefits by enhancing yield and improving nutritional profile. ICRISAT has developed a proso millet mini-core set of all accessions in its genebank. Likewise, a trait-specific working subset from the entire collection can be developed to capture the maximum diversity. Varieties across the world are developed through selections from landraces, recombining promising parents and selections from the segregating generation, and very few from mutation breeding. The inflorescence type and floral biology of proso millet do not allow for developing hybrids, which is a main limitation in realizing quantum jumps in yield. Nevertheless, biotic and abiotic stress resilience make it a good crop with less yield loss in the field. Breeding for high yield is slow paced, and by complementing with advanced molecular techniques, genetic gains can be enhanced. Gene editing in proso millet may be a potential technology for the rapid fixation of traits, but it must be supported by genomic resources such as genomics, proteomics, and metabolomics. Zou et al. (2019) sequenced the proso millet genome, which has led to more studies on the genomics and transcriptomics. Its genomic resources must be enriched so that advanced molecular techniques can be deployed to harness the potential of the crop for sustainable crop production, ensuring nutritional security, addressing the challenges of climate change, and depleting natural resources.

7.2 The Crop

7.2.1 Antiquity and Domestication of Proso Millet

Hunt et al. (2011) proposed that proso millet be domesticated at least 10,000 BC in China. Archaeological findings at the Cishan site in the Yellow River Valley provided the oldest evidence of proso millet cultivation. Substantial evidence based on phytoliths and biomolecular components has been collected to demonstrate that proso millet was the earliest crop domesticated by humankind as early as 10,300–8700 cal BP (Lu et al. 2009). A number of evidence was found in Eastern Europe, dating back to 7000 cal BP. These two locations were far away from each other, indicating either independent domestication events of proso millet across continents, as suggested by Harlan (1975). The researchers theorized that if domesticated proso millet dispersal has taken place across continents, then the event unparalleled any other crops in the ancient period between 10,000 and 7000 cal BP. This unprecedented dispersal of domesticated cereals over vast geographical areas was not documented until 5000 BP, and the exchange of material goods was not documented until after 4000 BP (Sherratt 2017). The origin and dispersal of proso millet has raised many interesting questions regarding the early agriculture and cultural exchange between humans across continents. Additional studies with microsatellite markers were not sufficient to map the route of dispersion to Eurasia, Central Asia, and the Americas, and the hypothesis of dispersal did not corroborate archaeobotanical evidence from various cites (Hunt et al. 2011) and Hu et al. (2008)

Pokharia et al. (2014) collated information from research articles and repository materials to discern the antiquity of millets in general, including proso millet in India. Archaeobotanical studies on crops and ethnobotany of early civilization in India have gathered evidence of its cultivation in the Indian subcontinent. The era and site of pros millet cultivation evidence are as follows: (i) Harappan time (Rojdi during 2500–1700 BC, Surkotada in Gujarat 2500–700 BC, Harappa in Punjab 2600–1700 BC, and Pirak in present-day Pakistan 2100–1500 BC); (ii) the Chalcolithic era was cultivated in Raja-Nal-ka-tila in Uttar Pradesh during 1600–1300 BC); and (iii) Iron Age comes from Jammu and Kashmir in Semthan (1–500 AD) and Uttar Pradesh Raja-Nal-ka-tila (1300–700 BC). Currently, proso millet is sporadically cultivated in parts of Madhya Pradesh, eastern Uttar Pradesh, Bihar, Tamil Nadu, Maharashtra, Andhra Pradesh, and Karnataka. Proso millet, foxtail millet, and ragi are cultivated in NE India, but this significant area is not accounted for in the total acreage. However, during 1980 and before, significant proso millet acreage was documented, of which nearly 80,000 ha was in Bihar alone. As the crop requires less water, it is best suited for summer cultivation in Bihar, thriving on residual moisture after the winter crop (*rabi*).

Proso millet has a wide adaptability and is cultivated in China, India, Nepal, Ukraine, Belarus, Africa, Russia, the Middle East, Turkey, and Romania. The proso millet cultivation in North America was attributed to people migrating from German Russian in 1875. The drought resilience of the crop is suitable for cultivation in the dry states of Nebraska, Colorado, and South Dakota in the United States. Presently, it is consumed as porridge/pudding in traditional food in Russia, Germany, India, and China Zohary & Hopf (2000). Some of the documented recipes for proso millet are (a) *awaokoshi*, candied millet puffs, a specialty in Osaka, Japan; (b) alcoholic beverage called *tongba* in Nepal and India; (c) white millet soup; and (d) millet with mushrooms (Santra and Rose (2013) published online http://extension.unl.edu/publications). Since proso millet was domesticated in China, its essence is obvious in Chinese culture, the past of the Chinese Spring Festival, and the wedding ceremonies. Its shiny and red-yellow grains are presented to newlyweds, marking fertility and wishing many children from the marriage (Wang et al. 2016).

7.2.2 Nutritional Benefits of Proso Millet

Proso millet has been an excellent food crop since the beginning of human settlements. The hunters and gatherers started practicing agriculture by utilizing available crops to meet food demands. Proso millet was among the first crops domesticated by humans. We now know that proso millet not only meets the carbohydrate requirements of early humans but also supplements proteins, vitamins, and minerals.

Table 7.1 Nutritional profile of important cereal crops

Cereals	Protein (%)	Fat (%)	Crude fiber (%)	Ash (%)	Starch (%)	Total dietary fiber (%)	Total phenol (mg/100 g)
Proso millet	**11**	**3.5**	**9**	**3.6**	**56.1**	**8.5**	**13.3**
Wheat	14.4	2.3	2.9	1.9	64	12.1	20.5
Rice	7.5	2.4	10.2	4.7	77.2	3.7	2.51
Maize	12.1	4.6	2.3	1.8	62.3	12.8	2.91
Sorghum	11	3.2	2.7	1.8	73.8	11.8	43.1
Finger millet	7.3	1.3	3.6	3	59	19.1	102
Pearl millet	14.5	5.1	2	2	60.5	7	51.4
Foxtail millet	11.7	3.9	7	3	59.1	19.1	106
Kodo millet	8.3	1.4	9	3.6	72	37.8	368

Kalinova and Moudry (2006) stated that proso protein is an exceptional source of protein due to its high levels of leucine, isoleucine, and methionine. Protein quality was found to be better than that of other cereals based on essential amino acid composition, viz., methionine, phenylalanine, tryptophan, and valine. Generally, cereals have low levels of lysine and tryptophan, whereas in proso millet, the prolamin fraction is high, as in the quality protein maize, where the delta zein fraction is greater than the alpha and beta fractions (Kalinová 2007). The waxy type is amylose-free proso millet, which is extensively used in China to make vermicelli Araki et al. (2012).

Proso millet contains various secondary metabolites having many health benefits. It is a rich source of fiber, phosphorus, magnesium, and traces of essential minerals such as calcium, potassium, sodium, manganese, iron, and zinc (Tables 7.1 and 7.2). Its consumption improves lipid and cholesterol metabolism. Proso millet is also gluten-free, which makes it suitable for gluten-sensitive subjects or subjects with celiac diseases, and is a good source of carbohydrates and protein for the elderly. Among the secondary metabolites, a high amount of lecithin is present, which is known to benefit the neural health system by repairing and regenerating myelin fibers and intensifying brain cell metabolism. Proso millet is also a source of niacin, B-complex vitamins, vitamin B6, and folic acid. Proso millet grain antioxidants and other phytochemicals, including phenolics, flavonoids, and anthocyanins, are known to prevent cancer (Amadou et al. 2013; Bangar et al. 2021; Das et al. 2019; Habiyaremye et al. 2017; Kalinová 2007; Kalinova and Moudry 2006; Shahidi and Chandrasekara 2013).

7.2.3 Origin and Distribution of Proso Millet

Based on genome sequence analysis, it is evident that *Seteria italica* and *Panicum miliaceum* shared a common ancestry 18 million years ago (Zou et al. 2019). Proso millet is a member of the family *Poaceae*, Paniceae tribe, and Poales order

Table 7.2 Micronutrient composition in major cereal crops

Cereals	Ca (%)	P (%)	K (%)	Na (%)	Mg (%)	Fe (%)	Mn (%)	Zn (%)	Thiamin (mg/100 g)	Riboflavin (mg/100 g)	Nicotinic acid (mg/100 g)
Proso millet	0.01	0.15	0.21	0.01	0.12	33.1	18.1	18.1	0.63	0.22	1.32
Wheat	0.04	0.35	0.36	0.04	0.14	40.1	40	30.9	0.57	0.12	7.4
Rice	0.02	0.12	0.1	0	0.03	19	12	10	0.07	0.03	1.6
Maize	0.03	0.29	0.37	0.03	0.14	30	5	20	0.38	0.14	2.8
Sorghum	0.04	0.35	0.38	0.05	0.19	50	16.3	15.4	0.46	0.15	4.84
Finger millet	0.33	0.24	0.43	0.02	0.11	46	7.5	15	0.48	0.12	0.3
Pearl millet	0.01	0.35	0.44	0.01	0.13	74.9	18	29.5	0.38	0.22	2.7
Foxtail millet	0.01	0.31	0.27	0.01	0.13	32.6	21.9	21.9	0.48	0.12	3.7
Kodo millet	0.01	0.32	0.17	0.01	0.13	7			0.32	0.05	0.7

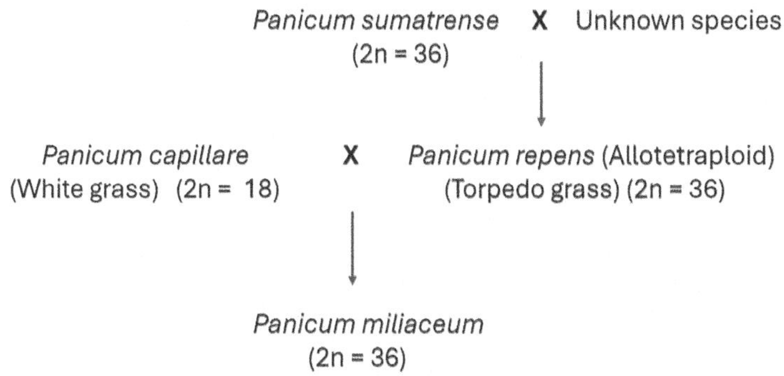

Fig. 7.1 Possible source of *Panicum miliaceum* progenitors

(Christenhusz and Byng 2016). It is a self-pollinating crop, with 10% outcrossing approximately (Baltensperger 2002). Proso millet is an allotetraploid species with $2n = 4x = 36$ chromosomes, which produces 18 regular bivalents. In situ hybridization of nuclear and chloroplast DNA sequences from *Panicum miliaceum* and various diploid and tetraploid relatives revealed that Torpedo grass (*P. repens*) and witch grass (*P. capillare*) could be potential progenitors of proso millet (Hunt et al. 2014) (Fig. 7.1). Among the two subspecies, *miliaceum* is cultivated, and subsp. *ruderale* encompasses the weedy forms (Williams et al. 2011).

Wild proso millet types exhibit narrow lemmas, loose panicles, and jointed spikelet stalks, whereas cultivated types display loose or compressed panicles, jointless spikelet stalks, and broader lemmas. The wild types observed in the temperate regions of the United States, Asia, and Europe differ from those found in China that indicated the domestication of proso millet independently across continents, potentially separated by temporal factors.

7.2.4 Description of Proso Millet

Proso millet is a short-duration (60–90 days to mature) annual crop, similar to other grass species. Typically, plant height ranges from 70 to 120 cm, but the range is from 30 to 120 cm. It has a fibrous and shallow root system, and most roots are confined to the top 15–30 cm soil depth, depending on the soil type. The leaves are simple initiating at the base of the internode usually 20–40 cm in length and 2–10 cm wide. Phyllotaxy is an alternate, ligule marked by dense hairs 2–3 mm long, and at the base is the leaf sheath, which has dense pubescence. Leaves are pubescent and dull green in color, which can be clearly distinguished from other millet crops.

The stem is hollow and pubescent; however, the density of pubescence varies significantly among varieties. Plant parts do not show pigmentation, and the observed pigmentation on the stem is mostly influenced by the environment and is not consistent across seasons or years. The crop shows variations in internode length, and the arrangement of internodes can be an important trait for varietal

Table 7.3 Various races of proso millet

Race	Inflorescence morphology	Characteristics
Compactum	Cylindrical-shaped erect inflorescences	Compact inflorescences and highly evolved cultivars of proso millet ""
Contractum	Compact and drooping inflorescences	
Miliaceum	The inflorescence is open and large, exhibiting slightly erect branches that are minimally divided. The morphology of the inflorescence structure resembles that of wild *P. miliaceum*.	These two races befall across the range of broomcorn millet cultivation, from Eastern Europe to Japan
Ovatum	Compact and a little curved inflorescences with ovate shape	
Patentissimum	It has slender and diffused panicle branches, which are quite similar to race *miliaceum*.	

identification. The stem can be straight or zigzag, which is a very stable trait. The crop also shows variation in tillering; some of the accessions are single culmed, which is a desirable trait for intensive cultivation and amenable for mechanical harvesting. The number of tillers may vary from 1 to 32, and the plant generates tillers at the base of the plant. Branching from the stem was also observed in proso millet. The branches may produce panicles and add to the yield; however, panicle maturation is not synchronized. The panicle in proso millet is covered in the flag leaf, and the length of the flag leaf may vary from 15 to 35 cm long by 2–10 cm wide.

Based on inflorescence morphology, proso millet was classified into five races (*miliaceum*, *patentissimum*, *contractum*, *compactum*, and *ovatum*) by Lyssov (1975). Panicle-type races have been described in proso millet (Reddy et al. 2007) (Table 7.3). They could be diffuse, open, pyramidal to compact, and cylindrical, reaching up to 45 cm in length. The florets on the inflorescence present on the rachis are borne singly at the ends of branches or on short pedicels; these rachises are highly branched. A spikelet consists of a fertile and sterile floret enclosed by two outer glumes. Floral biology is detailed in a later section. Soon after pollination, each floret will start becoming a seed and grain filling is clearly seen only after 15 days of pollination. The grain filling stage can be divided into milky, dough, and grain hardening stages. Maturity in proso millet initiates from the tip of the panicle, and the outer husk of the grains loses chlorophyll and dries. This drying process gradually proceeded to the base of the panicle. At 75% of panicle drying, the crops reached physiological maturity Gomashe (2017).

Similar to other cereals, proso millet produces caryopsis fruits. The grains were tightly packed inside the palea and lemma. The grains of proso millet were larger and heavier in comparison to foxtail millet. The test weight of the unhusked grains could be 4.0–7.0 g/1000 grains. Seed shattering is observed; hence, it must be harvested soon after physiological maturity (Kaume 2006).

7.2.5 Flower Biology

The proso millet inflorescence is characterized by drooping and broom-like structures ranging from 10 to 45 cm in length. The inflorescence has many branches, and each branch ends in one floret. Florets may also develop toward the end of the rachis or on the side with a short pedicel. Florets can be fertile or sessile. The floret opens in a basipetal pattern in the inflorescence (Habiyaremye et al. 2017). The flower opening at the apex may start even when the inflorescence is still emerging from the flag leaf. The latter florets, usually the latter 20%, remain unopened and do not produce seeds. Complete anthesis in the inflorescence may take up to 15 days; in a few genotypes, grain filling starts in the apex florets, and the bottom florets will still flower.

The spikelet is approximately 0.5 cm which comprises two glumes and two lemmas, accompanied by a single palea (Lu et al. 2009). The inner glume matches the spikelet length, whereas the outer glume is comparatively shorter. The lower lemma houses a sterile floret lacking stamens, whereas the upper lemma contains a fertile floret that is shorter in length. In contrast to the well-preserved palea of the upper lemma, the palea associated with the sterile floret are notably reduced (Seetharam et al. 2003). The upper lemma features three staminate structures and two feathery stigmas. As noted by Nanda and Agrawal (2008), the ovary is composed of a bifid style and a plumose stigma. Flowering typically occurs between 10:00 and 12:00, depending on the weather conditions. The flower opened for a short period of up to 7 min before closing. Environmental factors, such as cloud cover, may delay flowering, while intense sunlight and low humidity can expedite the process. Anther dehiscence coincides with stigma receptivity, and anthers undergo dehydration within minutes of floral opening. Usually, flower opening is accompanied by pollination and care must be taken during hand emasculation and pollination for hybridization. The entire flowering period from the onset of anthesis to the final floret on the panicle spans approximately 12–15 days.

Nelson (1984) described crossing procedures that employ manual emasculation techniques. Nandini and Srinathareddy (2019) outlined a modification of Nelson's procedure. Stepwise hand emasculation and pollination in proso millet followed by ICAR-IIMR is mentioned below and also depicted in Fig. 7.2:

(i) *Selection of parent plants*—Choose male and female parent plants with the desired traits. Ensure that the plants are healthy and at the appropriate reproductive (booting) stage. The synchronicity of flowering in male and female plants must be determined before attempting crossing to ensure pollen availability for crossing. Staggered sowing can be performed to synchronize flowering if the parents do not flower at the same time.

(ii) *Preparation for emasculation and pollination*
- Cold water container (temperature around 5–10 °C)
- Fine-tipped forceps
- Scissors
- Labels and tags

Fig. 7.2 Stepwise hand emasculation and pollination in proso millet. (**a**) Selection of panicle having right stage of florets for emasculation. (**b**) Hand emasculation of selected florets in proso millet panicle. (**c**) Emasculated florets in the panicle; rest of all the florets are excised from the panicle. (**d**) Pollination of emasculated florets by contact method

- Small parchment paper bags or cotton wool for covering emasculated flowers
- A clean workspace

(iii) *Emasculation*

- *Selection and time of emasculation:* Conduct emasculation early in the morning when the flowers are at the booting stage but before anthesis (anther dehiscence).
- *Cold water treatment:* The inflorescence was submerged in cold water (5–10 °C) for 2–3 min. Ready florets start opening in response to cold treatment.
- *Removing anthers:* As the florets open, the anther must be removed carefully without damaging the stigma using forceps. Anthers must be removed as quickly as possible because many florets will be opening simultaneously.

- *Retaining only the emasculated florets:* Emasculated florets must be retained in the inflorescence, usually 5–10 florets, and the rest of the florets must be excised carefully. This will ensure the recovery of crossed seeds and prevent the mixing of self-pollinated and crossed seeds in the panicle.
- *Covering the emasculated flower:* Cover the emasculated inflorescence with a small parchment paper bag. Label the plant with the emasculation date and parent details.

(iv) *Pollination*

- *Collection of pollen*: Collect mature anthers from the male parent in the morning when the pollen is fresh in small containers such as Eppendorf tubes. The anthers and pollen must be placed on the *stigma* of emasculated flowers using a small brush or by gently tapping the anthers over the stigma. For effective fertilization, ensure that the pollen grains are deposited evenly on the stigma.
- *Contact method:* Another method is to find the right panicle on the male parent and tie the emasculated panicle with the male parent panicle. The synchronicity of flowering must be determined prior to planning crossing in proso millet.

7.3 Germplasm Conservation Status

There are 29,308 proso millet germplasm accessions conserved in various genebanks worldwide as mentioned in Table 7.4. The Russian Federation maintained the highest number of accessions (8778), followed by China (8451). Ukraine, India, and the United States are other countries that conserve proso millet germplasm accessions (Upadhyaya et al. 2016). In India, approximately 3000 accessions are safeguarded by two national entities: the All-India Coordinated Research Project on Small Millets (AICRP-Small Millets) and the National Bureau of Plant Genetic Resources (NBPGR). The AICRP established a center in Bengaluru in 1976 (Rajasekaran and Francis 2021) and managed 920 accessions. The center has documented germplasm resources and has generated data on qualitative and quantitative traits. The National Genebank of ICAR-NBPGR was approximately 994. The ICRISAT database contains 849 accessions collected from 30 countries. The institute is actively engaged in the classification and assessment of proso millet germplasms. ICRISAT characterized and documented all accessions in their genebank and developed a mini-core set representing diversity in the entire set of germplasms (Upadhyaya et al. 2011). The establishment of a core collection and utilization of molecular markers play a pivotal role in the examination of genetic diversity (Wang et al. 2018).

The National Germplasm Resources Bank in China has 8331 accessions, augmented since 1982. Most (98%) accessions were indigenous landraces in China, and the remaining 2% were acquired from 14 countries.

Table 7.4 Proso millet accessions conserved in genebanks worldwide

Country	Institute	Germplasm accessions		
		Cultivated	Wild	Total
Australia	Australian Tropical Crops and Forages Genetic Resources Centre	225		228
Bangladesh	Plant Genetic Resources Centre, BARI	198		198
Bulgaria	Institute for Plant Genetic Resources "K. Malkov"	489		489
China	Institute of Crop Science, Chinese Academy of Agricultural Sciences	8451		8451
Czech Republic	Genebank Department, Division of Genetics and Plant Breeding, Research Institute of Crop Production	171		171
Germany	Genebank, Leibniz Institute of Plant Genetics and Crop Plant Research	165	1	166
Hungary	Institute for Agrobotany	243	1	244
India	AICRP on Small Millets	920	4	920
	International Crop Research Institute for the Semi-Arid Tropics (ICRISAT)	849		849
	National Bureau of Plant Genetic Resources	994		998
Japan	Department of Genetic Resources I, National Institute of Agrobiological Sciences	516		516
Mexico	Estacion ´ de Iguala, Instituto Nacional de Investigaciones Agrícolas	400		400
Poland	Botanical Garden of Plant Breeding and Acclimatization Institute	354		354
	Plant Breeding and Acclimatization Institute	359		359
Russia	N.I. Vavilov All-Russian Scientific Research Institute of Plant Industry	8778		8778
Ukraine	Institute of Plant Production n.a. V. Y. Yurjev of UAAS	1046	1	1046
	Ustymivka Experimental Station of Plant Production	3975		3976
USA	North Central Regional Plant Introduction Station, USDA-ARS	717	4	721

Adapted from Ref: Narciso and Nystrom (2023)

7.3.1 Genetic Variability in Proso Millet

Morphological characterization Proso millet accessions conserved in the ICRISAT genebank (842 accessions) were characterized by 9 agro-morphological traits. The results showed that Syrian accessions were early flowering and Indian were late; Mexican accessions were dwarf plant height, Sri Lankan was tall, good panicle exserting accessions were Australia and China, and shorter panicles were from former USSR, whereas the longest panicles were from Nepal (Reddy et al. 2007). Vetriventhan et al. (2019) published variability in proso millet accessions conserved in the ICRISAT genebank for all traits described by IBPGR for proso millet characterization. Eurasia was predominantly cultivated in proso millet, but with time, the crop was forgotten; however, to relaunch proso millet in Italy and Southern Austria, accessions from Austria, the United States, Poland, and Russia were evaluated using agro-morphological traits. Significant variability was found in growth habit, culm

Fig. 7.3 Inflorescence shape, panicle compactness, and seed coat color variability in proso millet

length, tillering, branching on the main stem, blade length, and width. The US bred lines were single-culmed and better suited to give higher yields in the tested location than others (Vischi et al. 2021). Variation for inflorescence shape, panicle compactness, and seed coat color variability in proso millet are depicted in Fig. 7.3.

The National Active Germplasm Site (NAGS) genebank in ICAR-Indian Institute of Millets Research holds more than 1000 accessions of proso millet augmented by the ICAR-National Bureau of Plant Genetic Resources (1007), All-India Coordinated

Table 7.5 List of promising proso millet genotypes

Character	Promising genotypes	Reference
Grain yield (g)	EC 24113, EC 24114, IC 28829, IC 41847, IC 52792, IPM 33, IPM 36, IPM 364	National Bureau of Plant Genetic Resource (NBPGR report, 2012)
Days to flower	JPM 44, IC345024, IPM 40, IPM 84, IPM 273, IPM 982, IPM 2125, IPM 2589	National Bureau of Plant Genetic Resource (NBPGR report, 2012)
Panicle length	Rampur Local, Shai Local	National Bureau of Plant Genetic Resource (NBPGR report, 2012)
Rust	CO 5, TNAU145, TNAU202, TNAU218, TNAU220	AICRP on Small Millets (2014)
Blast, sheath blight, leaf blight, grain smut	DHPRMV2769, TNAU148, TNAU164, TNAU218	Anonymous (2014)

Research Project on Small Millets (300), breeding lines from Tamil Nadu Agriculture University (439), and mini-core set from ICRISAT. Of these accessions, 634 were evaluated and characterized for 11 quantitative traits. Promising lines identified for each trait were documented and catalogued for further utilization in the proso millet breeding program (Elangovan et al. 2023) (Table 7.5).

Molecular characterization Molecular characterization has helped researchers address many challenging questions in proso millet, viz., phylogenetic studies, relating varieties to their ancient progenitors, reconstructing old varieties, assessing molecular diversity, and population genetics studies. Morphological and molecular characterizations were performed in Hungary with the aim of reconstructing 600-year-old varieties using modern cultivars. The 24 agro-morphological and ISSR, SSR, CAP, and mtDNA markers helped to relate the old excavated seeds to the present-day *Omszkoje* variety (Gyulai et al. 2006, 2012; Lágler et al. 2005). Molecular studies were able to discern diversity patterns with more confidence than the morphological markers. DNA-based markers, especially SSRs, were used in general for molecular studies of proso millet (Cho et al. 2010). Initially, heterologous SSRs from related species, such as rice, wheat, oats, and barley, were used (Hu et al. 2009). SSR developed in switch grass (*Panicum virgatum*) (Wang et al. 2011) has mostly been used in molecular studies in proso millet (Rajput et al. 2014, 2016). Transcriptome-based functional markers were also developed (Hou et al. 2017). Further, Zou et al. (2019) published the first genome sequence of proso millet, which opened a wide scope for molecular studies, including genotyping by sequencing (GBS), transcriptomics, proteomics, and metabolomics.

Protein content The genetic variability of protein content in Indian proso millet lines was in the ranged of 11.5% to 21%. The lines with high protein content (more than 20%) identified are IIMR-pm 56, IIMR-pm 143, and IIMR-pm 186, which are being used in metabolite studies. The protein quality, based on the Essential Amino

Acid Index (EAAI) in proso millet, is 51–67 which is higher than that of maize, wheat, rye, barley, and oat (Kalinová 2007).

Quality analysis Tannins and phytic acid are found as antinutritional factors in proso millet. In US bred proso millet lines and varieties, phytic acid ranged from 0.17% to 0.47%; however, hulling of grains reduced the phytic acid up to 30% (Lorenz 1983). The starch granules of proso millet using scanning electron microscope (SEM) revealed small spherical, large spherical (small numbers), and large polygonal granules (greater numbers) with 1.3 to 8.0 μm size (Krishna Kumari and Thayumanavan 1998). Graybosch and Baltensperger (2009) screened 650 accessions for waxy traits in the USDA-ARS proso millet germplasm to identify six waxy accessions: one was from Russia, PI 346946, and the rest were from China, PI 436622 (Lung Shu 5), PI 436623 (Lung Shu 7), PI 436624 (Lung Shu 14), PI 436625 (Lung Shu 16), and PI 436626 (Lung Shu 18). Further dissection of waxy traits in proso millet revealed the presence of two genes: L- and S-type *GBSS*1. "Waxy" is manifested due to nonfunctional L and S GBSS gene (Hunt et al. 2010).

7.4 Utilization of Germplasm in Crop Improvement

Similar to any other crop, economic yield is the most important trait for improvement, followed by biotic and abiotic stress tolerance. The major reason for the decline in the proso millet area is its low yield potential. It has built-in abiotic stress tolerance and harbors novel gene variants for many agronomic traits. Elucidating the adaptation mechanisms of proso millet and conducting gene mining could serve as viable approaches for investigating climate-resilience traits. The development of novel cultivars tailored to region-specific climatic factors, including soil composition, temperature, humidity, photoperiod, precipitation, and cropping patterns, is essential for producing genotypes of various maturity duration groups with specific breeding objectives. The huge collections of proso millet in the genebank were conserved in the medium- and long-term modules. Genetic characterization of accessions and breeding materials for economically important traits is necessary for their further utilization in crop improvement breeding programs. Several varieties have been bred utilizing germplasm resources, and most of them were selected from landraces.

The accessions collected and conserved in the genebank must be evaluated in the field for agronomic traits and other economically important traits such as disease and insect pest resistance. The formulation of a core set can be very useful for phenotyping or initiating a breeding program in proso millet. In some cases, the core set is also large; in such cases, a mini-core set can be formulated as in the ICRISAT (Upadhyaya et al. 2011). The entire set of proso millets in ICRISAT contained 833 accessions. These accessions were evaluated in the field for phenotypic characterization using 9 qualitative and 11 quantitative traits. A phenotypic distance matrix was calculated using Gower's distance and subjected to Ward's hierarchical clustering. Using the cluster information, 10% of the total accessions were selected to formulate a core capturing maximum diversity from the entire set with minimum redundancy. The core set was validated using Shannon-Weaver Diversity Index, mean difference percentage (MD%), variance difference percentage (VD%) of traits, and coincidence (CR%) and variable (VR%) rates. The National Active Germplasm Sites for millets (in ICAR, Indian Institute of Millets Research, Hyderabad, INDIA) also conducted field trials to formulate a subset of collections that included genotypes superior for one or more traits (Table 7.2). These accessions are used directly in breeding programs to improve agronomic traits.

7.4.1 Proso Millet Cultivar Development Using Germplasm Lines

In India, systematic breeding for small millets, including proso millet, started with the inception of the All-India Coordinated Research Project on Small Millet in 1962. Early proso millets were developed through mass or pure line selection in landraces. The varieties that developed through mass selection until 1970 were *Ram-cheena*, *Shyam-cheena*, V 27, and V 306 suitable for cultivation in the states of Andhra Pradesh and Koraput in Orissa. Most cultivars developed in early breeding programs have a short duration with a maturity of less than 80 days (Harinarayana 1989). In Bangladesh, three local landraces, Kumardhon (1.339 t/ha), Bagaikandi (1.288 t/ha), and Telipara (1.239 t/ha), and one exotic line called white millet (1.209 t/ha) were utilized in breeding programs and new varieties have been developed with higher-yield potential, such as BPm-40, BPm-14, BPm-34, and BPm-21 (Majid et al. 1989).

The breeding program in Russia (then the USSR) started in early 1930, and germplasm collection and selection followed to develop new varieties. Interestingly, the USSR practiced hybridization and selection to improve grains in proso millet. The variety Saratov 853 was developed with an increased test weight (7.5 g) through selection. Despite considerable selection work carried out between 1938 and 1968, the seed weight could not be increased. Later, through hybridization and selection, 1000 seed weight was improved significantly. A few selected cultures had test weights of even 10 g. Some of the early varieties developed in the USSR utilizing local landraces are Scorospeloe 66 (developed by crossing Aureum 1113 from Kazakhstan group with Saratov 853 form Volga group). The variety Volga 3 resulted from the cross Victoria 48 (Kazakhstan) with hybrid populations obtained from Orenburgskoe 42 and Sanquineum 75 from the Volga group (Ilyin and Zolotukhin 1989).

In the United States, proso millet is cultivated for bird feed; however, after realizing the prospects of the international market for amylose-free proso millet, the breeding objective included grain quality parameters too (Das et al. 2019). The Nebraska Agricultural Experiment Station released the amylose-free proso millet variety "Plateau," a cross between "Huntsman" (PI 578074) and the Chinese line Lung Shu 18 (PI 436626). Huntsman is a selection derived from a series of crosses using the previously released varieties of Cope, Dawn, and Common. The other cultivars in the United States include the white-seeded varieties "Sunrise" and "Earlybird" (Santra et al. 2015). In China, proso millet breeding relies on pure line selection and recombination. Several cultivars have been developed through crossing and selection. There have been 162 varieties released in China since 1950, mostly from landraces. Some proso millet varieties released in China at the national level are Neimi 5, Ningmi 9, Longmi 4, Ningmi 10, Longshu 16, Jinshu 4, and others.

7.4.2 Breeding Objectives in Proso Millet Improvement Program

Proso millet is an annual crop with a shallow root system that shows high resilience to abiotic stress. It is a potentially contingent crop alongside major cereals crops. Nevertheless, low productivity does not make the crop less important; cultivation of proso millet is more sustainable due to its short duration, less input requirement, and meager agronomic management. Researchers worldwide have set the following objectives for proso millet breeding:

Yield improvement Yield is the most important trait for breeding proso millets. In India, the acreage under proso millet has drastically reduced; at present, it is cultivated on 30,000 ha. Low yield is one of the major reasons for the reduced acreage in proso millet. The higher the yield, the greater the remuneration of farmers. With the increased demand for small millet, including proso millet, production must be enhanced per unit area of cultivation. The global demand for proso millet is likely to increase along with the demand for plant-based proteins.

Protein content Proso millet has a higher protein content than any other cereal crop. On an average, the Indian cultivars show 15–16% protein in the hulled grains. Protein content as high as 20% was identified in Indian germplasm collection; however, the yield levels of these lines must be enhanced. The test weight of the high-protein lines was lower than that of the released cultivars. Protein content estimation in an association mapping panel of 188 lines showed no negative correlation between the yield and protein content. However, the mechanism of high protein accumulation must be understood at various levels of expression to enhance the protein content in proso millet.

Breeding for shoot fly resistance Shoot fly (*Atherigona* spp.) is a devastating insect pest of proso millet (Fig. 7.4). Among the small millets, the shoot fly prefers the proso millet the most. Deadheart plants show stunted growth and trigger tillering, leading to unproductive plants. The yield loss estimated are up to 30–35%. Under heavy infestation, deadhearts can reach up to 50% under natural conditions. Early sowing is recommended in India to reduce shoot fly infestation. Shoot flies were screened using the fish-meal method or by releasing shoot flies in a net house. Proso millet genotypes show genetic variability in shoot fly tolerance. There is a potential for developing shoot fly-resistant lines using tolerant lines in breeding programs.

Fig. 7.4 Symptoms of shoot fly infestation at different growth stages

Breeding for biotic stress The reported plant diseases in proso millet include blast (*Pyricularia grisea*), sheath and leaf blight (*Rhizoctonia solani*), and grain smut (*Sorosporium paspali-thunbergii*). Enhancing host-plant resistance using resistance genes in breeding is the most viable and economic solution for disease management in crop species. In Russia, many smut-resistant varieties including Quartet, Soyuz, Sputnik, and Slavjanskoe have been bred (Konstantinov et al. 1989). *Pseudomonas avenae* subsp. avenae causes bacterial stripe disease (also known as bacterial leaf blight) in proso millets. Infection can cause stunting and seedling mortality (Liu et al. 2012). Few reports are available on the prevalence of this disease in proso millet. *Ustilago crameri* (Korn1874) causes a kernel smut in the proso millet Dyussibayeva et al. (2020). Infected plants transform grains into white-grayish sacs called smut sori, producing black spores. It is an external seed-borne pathogen. Five resistant variety identified was CO 5 (TNAU 143). The leaf spot of proso millet is a seed-transmitted disease, *Bipolaris panicimiliacei* (syn. *Helminthosporium panici-miliacei*). It produces brown rectangular lesions on affected leaves; under severe conditions, it causes seed rotting, coleoptile spots, and seedling blight (Lee 1997). The RAUM 7 variety developed by RAU Bihar was found to be resistant to leaf spot disease. GPUP 8 and GPUP 28 are resistant to the brown spot disease.

In addition to the primary concern of banded leaves and sheath blight, this crop is also susceptible to rust, melanosis, and blast, although these diseases are of comparatively minor importance. The current climate change scenario and the potential for increased disease occurrence underscore the critical need to evaluate germplasms for biotic stress tolerance, which could serve as valuable genetic resources in future breeding endeavors. Complementing phenotypic analyses, the utilization of markers for diversity studies, gene identification, and the development of novel molecular markers linked to disease resistance facilitated by available genomic sequences will accelerate the breeding process. Furthermore, the integration of omics approaches and high-throughput phenotyping techniques can contribute to more efficient and effective crop improvement strategies. The fall armyworm (*Spodoptera frugiperda*) has emerged as a major pest; however, the outbreaks are not regular. It is a polyphagous insect and has a wide host range; in proso millet, the entire whorls are fed by the fall armyworm larva. No studies have been conducted to understand the mechanisms of host-pest interactions.

Agronomic traits Earliness is an important trait in proso millet breeding. Proso millet maturing in 65-70 days yielding 1.5–1.8 tons per hectare will be highly rewarding to farmers, and such maturity groups will allow proso millet to fit into any cropping system. Shattering can cause significant losses in the field. These farm losses could be reduced by developing non-shattering cultivars. As proso millet matures, the culm strength is reduced, and the panicle weight lodges the plant. In locations where high wind prevails during the seed setting or grain maturation stage, a non-lodging crop will be highly beneficial. Enhancing yield by increasing plant density and maintaining a single culm will make crops amenable to mechanical harvesting. Some of these traits, viz., non-lodging and non-shattering, are difficult to breed because no standard procedure is available to screen germplasm lines for these traits.

Breeding for nutritional quality Proso millet is rich in protein; on average, proso millet contains 13–15% protein. Some genotypes were identified with improved protein content up to 21%. In addition to the protein content, the protein composition can be improved. In general, cereals are poor in lysine and tryptophan, and proso millet protein composition can be improved by increasing the protein prolamin fraction in the total protein. Proso millet is rich in phosphorus and magnesium. Iron and zinc are negligible; however, agronomic methods can be used to enhance grain Fe and Zn. Phytic acid is one of an antinutritional factor present in proso millet (up to 0.47% in unhulled whole grains) can be reduced (Lorenz 1983).

Breeding for abiotic stress tolerance Proso millet has a shallow root system and is cultivated in marginal lands in arid and semiarid regions under rainfed conditions (Upadhyaya et al. 2016; Rajasekaran and Francis 2021; Das et al. 2019). Intermittent drought adversely affects yield; nevertheless, proso millet shows considerable drought and heat tolerance. Molecular studies could identify novel variants of known genes in proso millet (Table 7.6). Salinity was strongly associated with drought, and comparable results were anticipated in drought and salinity experiments. Yue et al. (2016) conducted this study.

7.5 Detailed Account on Proso Millet Genetic Improvement in India

A systematic breeding program with specific objectives was followed for the proso millet. In India, postindependence, the state agriculture universities in Tamil Nadu, Andhra Pradesh, Karnataka, and Uttar Pradesh focused on developing millet cultivars using the available landraces. Research on finger millet was initiated as early as 1900 in Bangalore. In 1923, the Millet Research Station was established in

Table 7.6 Molecular markers developed in proso millet (*Panicum miliaceum* L.) for abiotic stress

Molecular markers	Total number of molecular markers	References
Expressed sequence tags (ESTs)	32 ESTs, homologous to recognized plant sequences articulated in response to abiotic or biotic stress factors	Lin et al. (2006b)
Simple sequence repeats	Development and characterization of SSR markers in proso millet based on switchgrass genomics	Rajput et al. (2014)
Differentially expressed genes (DEGs)	62,543 unigenes, functionally annotated from the de novo assembly and transcriptome description of the proso millet. 32 PmWRKY genes intricated in abiotic stress response	Yue et al. (2016)
Expressed sequence tags (ESTs)	211 ESTs, in response to drought stress	Saha et al. (2016)
DEGs	42,240 unigenes, 2301 SSRs, 1,447,148 SNPs under moisture stress	Wang et al. (2017)
DEGs	1695 DEGs in response to salt stress	Zhang et al. (2019)
Transcription factors	180 NAC (PmNAC) genes involved in drought response	Shan et al. (2020)
DEGs	For cell wall biosynthesis under salt stress	Yuan et al. (2021)

Coimbatore under the erstwhile Madras Presidency and then extended to Anakapalle (A.P.) and Hagari (Karnataka). Millet improvements in Kanpur and Gorakhpur in the state of Uttar Pradesh started in 1944. The first variety released from Coimbatore in proso millet was Co 1. With the establishment of the Project on Intensification of Research on Cotton, Oil Seeds, and Millets (PIRCOM) in Hyderabad during the 1950s, many centers were opened to work on millets.

The importance of germplasm was recognized, and countrywide germplasm collection was conducted even before the establishment of a coordinated research program in India. A whooping 3000 accessions of different millets were collected and conserved, including 250 accessions of proso millet. The All-India Coordinated Millets Improvement Project (AICMIP) was established in 1969. Five crop-specific lead centers were identified, and Dholi in Bihar was the lead center for proso millet improvement. This was because Bihar had the highest acreage in proso millet. In 1986, the AICMIP headquarters were established at the University of Agriculture Science Bangalore. The AICMIP focused on developing agro-technologies for millets, which were primarily cultivated by resource-poor farmers or tribals in drylands and hill regions. Improved high-yielding varieties and production technologies help secure food security for poor farmers in India.

The establishment of germplasm unit as National Active Germplasm Site (NAGS) for millet in Bangalore aided in conservation of millet germplasm; much later, another NAGS was established in the present ICAR-Indian Institute of Millets

Research Hyderabad (ICAR-IIMR). The aim was to collect, characterize, document, conserve, and distribute germplasms across countries for research and development. NAGS was enriched with more germplasm accessions, now with 939 accessions of proso millet. Based on field evaluation, the origin of the collection core set was formulated. By the end of 1986, 84 millet varieties had been released, of which 8 were proso millets. By the end of 2012, 24 proso millet varieties were released in India, mostly from pure line selection. The yield potential of proso millet was enhanced to 1800 kg per hectare.

Proso Millet in Twenty-First-Century India

The acreage under proso millet has gradually declined since the 1990s. In general, farmers prefer remunerative crops to millets. The major reasons for the decline in acreage were the low yield and the need for primary processing. Proso millet cultivation was reduced, and people reduced their consumption of proso millet. A complete void was created in society; people started forgetting ancient crops until 2010. In the year 2018, Nation Year of millets was celebrated in India, which became the precedence for "International year of millets" in 2023. Tamil Nadu Agricultural University developed variety K2, which is a non-lodging and non-shattering variety. Hybridization has emerged as a crucial breeding method for small millets, accounting for 29% of proso millet cultivar development (Vetriventhan et al. 2020). This approach has led to the creation of high-yielding cultivars in India, such as TNAU 202 and ATL 1 (Rajasekaran and Francis 2021). The current focus of proso millet research is to develop high-yielding varieties with high protein content. Rigorous selections were made to develop non-lodging and non-shattering varieties. High-protein proso millet is used in recombination breeding to simultaneously enhance yield and protein content. Basic research was performed to understand the mechanism of protein accumulation in the proso millet grains.

7.6 High-Throughput Phenotyping (HTP) in Proso Millet Breeding Program

High-throughput plant phenotyping (HTP) represents an advanced platform for data collection that employs sensors and imaging technologies to capture traits both above and below the soil surface in real time. The importance of highly accurate phenotyping cannot be overstated in genetic research, which spans traditional and cutting-edge areas of crop enhancement (Barbadikar et al. 2019). In a recent study, Zhao et al. (2022) introduced a novel approach for the automatic detection of proso millet panicles and estimation of heading percentage by utilizing standard RGB images acquired by unmanned aerial vehicles (UAVs). This method demonstrated a coefficient of determination of 0.728 and achieved 92.4% accuracy in identifying plots that had reached 50% heading. These results indicate the potential of this technique to enhance HTP protocols for proso millet breeding initiatives.

7.7 Genetic and Genomics Resources

The field of plant genomics encompasses a comprehensive analysis of entire plant genomes, including their sequencing, characterization, and examination of genetic composition, structure, organization, function, and interaction (Abdurakhmonov 2016). While early genomic research has concentrated on genomic size and mapping techniques, the introduction of next-generation sequencing (NGS) has significantly facilitated the sequencing, assembly, and analysis of plant species genomes. The application of genomic tools and resources for proso millet improvement has been extensively explored, as discussed in the subsequent sections of this work.

7.7.1 Genomic Resources

Rajput et al. (2016) developed the first genetic linkage map of proso millets. They used RIL population of 93 derived from Huntsman × Minsum cross to map quantitative trait loci (QTLs) for 9 morpho-agronomical traits. 833 SNPs were assigned to 18 linkage groups in proso millet. QTLs for heading date, 100 grain weight, grains per panicle, lodging, peduncle length, plant height, grain shattering, and panicle length were mapped. Wang et al. (2018) sequenced waxy genes locus (wx-L) to identify 6 SNPs linked with waxy gene. Johnson et al. (2019) used 1882 SNPs to study population structure in proso millet.

Zou et al. (2019) sequenced and assembled the genome of proso millet, resulting in a high-quality draft with an approximate size of 921 Mb. Genome annotation identified 55,930 protein-coding genes. Comparative genomic analysis with foxtail millet, rice, and maize revealed distinct evolutionary pathway for proso millet. Identification of genetic regions associated with domestication events have become possible through sequencing. These regions are linked with traits that are essential for the adaptation of crops to agricultural settings, including seed size, flowering time, and plant architecture. This breakthrough has considerable potential for gene mining applications (Shan et al. 2020) and GWAS (Boukail et al. 2021).

Hou et al. (2017) developed simple sequence repeat (SSR) markers from transcriptome of proso millet. Transcriptome analysis has enabled the discovery of evolutionary relationships, gene architectures, protein motifs, chromosomal distribution, duplication, expression patterns in various tissues, and drought stress responses of 180 NAC genes in proso millet (Shan et al. 2020). Liu et al. (2022) unraveled the mechanisms for achieving high nitrogen use efficiency (NUE) through both field and hydroponic experiments. Roy et al. (2017) conducted a complete proteomic analysis of seed proteins in four Korean proso millet cultivars using 2-DE and mass fingerprinting methodologies. Their investigation revealed 1152 differentially expressed proteins on a 2D gel, with 26 successfully identified by MALDI-TOF/MS. Interestingly, 2 proteins exhibited upregulation across all cultivars, whereas in 2 specific cultivars, 13 proteins showed increased expression and 11 displayed decreased expression. Yuan et al. (2022) conducted salt stress tolerance through multi-omics approach. Salt-unique proteins were identified in salt-sensitive and salt-tolerant genotypes suggesting posttranscriptional regulatory mechanisms in the response to salt stress and recovery.

Metabolomic studies offer a comprehensive view of gene expression at both spatial and temporal, when combined with transcriptomics and proteomics, and these investigations facilitate delineation of biological processes and mechanisms (Diola et al. 2014). Metabolomic research can illuminate the pathways and mechanisms underlying the accumulation of these compounds in proso millet (Shahidi and Chandrasekara 2013; Chandrasekara et al. 2012). However, there is a paucity of information regarding the quantification and characterization of these compounds in proso millet. Initial metabolomics research on proso millet employed gas chromatography-time-of-flight mass spectrometry (GCTOFMS) to evaluate the grain quality (Kim et al. 2013; Li et al. 2021a, b).

7.7.2 New Plant Breeding Techniques (NPBT)

Utilization of knowledge derived from various omics approaches in prominent crops has the potential to augment genetic advancements in proso millet. Key New Plant Breeding Techniques (NPBTs) utilized for crop enhancement include cisgenesis, intrageneric sequence-specific nuclease technology (SSN 1,2), oligo-directed mutagenesis, RNA-dependent DNA methylation, and reverse breeding. The Clustered Regularly Interspaced Short Palindromic Repeats (CRISPR)/CRISPR-associated protein 9 (Cas9) system, along with alternative gene editing technologies such as TALENS and ZFNS, shows promise in developing superior crops for the future. Moreover, the application of site-specific mutagenesis and high-precision genome-editing techniques may further expedite crop improvement initiatives in an environmentally sustainable manner.

7.8 Future Prospects

Panicum miliaceum L., commonly known as proso millet, exhibits remarkable characteristics, such as efficient water utilization, substantial nutritional content, and the capacity to thrive in marginal environments. These attributes make it a prime candidate for enhancing crop diversity and bolstering food security. The extensive genetic variability within proso millet germplasm, as demonstrated through DNA marker studies and the establishment of a core collection, serves as a crucial resource for breeding initiatives aimed at enhancing yield, nutritional profiles, and stress resilience. Notwithstanding its potential, proso millet remains inadequately explored and underutilized, with certain regions displaying limited genetic diversity and insufficient genomic resources, thus hindering genetic advancement. The integration of various omics technologies and the creation of innovative molecular markers are projected to expedite breeding processes. The development of a core collection and identification of germplasm accessions with specific traits represents significant genetic assets for augmenting the utilization of proso millet germplasms. The preservation of broad genetic diversity in genebanks, coupled with the application of cutting-edge genomic tools, is anticipated to facilitate the creation of superior proso millet cultivars. These improved varieties are expected to exhibit enhanced yield potential, elevated nutrient content, and increased resilience to environmental stressors, thereby contributing to the advancement of sustainable agricultural practices for an improved future.

References

Abdurakhmonov IY (2016) Genomics era for plants and crop species—advances made and needed tasks ahead. Plant Genomics

Amadou I, Gounga ME, Le GW (2013) Millets: nutritional composition, some health benefits and processing – a review. Emir J Food Agric 25(7):501–508. https://doi.org/10.9755/ejfa.v25i7.12045

Baltensperger DD (2002) Progress with proso, pearl, and other millets. Trends New Crops New Uses:100–103

Bangar SP, Ashogbon AO, Dhull SB, Thirumdas R, Kumar M, Hasan M, Chaudhary V, Pathem S (2021) Proso-millet starch: properties, functionality, and applications. Int J Biol Macromol 190(September):960–968. https://doi.org/10.1016/j.ijbiomac.2021.09.064

Barbadikar KM, Balakrishnan D, Gireesh C, Kardile H, Bosamia TC, Mishra A (2019) High throughput phenotyping: potential tool for genomics. OMICS based approaches in plant biotechnology, 303-321

Boukail S, Macharia M, Miculan M, Masoni A, Calamai A, Palchetti E, Dell'Acqua M (2021) Genome-wide association study of agronomic and seed traits in a world collection of proso millet (*Panicum miliaceum* L.). BMC Plant Biol 21(1):1–12

Chandrasekara A, Naczk M, Shahidi F (2012) Effect of processing on the antioxidant activity of millet grains. Food Chem 133(1):1–9

Christenhusz MJ, Byng JW (2016) The number of known plants species in the world and its annual increase. Phytotaxa 261(3):201–217

Dai HP, Zhang PP, Lu C, Jia GL, Song H, Ren XM, Chen J, Wei AZ, Feng BL, Zhang SQ (2011) Leaf senescence and reactive oxygen species metabolism of broomcorn millet (Panicum miliaceum L.) under drought condition. Aust J Crop Sci 5(12):1655–1660

Das S, Khound R, Santra M, Santra DK (2019) Beyond bird feed: proso millet for human health and environment. Agriculture 9(3):2–19. https://doi.org/10.3390/agriculture9030064

Diola V, de Menezes Daloso D, Antunes WC (2014) Metabolomics. In: *Omics plant breeding*, pp 81–101

Elangovan M, Singode A, Venkatesh K, Amasiddha B, Karthik AS, Pandey S, Pandey CD (2023) Agro-morphological characterization of proso millet germplasm for utilization. Int J Plant Soil Sci 35(16):381–392. https://doi.org/10.9734/ijpss/2023/v35i163165

Goron TL, Raizada MN (2015) Genetic diversity and genomic resources available for the small millet crops to accelerate a New Green Revolution. Front Plant Sci 6(MAR). https://doi.org/10.3389/fpls.2015.00157

Graybosch RA, Baltensperger DD (2009) Evaluation of the waxy endosperm trait in proso millet (*Panicum miliaceum*). Plant Breed 128(1):70–73. https://doi.org/10.1111/j.1439-0523.2008.01511.x

Gyulai G, Humphreys M, Lagler R, Szabo Z, Toth Z, Bittsanszky A, Gyulai F, Heszky L (2006) Seed remains of common millet from the 4th (Mongolia) and 15th (Hungary) centuries: AFLP, SSR and mtDNA sequence recoveries. Seed Sci Res 16(3):179–191. https://doi.org/10.1079/ssr2006251

Gyulai G, Lágler R, Holly L, Horváth L (2012) The Hungarian Genebank collections of common millet (*Panicum miliaceum*) and the application to conservation genetics. Europ J Plant Sci Biotechnol 6(2):96–102

Habiyaremye C, Matanguihan JB, D'Alpoim Guedes J, Ganjyal GM, Whiteman MR, Kidwell KK, Murphy KM (2017) Proso millet (*Panicum miliaceum* L.) and its potential for cultivation in the Pacific Northwest, U.S.: a review. Front Plant Sci 7. https://doi.org/10.3389/fpls.2016.01961

Harinarayana G (1989) Breeding and Varietal Improvement of Small Millets in India G. In: Seetharam A, Riley KW, Harinarayana G (eds) Small millets in global agriculture. Proceedings of 1st international small millets workshop, Bangalore, India, 1986. IDRC. Pub. by Mohan Primlani for Oxford and IBH Publishing Co. Pvt. Ltd., New Delhi, India, pp 59–70

Harlan JR (1975) Crops and man. American Society of Agronomy, Madison, Wisconsin

Hou S, Sun Z, Li Y, Wang Y, Ling H, Xing G, Han Y, Li H (2017) Transcriptomic analysis, genic SSR development, and genetic diversity of proso millet (*Panicum miliaceum*; Poaceae). Appl Plant Sci 5(7):1600137. https://doi.org/10.3732/apps.1600137

Hu X, Wang J, Lu P, Zhang H (2009) Assessment of genetic diversity in broomcorn millet (*Panicum miliaceum* L.) using SSR markers. J Genet Genomics 36(8):491–500. https://doi.org/10.1016/S1673-8527(08)60139-3

Hunt HV, Denyer K, Packman LC, Jones MK, Howe CJ (2010) Molecular basis of the waxy endosperm starch phenotype in broomcorn millet (*Panicum miliaceum* L.). Mol Biol Evol 27(7):1478–1494. https://doi.org/10.1093/molbev/msq040

Hunt HV, Campana MG, Lawes MC, Park Y-J, Bower MA, Howe CJ, Jones MK (2011) Genetic diversity and phylogeography of broomcorn millet (*Panicum miliaceum* L.) across Eurasia. Mol Ecol 20(22):4756–4771. https://doi.org/10.1111/j.1365-294X.2011.05318.x

Hunt HV, Badakshi F, Romanova O, Howe CJ, Jones MK, Heslop-Harrison JP (2014) Reticulate evolution in *Panicum* (Poaceae): the origin of tetraploid broomcorn millet, P miliaceum. J Exp Bot 65(12):3165–3175

Ilyin VA, Zolotukhin EN (1989) Breeding of proso millet (*Panicum miliaceum* L.) in Volga region of USSR. In: Seetharam A, Riley KW, Harinarayana G (eds) Small millets in global agriculture. Proceedings of 1st international small millets workshop, Bangalore, India, 1986. IDRC. Pub. by Mohan Primlani for Oxford and IBH Publishing Co. Pvt. Ltd, New Delhi, India, pp 105–114

Johnson M, Deshpande S, Vetriventhan M, Upadhyaya HD, Wallace JG (2019) Genome-wide population structure analyses of three minor millets: Kodo millet, little millet, and proso millet. Plant Genome 12(3):1–9

Kalinova J, Moudry J (2006) Content and quality of protein in proso millet (*Panicum miliaceum* L.) varieties. Plant Foods Hum Nutr 61(1):45–49. https://doi.org/10.1007/s11130-006-0013-9

Kaume RN (2006) Panicum miliaceum L. In: Brink M, Belay G (eds) PROTA 1: Cereals and Pulses/Céréales et légumes secs (CD-ROM). PROTA, Wageningen

Kim JK, Park SY, Yeo Y, Cho HS, Kim YB, Bae H, Park CH, Lee JH, Park SU (2013) Metabolic profiling of millet (*Panicum miliaceum*) using gas chromatography-time-of-flight mass spectrometry (GC-TOFMS) for quality assessment. Plant Omics 6(1):73–78

Konstantinov SI, Linnik VM, Ya SL (1989) Use of smut-resistant induced mutants in breeding proso millet. Selektsiyai Semenovodstvo (Kiev) 66:25–28

Krishna Kumari S, Thayumanavan B (1998) Characterization of starches of proso, foxtail, barnyard, kodo, and little millets. Plant Foods Hum Nutr 53(1):47–56. https://doi.org/10.1023/A:1008083020810

Lágler R, Gyulai G, Humphreys M, Szabó Z, Horváth L, Bittsánszky A, Kiss J, Holly L, Heszky L (2005) Morphological and molecular analysis of common millet (P. miliaceum) cultivars compared to an aDNA sample from the 15th century (Hungary). Euphytica 146(1–2):77–85. https://doi.org/10.1007/s10681-005-5814-7

Lee DH (1997) Morphological characters and seed transmission of *Bipolaris panici-miliacei* causing leaf spot of common millet. Korean J Plant Pathol 13(1):18–21

Li J, Wang Y, Wang L, Zhu J, Deng J, Tang R, Chen G (2021a) Integration of transcriptomic and proteomic analyses for finger millet [*Eleusine coracana* (L.) Gaertn.] in response to drought stress. PLoS One 16(2):e0247181. https://doi.org/10.1371/journal.pone.0247181

Li W, Wen L, Chen Z, Zhang Z, Pang X, Deng Z, Liu T, Guo Y (2021b) Study on metabolic variation in whole grains of four proso millet varieties reveals metabolites important for antioxidant properties and quality traits. Food Chem 357:129791

Liu H, Qiu H, Zhao W, Cui Z, Ibrahim M, Jin G, Li B, Zhu B, Xie GL (2012) Genome sequence of the plant pathogen *Pseudomonas syringae* pv. *panici*. LMG 2367:5693–5694
Liu C, Yuan Y, Liu J, Wang H, Ma Q, Zhou Y, Liu C, Gong X, Feng B (2022) Comparative transcriptome and physiological analysis unravel proso millet (*Panicum miliaceum L.*) source leaf adaptation to nitrogen deficiency with high nitrogen use efficiency. Environ Exp Bot 199:104891
Lorenz K (1983) Tannins and Phytate Content in Proso millets (*Panicum miliaceum*). Cereal Chem 60(6):424–426
Lu H, Zhang J, Liu K, Wu N, Li Y, Zhou K, Ye M, Zhang T, Zhang H, Yang X, Shen L, Xu D, Li Q (2009) Earliest domestication of common millet (*Panicum miliaceum*) in East Asia extended to 10,000 years ago. Proc Natl Acad Sci 106(18):7367–7372. https://doi.org/10.1073/pnas.0900158106
Lyssov BH (1975) Proso (Panicum L.). In: Krotov AS (ed) The USSR Flora of Cultivated Plants, vol 3. USSR, Kolos, Leningrad
Majid MA, Hamid MA and Mannujan Importance, Genetic Resources and Breeding of Small. In: Seetharam A, Riley KW, Harinarayana G (eds) Small millets in global agriculture. Proc. 1st International small millets workshop, Bangalore, India, 1986. IDRC. Pub. by Mohan Primlani for Oxford and IBH Publishing Co. Pvt. Ltd., New Delhi, India, pp 71–76
Nanda JS, Agrawal PK (2008) Botany of field crops, vol I. Kalyani Publisher, India
Nandini C, Srinathareddy J (2019) Modified crossing (SMUASB) method for artificial hybridization in proso millet (*Panicum miliaceum L.*) and little millet (*Panicum sumatrense*). Electron J Plant Breed 10(3):1161–1170
Nelson LA (1984) Technique for crossing proso millet[1]. Crop Sci 24(1):205–206
Pokharia AK, Kharakwal JS, Srivastava A (2014) Archaeobotanical evidence of millets in the Indian subcontinent with some observations on their role in the Indus civilization. J Archaeol Sci 42(1):442–455. https://doi.org/10.1016/j.jas.2013.11.029
Rajasekaran R, Francis N (2021) Genetic and genomic resources for improving proso millet (*Panicum miliaceum* L.): a potential crop for food and nutritional security. Nucleus (India) 64(1):21–32. https://doi.org/10.1007/s13237-020-00331-2
Rajput SG, Plyler-Harveson T, Santra DK (2014) Development and characterization of SSR markers in proso millet based on switchgrass genomics. Am J Plant Sci 5(1):175–186. https://doi.org/10.4236/ajps.2014.51023
Rajput SG, Santra DK, Schnable J (2016) Mapping QTLs for morpho-agronomic traits in proso millet (*Panicum miliaceum* L.). Mol Breed 36(4):37. https://doi.org/10.1007/s11032-016-0460-4
Reddy VG, Upadhyaya HD, Gowda CLL (2007) Morphological characterization of world's proso millet germplasm. ICRISAT EJ 3(1):1–4
Roy SK, Kwon SJ, Yu JH, Sarker K, Cho SW, Moon YJ, Jung TW, Park CH, Woo SH (2017) Comparison of protein profiles of proso millet (*Panicum miliaceum*) seeds of various Korean cultivars. Korean J Crop Sci 62(1):40–50
Santra DK (2013) Proso millet varieties for western Nebraska. University of Nebraska-Lincoln NebGuide, G2219
Santra DK, Heyduck RF, Baltensperger DD, Graybosch RA, Nelson LA, Frickel G, Nielsen E (2015) Registration of 'Plateau' Waxy (Amylose-Free) Proso Millet. J Plant Regist 9(1):41–43. https://doi.org/10.3198/jpr2013.11.0067crc
Seetharam A, Gowda J, Halaswamy JH (2003) Small millets. In: Chowdhury SK, Lal SK (eds) Nucleus and breeder seed production manual. Indian Agricultural Research Institute, New Delhi, pp 54–67
Shahidi F, Chandrasekara A (2013) Millet grain phenolics and their role in disease risk reduction and health promotion: a review. J Funct Foods 5(2):570–581. https://doi.org/10.1016/j.jff.2013.02.004
Shan Z, Jiang Y, Li H, Guo J, Dong M, Zhang J, Liu G (2020) Genome-wide analysis of the NAC transcription factor family in broomcorn millet (*Panicum miliaceum L.*) and expression analysis under drought stress. BMC Genomics 21(1):1–13

Sherratt A (2017) 2. The Trans-Eurasian exchange: the prehistory of Chinese relations with the West. In: Contact and exchange in the ancient world. University of Hawaii Press, pp 30–61. https://doi.org/10.1515/9780824841676-005

Upadhyaya HD, Sharma S, Gowda CLL, Gopal Reddy V, Singh S (2011) Developing proso millet (*Panicum miliaceum* L.) core collection using geographic and morpho-agronomic data. Crop Pasture Sci 62(5):383–389. https://doi.org/10.1071/CP10294

Upadhyaya HD, Vetriventhan M, Dwivedi SL, Pattanashetti SK, Singh SK (2016) Proso, barnyard, little, and kodo millets. In: Singh M, Upadhyaya HD (eds) Genetic and genomic resources for grain cereals improvement. Academic Press, Cambridge, MA, pp 321–343

Vetriventhan M, Azevedo VCR, Upadhyaya HD, Naresh D (2019) Variability in the global Proso millet (*Panicum miliaceum* l.) Germplasm collection conserved at the Icrisat Genebank. Agriculture (Switzerland) 9(5):1–16. https://doi.org/10.3390/agriculture9050112

Vetriventhan M, Azevedo VC, Upadhyaya HD, Nirmalakumari A, Kane-Potaka J, Anitha S, Ceasar SA, Muthamilarasan M, Bhat BV, Hariprasanna K, Bellundagi A (2020) Genetic and genomic resources, and breeding for accelerating improvement of small millets: current status and future interventions. Nucleus 63(3):217–239

Vischi M, Zorzin N, Bernhart M, Winkler J, Santra D, Pappalardo C, Marchetti S (2021) Morphophysiological and molecular characterization of millet (*Panicum miliaceum* L.) varieties for crop improvement in Western Europe. CABI Agric Biosci 2(1):1–12. https://doi.org/10.1186/s43170-021-00047-6

Wang YW, Samuels TD, Wu YQ (2011) Development of 1,030 genomic SSR markers in switchgrass. Theor Appl Genet 122(4):677–686. https://doi.org/10.1007/s00122-010-1477-4

Wang R, Hunt HV, Qiao Z, Wang L, Han Y (2016) Diversity and cultivation of Broomcorn Millet (*Panicum miliaceum* L.) in China: a review. Econ Bot 70(3):332–342. https://doi.org/10.1007/s12231-016-9357-8

Wang R, Wang H, Liu X, Ji X, Chen L, Lu P, Liu M, Teng B, Qiao Z (2018) Waxy allelic diversity in common millet (*Panicum miliaceum L.*) in China. Crop Journal 6(4):377–385

Williams WM, Williamson ML, Real D (2011) Bromus L. wild crop relatives: genomic and breeding. Resources 9(Jauhar 1981):197–216. https://doi.org/10.1007/978-3-642-14255-0

Yuan Y, Liu J, Ma Q, Gao Y, Yang Q, Gao X, Feng B (2022) Cleaner production of proso millet (*Panicum miliaceum L.*) in salt-stressed environment using re-watering: from leaf structural alleviations to multi-omics responses. J Clean Prod 334:130205

Yue H, Wang L, Liu H, Yue W, Du X, Song W, Nie X (2016) De novo assembly and characterization of the transcriptome of broomcorn millet (*Panicum miliaceum L.*) for gene discovery and marker development. Front Plant Sci 7:1083

Zhao B, Khound R, Ghimire D, Zhou Y, Maharjan B, Santra DK, Shi Y (2022) Heading percentage estimation in proso millet (*Panicum miliaceum L.*) using aerial imagery and deep learning. Plant Phenome J 5(1):e20049

Zou C, Li L, Miki D, Li D, Tang Q, Xiao L, Rajput S, Deng P, Peng L, Jia W, Huang R, Zhang M, Sun Y, Hu J, Fu X, Schnable PS, Chang Y, Li F, Zhang H et al (2019) The genome of broomcorn millet. Nat Commun 10(1):436. https://doi.org/10.1038/s41467-019-08409-5

Kodo Millet

8

C. Deepika, K. Venkatesh, K. N. Ganapathy, B. Amasiddha, S. Avinash, Sooganna, and C. Tara Satyavathi

Abstract

Kodo millet (*Paspalum scrobiculatum L.*), known for its resilience to harsh climatic conditions, is a valuable crop for ensuring food and nutritional security. Kodo millet is one of the most important nutri-cereals cultivated in tropical and subtropical regions worldwide. In India, it is predominantly cultivated in Madhya Pradesh and Chhattisgarh, and among small millets, kodo millet occupied largest area in India. It is nutritively potential crop compared to rice, wheat, and barley. The crop possesses valuable characteristics like high herbage, branched ears, numerous seeds per raceme, high fertility, and unique storage ability, making it one of the hardiest among small millets. Its production potential is particularly noteworthy in marginal, low-fertility soils and areas with chronic moisture deficits, contributing to its survival in modernized agricultural systems. Kodo millet production can be further improved by using potential germplasm in the breeding programs. Efforts to conserve and characterize germplasm collections preserved in various genebanks globally are ongoing to know the genetic diversity and identify suitable donors. These genetic resources are crucial for improving traits such as drought resistance, pest tolerance, and nutritional quality. Breeding programs aim to develop improved varieties with better yields, adaptability, and nutrient content by utilizing both conventional breeding methods and advanced technologies.

Keywords

Kodo millet · Plant genetic resources · Germplasm · Breeding · Utilization · Pre-breeding · Trait-specific germplasm

C. Deepika (✉) · K. Venkatesh · K. N. Ganapathy · B. Amasiddha · S. Avinash · Sooganna
C. T. Satyavathi
ICAR-Indian Institute of Millets Research, Hyderabad, Telangana, India
e-mail: deepika@millets.res.in

K. Tripathi et al. (eds.), *Plant Genebank Utilization for Trait Discovery in Millets*, https://doi.org/10.1007/978-981-96-4043-0_8

8.1 Introduction

Kodo millet (*Paspalum scrobiculatum* L.) thrives in damp habitats across tropical and subtropical regions worldwide. It serves as a pasture crop in arid regions of Asia, New Zealand, and the United States. However, it holds indigenous status in India, predominantly cultivated in Madhya Pradesh, Chhattisgarh, Uttar Pradesh, Tamil Nadu, Maharashtra, Karnataka, and certain parts of Andhra Pradesh (Yadava and Jain 2006). It is known by various names across India—such as *kodo* (Hindi), *Varagu* (Tamil), *Arika* (Telugu), *Harka* (Kannada), *Kodra* (Gujarati, Marathi, and Punjabi), and *Kodua* (Oriya) (Ayyangar and Wariar 1934; Deshpande et al. 2015). It also carries several English names, including bastard millet, creeping paspalum, ditch millet, Indian paspalum, koda grass, water couch, and Kodohirse (German) (Knees and Gupta 2013). In India, kodo millet covers about 1.39 lakh hectares (2023), primarily in southern, western, central, and northern states, yielding approximately 0.82 lakh tons of grains with a productivity of around 801 kg/ha. Madhya Pradesh and Chhattisgarh jointly account for nearly 80% of the area under cultivation, whereas Tamil Nadu, Maharashtra, Uttar Pradesh, and Gujarat are also significant contributors (Table 8.1).

Kodo millet boasts nutritive potential comparable to popular cereals like rice, wheat, barley, or bajra, offering 8–9% dietary fiber, the highest among cereals, along with higher levels of polyunsaturated fatty acids (PUFA) and minerals compared to rice or wheat. It is easily digestible and rich in B vitamins (especially niacin, B6, and folic acid) as well as minerals such as calcium, iron, potassium, magnesium, and zinc (Table 8.2). Moreover, being gluten-free, it's an ideal choice for individuals who are intolerant to gluten.

Regular consumption of kodo millet offers significant health benefits, particularly for postmenopausal women combating cardiovascular diseases, high blood pressure, and elevated cholesterol levels. It possesses valuable characteristics like high herbage, numerous seeds per raceme, high fertility, and unique storage ability, making it one of the hardest among small millets. Its production potential is particularly noteworthy in marginal, low-fertility soils and areas with chronic moisture deficits, contributing to its survival in modernized agricultural systems (Yadava and Jain 2006).

Table 8.1 Distribution of kodo millet in India

States	Area (Lha)	Production (LTonne)	Productivity(kg/ha)
Madhya Pradesh	1.005	0.431	429
Chhattisgarh	0.599	0.136	226
Tamil Nadu	0.045	0.083	1833
Uttar Pradesh	0.045	0.035	777
Gujarat	0.296	0.270	913
Maharashtra	0.123	0.062	509
Andhra Pradesh	0.012	0.009	723
Total	2.12	1.026	419

Source: Bhat et al (2018)

Table 8.2 Nutritive value of kodo millet vs. rice and wheat (per 100 g)

Item	Kodo millet	Wheat (whole)	Rice (raw, milled)
Carbohydrates (g)	65.9	71.2	78.2
Protein (g)	8.3	11.8	6.8
Fat (g)	1.4	1.5	0.5
Energy (KCal)	309	346	345
Crude fiber (g)	9	1.2	0.2
Mineral matter (g)	2.6	1.5	0.6
Amylose (%)	24	25	12–19
Amylopectin (%)	76	75	88–81
Ca (mg)	27	41	10
P (mg)	188	306	160
Vitamins (mg/100 g)			
Thiamin	0.15	0.41	0.41
Niacin	2.0	5.1	4.3
Riboflavin	0.09	0.1	0.04
Folic acid	23.1	36.6	8.0
Micronutrients (mg/100 g)			
Fe	0.5	5.3	0.7
Zn	0.7	2.7	1.4
Mg	147	138	90
Na	4.6	17.1	–
K	144	284	–
Cu	1.6	0.68	0.14
Mn	1.1	2.29	0.59

Source: Nutritive value of Indian food (2007), National Institute of Nutrition, India; Hariprasanna (2016)

8.2 The Crop

Kodo millet (*Paspalum scrobiculatum*) is an annual tufted grass with plant heights of 60–90 cm, characterized by glabrous stems with swollen nodes that turn purple upon maturity. The leaves are simple, alternate, and erect or suberect, extending up to 40 cm in length, with surfaces that are either glabrous or occasionally soft-hairy. The inflorescence comprises 3–5 alternately arranged racemes, each up to 13 cm long, bearing spikelets in 2–3 rows along one side of a flattened rachis. The grains are elliptic-orbicular, measuring 1.5–2.5 mm in length; are enclosed in hard, corneous husks; and vary in color from light red to dark gray, with a thousand-grain weight of approximately 6 grams.

Kodo millet matures within 3–4 months, yielding between 250 and 1000 kg/ha on average, with potential yields reaching up to 2000 kg/ha. The crop prefers full sunlight but can tolerate partial shading. Kodo millet is predominantly self-pollinated, with cleistogamous flowers. Flower opening occurs between 2:30 am and 11:30 am, varying by region, with some lines exhibiting protogyny (Youngman and Roy 1923; Rangaswami Ayyangar and Panduranga Rao 1934; Verma 1989; Yadava 1997).

8.2.1 History, Origin, Taxonomy, and Domestication

Yadava and Jain (2006) described kodo millet as a native crop of India, being cultivated since ancient times. Mentioned in the Brihad Samhita, it was believed that during years when Jupiter led, famine would trouble most, and only grains like *kodra* (kodo) and *samai* (*kutki*) would survive. The crop's drought stress tolerance has been recognized since long time. It was domesticated in southern Rajasthan and Maharashtra around 3000 years ago (Kajale 1977). India, with its rich biodiversity, is considered to be the center of origin for this crop, which spread throughout tropical regions globally.

Taxonomically, kodo millet belongs to the family Poaceae, subfamily Panicoideae, tribe Paniceae, genus *Paspalum*. The genus *Paspalum* comprises about 250 species, including *Paspalum scrobiculatum* L, *P. notatum, P. conjugatum*, *P. compactum*, and *P. dilatatum*, cultivated for grain or fodder. The wild progenitor of paspalum species is *P. sangunale*. Further, historical records indicate the presence of *P. compactum*, *P. conjugatum*, *P. orbiculare*, and *P. scrobiculatum* in Hasan district of Karnataka during 1772–1782 (Yadav 2005). Kodo millet is divided into three races (regularis, irregularis, and variabilis) based on panicle morphology. It is tetraploid in nature with the chromosome number as $2n = 4\times = 40$ based on the cytological studies (Hariprasanna 2015; Upadhyaya et al. 2016).

8.2.2 Production, Use, and Constraint

Kodo millet is nutritious millet with multiple uses. It is rich in fiber, protein, and essential minerals, hence beneficial for managing weight, controlling diabetes, and improving overall health. It is widely used in preparing porridge, upma, khichdi, and gluten-free baked goods, making it suitable for those with gluten intolerance. It also serves as fodder for livestock, highlighting its importance in both food and agriculture. Kodo millet is predominantly cultivated in marginal and shallow soils characterized by slow native fertility and limited moisture retention capacity. Given its status as a rainfed crop, recurrent droughts are common due to the erratic behavior of rainfall patterns. Cultivation of kodo millet is primarily restricted to hilly terrain where altitude, rainfall, and soil characteristics vary significantly, often exacerbated by severe soil erosion that diminishes fertility levels. Farming practices associated with kodo millet tend to be at subsistence levels with minimal management interventions. Traditional low-yielding landraces are commonly grown, and the adoption rate of high-yielding varieties has been sluggish. Fertilizer application is generally absent, and weed competition poses persistent challenges in kodo millet fields. Additionally, various biotic stressors such as sheath blight, head smut, and shoot fly can significantly reduce yields, especially in susceptible genotypes.

Given that kodo millet is traditionally cultivated by tribal and hill farmers, resource constraints are a prevalent issue. The adherence of tribal farmers to ancestral farming practices, coupled with limited literacy levels, often hampers the adoption of modern agricultural techniques. Furthermore, a lack of remunerative markets and inadequate recognition of its nutritive value contribute to the challenges faced by kodo millet growers. While kodo millet's contribution to total food grain production remains relatively low, it serves as a staple food source within agroecosystems. However, the persistent imbalance in cereal production exacerbates food insecurity, particularly among vulnerable populations. Recognizing the critical role of kodo millet in food security, there is a pressing need to prioritize efforts to enhance its production. This entails targeted interventions aimed at improving farming practices, promoting the adoption of high-yielding varieties, strengthening market linkages, and raising awareness about its nutritional benefits. Such initiatives are essential for bolstering the resilience of kodo millet farming communities and ensuring food security for vulnerable populations.

Another concern is toxicity in kodo millet due to fungal contamination producing cyclopiazonic acid, a toxic fungal secondary metabolite. This infection is common in farmers' fields, and some cultivars seem to be particularly susceptible to the disease. By effective adoption of both pre- and postharvest management, the kodo millet grains can be made safe for human consumption and can be popularized as a nutritious grain (Deepika et al. 2022).

8.3 Status of Germplasm Collection

In any crop improvement program, the initial and critical step involves the exploration of naturally available germplasm resources. The collection, characterization, and conservation of these germplasms provide breeders with essential material to integrate into breeding programs aimed at developing improved varieties. Ex situ genebanks worldwide have conserved a substantial number of Kodo millet accessions, totaling approximately 8587 (Goron and Raizada 2015; Padulosi et al. 2009), with the majority housed in India.

India's efforts to comprehend the gene pool of Kodo millet commenced in 1961 under the PL 480 project, resulting in the collection of approximately 584 germplasms. Subsequent endeavors aimed at amassing a diverse array of germplasms, including landraces and improved varieties, from various regions across the country. Notably, the National Bureau of Plant Genetic Resources (NBPGR) in New Delhi, India, conserves about 2273 germplasm accessions of kodo millet. Furthermore, the genebank at Tamil Nadu Agricultural University (TNAU), Coimbatore, India, safeguards 427 accessions. The establishment of the National Active Germplasm Site (NAGS) at the University of Agricultural Science (UAS) in Bangalore, India, in 1992 focused on small millets and conserve 1537 accessions. A comprehensive overview of organizations conserving the kodo millet gene pool is provided in Table 8.3 (Ravikesavan et al. 2023; Hariprasanna 2017; Johnson et al. 2019; Upadhyaya et al. 2014, 2016).

Table 8.3 Kodo millet and related species germplasm conserved in various genebanks across the globe

Organization	Location	No. of accessions	Resource
ICAR-NBPGR	New Delhi, India	2273	Chandra U et al. (1990)
ICAR-AICRP Small millet centre UAS, Bengaluru	Bengaluru, India	1537	http://www.smallmillets.res.in/
ICRISAT	Patancheru, India	656	Upadhyaya et al. (2014)
TNAU	Coimbatore, India	427	www.tnau.ac.in/
ICAR-IIMR, Hyderabad	Hyderabad, India	1927	
PGRCU of the USDA-ARS	Griffin, United States	336	http://www.arsgrin.gov/npgs/ index.html
EMBRAPA	Brazil	746	Acuña et al. (2019)
Germplasm bank of FCA-UNNE and IBONE	Argentina	434	Acuña et al. (2019)
CIAT	Columbia	156	Acuña et al. (2019)
Australian Pastures Genebank	Australia	301	Acuña et al. (2019)
Margot Forde Forage Germplasm	New Zealand	373	Acuña et al. (2019)

8.4 Use of Germplasm in Kodo Millet Crop Improvement

The fundamental understanding of variability is pivotal for the success of any plant breeding program. Selection of superior varieties is effective only when variability exists within the chosen materials for the improvement (Nirubana et al. 2019). Plant genetic diversity plays a crucial role in selecting diverse parents for exploiting heterosis or obtaining desirable recombinants and transgressive segregants. Dhagat (1971) conducted the initial study on genetic diversity estimation among 6 exotic and 90 indigenous germplasm of kodo millet. Significant efforts have been directed toward screening germplasm for grain yield, smut resistance, nutritional traits, and adaptation to biotic and abiotic stresses (Vetriventhan and Upadhyaya 2019). In another study, higher genetic variability has been observed in kodo millet for qualitative traits such as leaf characteristics, panicle attributes, and grain color, offering opportunities for genotype identification. Conversely, traits like growth habit and leaf pigmentation exhibited limited variation and may not be effective for genotype distinction. Specific germplasm lines with diverse characteristics have been identified for future breeding programs (Nirubana et al. 2019).

Considering the nutritional significance of kodo millet, studies have been conducted to analyze its nutritional composition alongside agronomical traits. Genotypes rich in multiple nutrients have been identified, aiding in the selection of high-yielding lines with enhanced nutritional content for release as cultivars after multilocation testing (Koutu 1989; Kulkarni and Naik 2000; Kewat 2016; Nirubana et al. 2017; Vetriventhan and Upadhyaya 2019). The development of varieties in kodo millet has predominantly involved mass or pure line selections due to challenges encountered in hybridization techniques, primarily attributed to tiny spikelets on brittle pedicels and cleistogamous flower (Nirmalakumari et al. 2007). However, mutagenesis has been used to create variability and the variety JK 76 was gamma-irradiated to develop a protogynous mutant with two rows of spikelets to be used in recombination breeding to identify superior varieties (Nagaraja et al. 2023).

Various studies have been conducted across India to enhance kodo millet production through morphological traits, yield, and stress tolerance using available germplasm resources. Association studies between yield, biochemical traits, and agronomic characteristics have elucidated important selection criteria. Traits such as the number of productive tillers per plant, fodder yield, and harvest index have shown positive associations with grain yield, highlighting their significance in crop improvement (Dalsaniya et al. 2023).

8.4.1 Characterization and Evaluation of Kodo Millet Germplasm

Characterization efforts have been undertaken for kodo millet germplasm, cataloguing 1038 germplasm lines for 16 qualitative traits and 11 quantitative traits at UAS, Bangalore (Hariprasanna 2016). Additionally, more than 2418 accessions are conserved in Consultative Group for International Agricultural Research (CGIAR) genebanks, with 665 germplasms held by the International Crops Research Institute for the Semi-Arid Tropics (ICRISAT). The Svalbard genebank, often referred to as the "doom's day vault," has safeguarded 654 kodo millet accessions from various countries worldwide. Moreover, the US Department of Agriculture holds 336 germplasms within the National Plant Germplasm System (GRIN). The representative pictures showing the growth stages of kodo millet crop are presented in Fig. 8.1.

In few other studies, phenotypic evaluation of kodo millet germplasm revealed diverse morphological traits and identified lines with desirable characteristics for varietal improvement. Plant height, number of productive tillers, and days to maturity are key traits influencing yield potential and adaptation to specific environments (Yadav et al. 2020; Jayaram 2007). Further, studies on lodging tolerance and associated grain loss have identified diverse germplasm lines and landraces with varying culm strengths and duration. High grain-yielding lines with medium culm strength and short stature could mitigate lodging-induced grain loss (Ragini et al. 2015). Also, genetic parameter estimation studies have provided insights into yield-related traits and heritability, facilitating early and simple selection methods for desirable traits (Sao et al. 2017; Subramanian et al. 2010; Thakur et al. 2018; Suthediya et al. 2021). The representative picture showing the panicle diversity in kodo millet germplasm is presented in Fig. 8.2.

Fig. 8.1 Collage showing the growth stages of kodo millet crop

8.4.2 Core Set Development in Kodo Millet

Upadhyaya et al. (2014) developed core set for kodo millet by characterizing the entire germplasm collections of kodo (656 accessions) conserved at ICRISAT for 20 morpho-agronomic traits. The data of quantitative traits were subjected to residual (or restricted) maximum likelihood analysis and the best linear unbiased predictors followed by Gower distance matrix and hierarchical clustering by Ward method. Minimum one accession or 10% accessions from each cluster were selected to establish a core set comprising 75 diverse accessions.

Fig. 8.2 Panicle diversity in kodo millet germplasm

8.4.3 Varieties of Kodo Millet Released Using Germplasm

Several high-yield potential varieties have been released for different states as outlined in Table 8.4. To date, the Government of India has notified 27 kodo millet varieties, including 13 by central variety release committee and others designated by various states.

Under the All India Coordinated Small Millets Improvement Project, AICSMIP (2014), kodo millet varietal evaluation trials are conducted at designated testing centers across the country, selected based on kodo millet growing regions. These centers are located in Chhattisgarh (Jagdalpur), Gujarat (Waghai), Karnataka (PC unit, Bengaluru), Madhya Pradesh (Dindori and Rewa), and Tamil Nadu (Athiyandal). Now, variety release system is known as All India Coordinated Research Project on Small Millets (AICRP-Small Millets). The latest national checks included in the trail are GPUK 3 (selected from germplasm GPLM 826, Bengaluru), RK 390–25 (a mutant of RK 390, Rewa), and TNAU 86 (selected from IPs 85, Coimbatore) (Hariprasanna 2016).

8.5 Genetic and Genomic Resources

Despite its high dietary fiber content, antioxidant activity, nutritional value, and resilience to various biotic and abiotic stresses, kodo millet lags behind in the development of linkage maps and molecular markers compared to other small millets such as foxtail millet, proso millet, and barnyard millet, where genomic and marker information is readily available. Nevertheless, several preliminary molecular

Table 8.4 List of kodo millet varieties developed using germplasm

S. no	Name	Pedigree	Year	Days to maturity	Yield (q/ha)	Recommended area
1	Dahod Kodo 1 (CKMV 3)	Pure line selection from locally collected germplasm	2023	108	30	Andhra Pradesh, Chhattisgarh, Jharkhand, Madhya Pradesh, Tamil Nadu
2	CKMV 1 (ATL 2)	Pure line selection from DPS 63/58	2021	106–110	28	Andhra Pradesh, Chhattisgarh, Gujarat, Jharkhand, Karnataka, Madhya Pradesh, Tamil Nadu, Telangana
3	CKMV 2 (GAK 3/ Dahod Kodo)	Pure line selection from germplasm RK 286	2020	105–110	25	Andhra Pradesh, Chhattisgarh, Gujarat, Madhya Pradesh, Tamil Nadu
7	TNAU 86	Pure line selection from IPS 85	2012	95–110	27–30	All states
8	INDIRA KODO1	Pure line selection	2012	100–105	22–25	Chhattisgarh
9	DPS 9–1	Pure line selection	2011	105–110	26–28	Madhya Pradesh
10	JK 98	Pure line selection from 317	2010	100–105	25–30	Madhya Pradesh
11	JK 36	Pure line selection	2009	75–80	10–12	Madhya Pradesh
12	JK 65	Pure line selection	2009	105–110	25–30	All states
13	JK 106	Pure line selection	2009	100–105	19–20	Madhya Pradesh
15	JK 439	Pure line selection	2004	100–110	22–23	Chhattisgarh and Madhya Pradesh
16	KK 2	Pure line selection	2002	80–85	20–23	Uttar Pradesh
17	JK 48	Pure line selection	2001	95–100	26–27	All states
18	JK 155 (RBK 155)	Selection from GPLMP 251	2000	80–100	20–22	Karnataka and Madhya Pradesh
19	KK1 (ICK 7125)	Pure line selection from Deoria district of UP	1999	80–85	18–20	Uttar Pradesh

(continued)

Table 8.4 (continued)

S. no	Name	Pedigree	Year	Days to maturity	Yield (q/ha)	Recommended area
20	KMV20 (VAMBAN 1/ VARGU 1)	Pure line selection from Pali	1996	100–110	16–18	Tamil Nadu
21	GK2 (GUJARAT KODRA 2)	Pure line selection	1993	110–115	15–18	GAU, Waghai, Gujarat
22	APK1 (VARAGU)	Pure line selection from PSC 5	1993	100–105	18–20	Tamil Nadu
23	GPUK 3	Pure line selection from GPLM 826	1991	100–105	18–20	All states
24	JK 76	Pure line selection from local germplasm of Sidhi	1991	85–90	18–22	Madhya Pradesh
25	JK 62	Pure line selection	1986	90–100	20–22	All states
26	JK 41	Pure line selection	1986	105–110	20–22	Plains of Madhya Pradesh, Gujarat, and South India
27	PSC 1	Pure line selection from IPS 115	1986	100–105	18–20	All states
28	JK 2 (DINDORI 73)	Pure line selection	1982	110–120	20–22	Madhya Pradesh
30	CO 3	Pure line selection from PS175	1980	115–120	12–13	Tamil Nadu
31	K 1	Pure line selection from KPS 209 (IPM 600)	1982	100–105	12–14	Tamil Nadu
32	JNK 364	Pure line selection	1977	95–100	18–20	Madhya Pradesh
33	GK 1 (GUJARAT KODRA 1)	Pure line selection from IPS 19	1977	115–120	20–25	Western Ghats of Gujarat

studies on kodo millet have been conducted during recent years, which are summarized in this section.

M'Ribu and Hilu (1996) developed RAPD markers to analyze genetic diversity and geographical variation in *Paspalum scrobiculatum* (kodo millet) and *P. polystachyum*. They observed high polymorphism and genetic diversity, clustering the accessions into three groups, with one representing an African type and the other two representing Indian accessions. Arya et al. (2014) examined SSR markers from finger millet and pearl millet for cross-species amplification in kodo millet. Twenty SSR markers (15.15%) were successfully transferred to kodo millet, with markers from pearl millet showing a higher transferability rate compared to finger millet, likely due to both belonging to the same tribe, Paniceae. Kushwaha et al. (2015) developed gene-specific primers called DNA binding with one finger (DoF) to analyze the genetic diversity of cereals and millets. Multiple bands of size 100–3000 bp were observed, with genetic diversity ranging from 0.0601 to 0.2118 among cereals and millets. Two major clusters were constructed based on Jaccard's similarity matrix. Yadav et al. (2016) conducted a diversity study on 96 Kodo millet germplasm accessions using Neutral-ISSR (Inter Simple Sequence Repeat), Functional-SCoT (Start Codon Targeted), and SRAP (Sequence-Related Amplified Polymorphism) markers. Nei's gene diversity and Shannon's information index indicated genetic diversity values of 0.15 and 0.13 for neutral markers and 0.26 and 0.22 for functional markers, respectively. Forty-two landraces of kodo millet collected from five regions of Madhya Pradesh were used for genetic diversity analysis using ISSR markers. The highest number of alleles (07) was observed with primer UBC-884, and the highest Polymorphism Information Content value was 0.58. The Dindori region exhibited the highest genetic diversity (Lalit et al. 2019).

These studies demonstrated that ISSR markers can be effectively used for diversity analysis until complete genomic information of kodo millet becomes available in the public domain. Through genotyping-by-sequencing (GBS), 3461 single-nucleotide polymorphisms (SNPs) were identified for kodo millet. Population structure analysis revealed seven putative subpopulations, with no correlation between genetic groups and designated races of kodo millet. It was suggested to improve core collections by incorporating population and genomic data along with race designations (Johnson et al. 2019).

8.6 Pre-breeding

Pre-breeding in kodo millet involves identifying and utilizing wild or unadapted germplasm to introduce novel traits into cultivated varieties. This process enhances the genetic diversity and resilience of the crop, making it more adaptable to changing environmental conditions and pests. The primary goal of pre-breeding in kodo millet is to improve traits such as yield, nutritional quality, disease resistance, and stress tolerance. Addressing challenges such as climate change, pre-breeding incorporates genes from wild relatives that confer tolerance to stresses like drought, heat, and pests. This equips kodo millet varieties to withstand environmental challenges. By exploring the genetic resources in wild relatives, breeders can identify genes that confer these desirable traits and integrate them into breeding programs. This enhances productivity and sustainability, benefiting both farmers and consumers.

Preservation of genetic diversity is another critical aspect. As modern varieties replace traditional landraces, valuable genetic traits face risk of depletion. Exploring genetic resources in wild relatives helps conserve unique genes, and utilizing them for broadening the genetic base of cultivated varieties. Pre-breeding efforts can also focus on improving postharvest processing and consumer preferences. Varieties with enhanced milling and cooking qualities increase market value, ensuring economic viability for farmers. In summary, pre-breeding expands the genetic diversity of kodo millet, enhances resilience to environmental stresses, and improves market value. Leveraging genetic resources from wild relatives unlocks the full potential of this nutritious and climate-resilient crop, benefiting farmers and consumers alike.

8.7 Future Prospects

There is ample possibility to enrich the variability for economic traits through vigorous collection from unexplored kodo millet growing areas. These economic traits may be utilized as donor for improvement in kodo millet for earliness, non-shattering, and higher grain yield coupled with in-built resistance against important biotic stresses. Though rich in nutrient and traditionally been taken as poor man's staple, thorough research is needed regarding their bioavailability and reduction of anti-nutritional factors. Limited availability of genomic resources is the major challenge in this crop too, like other millets except foxtail and finger millet; however, this could be overcome through use of genomic resources available in taxonomically closest species and high-throughput genotyping technologies. The other problem of limited use of kodo millet is milling for the grains, milling drudgery associated with upper husk sticking endosperm tightly that reduces the efficiency of grain recovery from each spikelet. In addition, mechanization in the area of cultivation, harvesting, and processing needs to be taken up. Finally, farmers should be educated against kodo poisoning so that suitable precautionary measures can be taken up for risk-free production in large area, instead of small pockets.

References

Acuña CA, Martínez EJ, Zilli AL, Brugnoli EA, Espinoza F, Marcón F, Urbani MH, Quarin CL (2019) Reproductive systems in *Paspalum*: Relevance for germplasm collection and conservation, breeding techniques, and adoption of released cultivars. Front Plant Sci 10:1377. https://doi.org/10.3389/fpls.2019.01377

AICSMIP (2014) Report on compendium of released varieties in small millets

Arya L, Chauhan D, Yadav Y, Verma M (2014) Transferability of simple sequence repeat (SSR) markers developed in finger millet and pearl millet to kodo millet and barnyard millet. In: Innovative approach in stem cell research, cancer biology, and applied biotechnology, pp 60–64

Ayyangar RG, Wariar UA (1934) Anthesis and pollination in ragi, *Eleusine coracana (Gaertn.)*: The finger millet part II. Indian J Agric Sci 4:386

Bhat BV, Dayakar Rao B, Tonapi VA (eds) (2018) The story of millets, Karnataka State Department of Agriculture, Benguluru, India & ICAR-Indian Institute of Millets Research, Hyderabad, India. 58

Chandra U, Koppar MN (1990) Diversity and domestication of minor millet species in Indian subcontinent. IJPGR 3(02):47–58.

Dalsaniya D, Das A, Patel K, Macwana S, Parmar DJ, Sonagara M, Patel DA, Pali V (2023) Character association and path analysis in kodo millet (*Paspalum scrobiculatum L.*) with relation to yield and its attributing traits. Electron J of Plant Breed 14(2):730–736

Deepika C, Hariprasanna K, Das IK, Jacob J, Ronanki S, Ratnavathi CV, Bellundagi A, Sooganna D, Tonapi VA (2022) 'Kodo poisoning': Cause, science, and management. J Food Sci Technol 59(7):2517–2526. https://doi.org/10.1007/s13197-021-05141-1

Deshpande SS, Mohapatra D, Tripathi MK, Sadvatha RH (2015) Kodo millet—Nutritional value and utilization in Indian foods. J Grain Process Storage 2(2):16–23

Dhagat NK, Joshi RC, Sharma D (1971) Correlation and heritability in kodo millet (Paspalum scrobiculatum L.). Indian J Agric Sci 41:682–684.

Goron TL, Raizada MN (2015) Genetic diversity and genomic resources available for the small millet crops to accelerate a new Green Revolution. Front Plant Sci 6:157. https://doi.org/10.3389/fpls.2015.00157

Hariprasanna K (2015) Genetic improvement in kodo millet. In: Tonapi VA, Patil JV, editors. Millets: ensuring climate resilience and nutritional security. New Delhi, India: Daya Publishing House. pp. 305–317

Hariprasanna K (2016) Kodo millet, *Paspalum scrobiculatum L.* In: Patil JV (ed) Millets and sorghum: biology and genetic improvement. Wiley, pp 199–225

Hariprasanna K (2017) Kodo millet, Paspalum scrobiculatum L. In: Patil JV (ed) Millets and sorghum: biology and genetic improvement. UK: John Wiley & Sons. pp. 199–225

Johnson M, Deshpande S, Vetriventhan M, Upadhyaya HD, Wallace JG (2019) Genome-wide population structure analyses of three minor millets: Kodo millet, little millet, and proso millet. Plant Genome 12(3):190021

Kajale MD (1977) Ancient grains from excavations at Nevassa, Maharashtra. Geophytology, 7:98–106.

Knees SG, Gupta AK (2013) *Paspalum scrobiculatum*. The IUCN Red List of Threatened Species

Kushwaha H, Jillo KW, Singh VK, Kumar A, Yadav D (2015) Assessment of genetic diversity among cereals and millets based on PCR amplification using Dof (DNA binding with One Finger) transcription factor gene-specific primers. Plant Syst Evol 301:833–840

Lalit PSR, Keerti T, Sajjan KP, Koji T (2019) Genetic diversity among the landraces of kodo millet (*Paspalum scrobiculatum* L.) using ISSR markers. Int J Chem Stud 7(6):2529–2534

M'Ribu HK, Hilu KW (1996) Application of random amplified polymorphic DNA to study genetic diversity in *Paspalum scrobiculatum* L. (Kodo millet, Poaceae). Genet Resour Crop Evol 43:203–210

Nagaraja TE, Nandini C, Sujata B, Parveen GS (2023) Artificial hybridization techniques in small millets—A review. Front Plant Sci 14. https://doi.org/10.3389/fpls.2023.1112117

Nirmalakumari A, Arulselvi S, Ganapathy S, Souframanian J, Senthil N, Devan P (2007) Gamma ray induced variation for lodging resistance and its associated characters in little millet (Panicum sumatrense). Madras Agric J 94(7-12):151–155

Nirubana V, Ganesamurthy K, Ravikesavan R, Chitdeshwari T (2017) Genetic diversity studies in kodo millet (Paspalum scrobiculatum L.) germplasm accessions based on biometrical and nutritional quality traits. Int J Curr Microbiol App Sci 6(10):832–839

Nirubana V, Ravikesavan R, Ganesamurthy K (2019) Characterization and clustering of kodo millet (*Paspalum scrobiculatum* L.) genotypes based on qualitative characters. Electron J Plant Breed 10(1):101–110

Padulosi S, Mal B, Bala Ravi S et al (2009) Food security and climate change: role of plant genetic resources of minor millets. Indian J Plant Genet Resour 22:1–16.

Ragini S, Subramanian A, Nirmalakumari A, Bapu JRK (2015) Genetic diversity in kodo millet (*Paspalum scrobiculatum*) relative to culm strength and yield. J Crop Improv 29:420–431

Rangaswami Ayyangar NG, Panduranga Rao V (1934) Studies in *Paspalum scrobiculatum* L.: the Kodo millet. Madras Agric J 22(12)

Ravikesavan RG, Jeeva J, Poornima Jency M, Muthamilarasan M, Francis N (2023) Kodo Millet (*Paspalum scrobiculatum* L.). In: Neglected and underutilized crops. Academic, pp 279–304

Sao A, Singh P, Kumar P, Pradhan A (2017) Estimates of genetic parameters for yield and contributing traits in kodo millet (*Paspalum scrobiculatum* L.). Res J Agric Sci 8(1):120–122

Subramanian A, Nirmalakumari A, Veerabadhiran P (2010) Trait-based selection of superior kodo millet (*Paspalum scrobiculatum* L.) genotypes. Electron J Plant Breed 1(4):852–855

Suthediya VR, Desai SS, Pethe UB, Naik KV, Mahadik SG, Pendyala S (2021) Genetic diversity studies in kodo millet (*Paspalum scrobiculatum* L.). Electron J Plant Breed 12(4):1337–1344

Thakur J, Kanwar RR, Kumar P, Salam JL, Kar S (2018) Studies of genetic parameters for yield and yield attributing traits of kodo millet (*Paspalum scrobiculatum* L.). Int J Curr Microbiol App Sci 7(9):278–287

Upadhyaya HD, Vetriventhan M, Dwivedi SL, Pattanashetti SK, Singh SK (2016) Proso, barnyard, little, and kodo millets. In: Genetic and genomic resources for grain cereals improvement. Academic, pp 321–343

Verma SNP (1989) Research on small millets at JNKVV. College of Agriculture, JNKVV, Rewa, p 126

Vetriventhan M, Upadhyaya HD (2019) Variability for productivity and nutritional traits in germplasm of kodo millet, an underutilized nutrient-rich climate-smart crop. Crop Sci 59(3):1095–1106

Yadav Y, Lavanya GR, Pandey S, Verma M, Ram C, Arya L (2016) Neutral and functional marker-based genetic diversity in kodo millet (*Paspalum scrobiculatum* L.). Acta Physiol Plant 38:1–12

Yadav Y, Lavanya GR, Arya L, Verma M, Rajput R (2020) Phenotype-based selection in kodo millet (*Paspalum scrobiculatum* L.) to identify elite accessions. Agric Sci Dig 40(4):357–363

Yadava HS (1997) Gamma rays induced variation in kodo millet. Adv Plant Sci 9(II Suppl):165–169

Yadava HS, Jain AK (2006) Advances in kodo millet research. New Delhi, India: DIPA, Indian Council of Agricultural Research. p. 95

Youngman W, Roy SC (1923) Pollination methods amongst the lesser millets. Agric J India 18:580–583

Brown Top Millet

9

P. Pranusha, Maggiga Revanth Kumar,
Vaidurya Pratap Sahi, and Sneka Chinna Durai

Abstract

With its hardiness and underutilization, brown top millet (*Brachiaria ramosa*) is promising in alleviating global food insecurity and making for sustainable agricultural practice. This chapter deals with the diverse aspects of brown top millet, ranging from introduction and historical relevance to its domestication and adaptation to different agroecological regions. Production systems are discussed to highlight the suitability of this plant for arid and semiarid environments, low input requirement, and adaptation to marginal soils. Features of nutritional value and potential food industry use through traditional cooking are enumerated for brown top millet. Notwithstanding this potential, there are several constraints such as limited awareness on brown top millet, low productivity, poor research investment, etc. The status of germplasm collection is examined through the study, and gaps are defined on genetic diversity and conservation. Some advancements in gene mapping and genomic resources are mentioned to focus on linear approaches on unravelling genetic traits in concern to increase quantity, stress tolerance, and nutritional quality. Related literature in conjunction with research priorities will shed light on the fact that brown top millet can turn out to be a climate-resilient crop with potential. They advocate strategic investments in research, breeding programs, and awareness-building programs for unlocking benefits for sustainable food systems and livelihoods.

P. Pranusha (✉)
ICAR-National Bureau of Plant Genetic Resources, Regional Station,
Hyderabad, Telangana, India
e-mail: p.pranusha@icar.gov.in

M. R. Kumar · V. P. Sahi
Department of Genetics and Plant Breeding, Sam Higginbottom University of Agriculture Technology and Sciences, Prayagraj, Uttar Pradesh, India

S. C. Durai
Siddha Central Research Institute, Chennai, India, India

K. Tripathi et al. (eds.), *Plant Genebank Utilization for Trait Discovery in Millets*, https://doi.org/10.1007/978-981-96-4043-0_9

Keywords

Brown top millet · Plant Genetic Resources · Genebank · Landraces · Germplasm · Utilization

9.1 Introduction

A good crop should be highly adaptive to changes in the environment without sacrificing its nutritional value. To address the widespread problem of chronic malnutrition, it is important to take into consideration the nutritional quality of food to ensure the overall improvement of human health and fitness (Singh and Raghuvanshi 2012). Many agricultural products aren't used as the main resource for food because people are unaware of them; millet is one among them. They are only used to feed cattle and birds (Yang et al. 2012). Millets, a type of cereal belonging to the Poaceae grass family, are among the first crops to be grown (Garí 2002). Millets exhibit extraordinary resistance to adverse climatic conditions, displaying flexibility in the face of many soil stressors such as variations in pH, moisture content, temperature fluctuations, and soil fertility. They grow best in sandy, infertile soils and are especially adapted to even desert settings with temperatures as high as 64 °C and little to no yearly precipitation (Chivenge et al. 2015). They are resilient and advantageous to grow in wider climatic conditions due to all of these qualities. Owing to adaptable and resilient features, millet grains are becoming more and more popular as food in countries such as China, India, and several African countries. Additionally, certain developed nations are also recognizing the potential of millet grains, particularly in the production of bioethanol and biofilms (Lei and Jakobsen 2004).

Further, it is mainly preferred due to its nutritional qualities in addition to its hardiness. A multitude of vital nutrients are found in millets, such as lipids, resistant starch, oligosaccharides, and antioxidants such as flavonoids, lignans, phenolic acids, and avenanthramides. Several health advantages are linked to these substances (Edge et al. 2005). Important vitamins such as folic acid, riboflavin, thiamine, and niacin are also abundant in millet grains (Dayakar Rao et al. 2017). It is commonly acknowledged that millet is a high-energy, nutrient-dense meal that is necessary to avoid and treat diseases including obesity, diabetes, and cardiovascular disease (CVD), in addition to feeding normal people and battling malnutrition. Moreover, millet is an excellent substitute for those suffering from celiac illness, because it is devoid of gluten (FAO 2012).

They sustain farming communities' year-round livelihoods and supply food as well as fodder, better nutrition, and better health benefits. Since certain crops are harvested at particular periods of the year, such as wheat and rice, they are principally in charge of ensuring food security. On the other hand, millets, an all-season crop, provide advantages beyond only ensuring food security. Millets are a beneficial addition to agricultural systems because of their multifaceted contribution, especially in areas where traditional crop production is impacted by seasonal unpredictability and climatic problems (Shivananjappa 2018).

Millets are commonly used in Indian cuisine for a variety of purposes. They are frequently used in dosa, a kind of flatbread prepared by combining millet with other grains. Moreover, couscous, biscuits, sushi, no-yeast pizza, and roti are made with millet (Dias-Martins et al. 2018). In fact, millets are a common ingredient in many different types of beverages worldwide. "Madua" is a popular finger millet-based beverage in India that is praised for both its distinct flavor and healthful properties. Similar to this, "Oshikundu" is a traditional millet beverage from Namibia that comes in both alcoholic and nonalcoholic varieties. These illustrations show the various culinary uses of millets in various societies and locales (Kumar et al. 2018a).

9.2 The Crop

Millets vary in morphological characteristics, maturity, grain type, appearance, and plant structure. Millets can be divided into two main categories: major and minor. Pearl millet (*Pennisetum glaucum)* is mostly utilized for human consumption (Ibidapo et al. 2019). Three more important millets include finger millet (*Eleusine coracana*), proso millet (*Panicum miliaceum*), and foxtail millet (*Setaria italica*) (Yang et al. 2012). Little millet (*Panicum sumatrense*), guinea millet (*Brachiaria deflexa*), brown top millet (*Urochloa ramosa*), barnyard millet (*Echinochloa* spp.), fonio (*Digitaria exilis*), sorghum (Sorghum spp.), Job's tears (*Coix lacryma-jobi)*, and kodo millet (*Paspalum scrobiculatum*) are examples of minor millets (Adekunle et al. 2013).

Out of all the millets listed above, brown top millet was the least popular. The significance of brown top millet (BTM), also known as *Panicum ramosum* L., *Brachiaria ramosa* L. Stapf, and *Urochloa ramosa*, has just now come to light. Its huge potential to yield significantly in locations with limited resources and delicate ecological conditions has been recognized. According to Maitra (2020), BTM plays a pivotal role in bolstering smallholder farming's sustainability by achieving economic and nutritional stability. Being a crop that grows well in less fertile soils like sandy loam, brown top is one of the most economical crops. High levels of protein (11.5 g per 100 g of grains), dietary fiber (12.5 g per 100 g of grains), and minerals (4.2 g $100\ g^{-1}$ of grains) are found in these grains according to the Indian Institute of Millets Research (2019).

Brown top millet is known by numerous local names in India, which translate to "illegitimate spouse of little millet [*Panicum sumatrense*]." This refers to its tendency to grow in fields of little millet, frequently seeming like weeds (Sakamoto 1987). The growth features of brown top millet are variable, displaying the capacity to form compact or open panicles and either indehiscent or shattering spikelets. But the cultivated type, with its compact panicles and partly indehiscent spikelets, usually looks like other domestic cereals (Klmata et al. 2000). When brown top millet appears as a weed in millet fields, it's frequently considered an insurance crop. The current investigation was started since there isn't much information available about brown top millet. The specific objective of this investigation is to explore the utilization of brown top millet. It is additionally known as "korale" or "karlakki" in

Karnataka and "andakorra" in Andhra Pradesh. According to Bhat et al. (2018), brown top millet is grown mostly in the districts of Anantapur in Andhra Pradesh, Tumakuru, Chitradurga, Chikkaballapur, and Mandya in Karnataka and where it is utilized in traditional recipes. In addition, a few regions in northern India and Tamil Nadu grow it. Despite its easy storage and excellent nutritional content, its use is disregarded. Currently, in addition to India, it is grown in China, Australia, Arabia, Western Asia, and Africa (Clayton 2008).

9.2.1 History and Origin

Archaeobotanical study at Neolithic sites in southern India provides solid evidence that brown top millet (BTM) was present and possibly domesticated throughout the ancient period. It is thought that this crop first appeared in the southern Indian Deccan region before spreading to other parts of the country. These results underline the significance of BTM in the agricultural history of the subcontinent of India and provide insight into its early occurrence and potential domestication (Kingwell-Banham and Fuller 2014).

According to Fuller (2006), brown top millet was initially identified as a weed in southern India's staple crop fields as early as the third century BCE. By the second millennium BCE, it had spread to Gujarat and Tamil Nadu (Cooke et al. 2005). According to archaeological data, brown top millet grains have also been discovered at several sites across Odisha and the Gangetic plains, indicating that the grain was widely available in antiquity. Numerous archaeological findings indicate that brown top millet (BTM) was grown in antiquity; nonetheless, there is a glaring dearth of literature outlining its commercial agricultural practices. Up to the seventh century, the history of brown top millet cultivation in Maharashtra—especially in places like Paithan—provides a useful window into the agricultural practices and crop preferences of the area. Brown top millet was initially recognized as a crop in these areas because of its importance in the diet and agricultural methods of the time (Kingwell-Banham and Fuller 2014). The origins of brown top millet (*Urochloa ramosa* L.) can be traced back to southern Asia, namely, to southern India (K Santhi Sirisha et al. 2022). Brown top millet, however, progressively lost popularity as farming methods changed and other millets with greater production and perhaps better adaptability were developed. Although it was once a staple crop in many areas of the world, especially southern India, there isn't much information in academic literature about its commercial cultivation practices, market trading, or economic importance (Harvey 2007).

9.2.2 Taxonomy

The taxonomy of brown top millet (*Urochloa ramosa*) is as follows:

Kingdom: Plantae.

Phylum: Angiosperms.
Class: Monocots.
Order: Poales.
Family: Poaceae (grasses).
Genus: Urochloa.
Species: *Urochloa ramosa.*

9.2.3 Domestication

Brown top millet was probably domesticated in South India, in the Deccan region, and then spread to other parts of India in prehistoric times. Researchers studying the archaeobotany of Neolithic sites in South India have discovered burned grains that have been classified as "*Brachiaria ramosa* type," which suggests that brown top millet was there. High ubiquity and relative frequency of this crop are seen at these areas. Although there is little archaeobotanical evidence from Mesolithic sites in the early Neolithic or preceramic periods to date the exact time of domestication, what is known suggests that brown top millet and other native South Indian crops such as *Vigna radiata, Macrotyloma uniflorum*, and *Setaria verticillata* emerged from wild populations around the start of the third millennium BCE (Fuller 2006).

Brown top millet is well known for its remarkable ability to withstand drought and adapt to semiarid environments. It is ideally suited for cultivation in arid or semiarid climates since it grows well in regions with restricted water availability. Furthermore, brown top millet grows well at moderate elevations, usually in the range of 2000–2500 m above sea level. It can grow in areas where the yearly precipitation varies from 75 to 150 cm (Roecklein and Leung 1987). Distribution map of archaeological finds of *Brachiaria ramosa* is given in Fig. 9.1.

Brown top millet with a shorter growth cycle than many other millet kinds matures in around 90 days. It is mostly cultivated as a monoculture instead of being combined with other crops in mixed-field systems. Harvesting is frequently done early in the morning, while the crop still has dew on it to reduce grain loss due to panicle shattering (Kingwell-Banham and Fuller 2014). Brown top millet has less shattering than wild millet, but some shattering still occurs. Cutting the crop at the base, winnowing, dehusking, and polishing are the steps in the harvesting process. Because it is semi-shattering, the grains may break loose when drying, minimizing the amount of threshing that is required. But like other millet cultivars, dehusking is usually necessary (Klmata et al. 2000).

9.2.4 Production

China and Nigeria have been the top producers of millet, after India (Perumal, 2018). Between 2016 and 2020, India's millet production increased from 10.28 million tons to 12.49 million tons. The proportion of millet produced worldwide that came from India increased from 37.34% in 2016 to about 41% in 2020. India

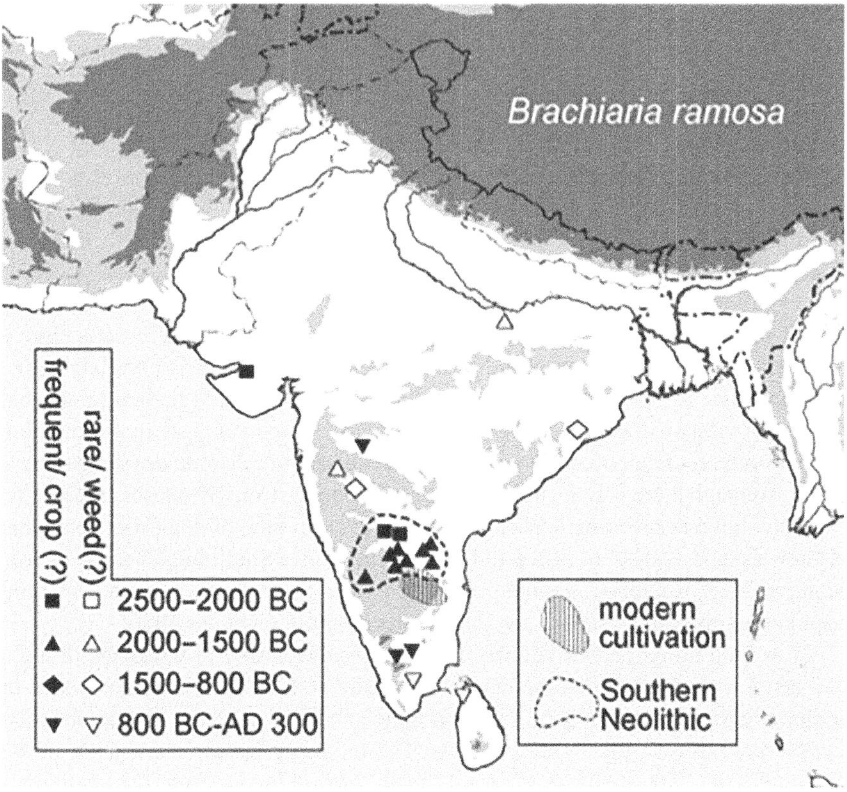

Fig. 9.1 Distribution map of archaeological finds of *Brachiaria ramosa* in relation to the southern Indian Neolithic and modern cultivation of brown top. (Kingwell-Banham and Fuller 2014)

contributed 80.46% of Asia's millet production in 2016 and 79.36% in 2020, which is a significant amount (FAOSTAT 2022). According to Niharika et al. (2020), the IC 617961 variety produces the highest grain output of 20.38 g per plant, whereas the average BTM grain production is 12.13 g per plant. The development and yield characteristics of brown top millet (*Brachiaria ramosa*) on the red sandy soils of Dharwad, Karnataka, were examined. According to the findings, grain yield was only 3.95 q/ha when fertilizer was not used. However, using a combination of organic manure and inorganic fertilizer increased grain output to 7.38 q/ha (Ashoka and Rajkumar 2020).

Brown top millet is being replaced by alternative, more productive millets, which is causing its use to diminish. This transformation was probably greatly influenced by millets like foxtail millet (*Setaria italica*), *Eleusine coracana* (finger millet), and *Sorghum bicolor* (sorghum). Brown top millet's cultivation gradually decreased as a result of these substitute millets' greater yields or other advantageous qualities (Kingwell-Banham and Fuller 2014). Brown top millet continues to be cultivated as a relict cultivar, which means it is far less widely grown than it formerly was. Brown top millet still has some cultural and ceremonial significance, despite its declining agricultural relevance. Its traditional worth and cultural history are preserved within certain communities through its continued use in rites and celebrations.

9.2.5 Consumption and Use

As a versatile grain, brown top millet is utilized as a raw material for several industries as well as for human consumption, animal feed and bird feed (Sheahan 2014). In regions such as Africa and India, little millets are consumed whole or processed into flour, primarily for culinary purposes. In contrast to other nations, tiny millets are widely used as bird feed in Japan (Asharani et al. 2010). Value addition for human consumption can be achieved in large quantities by making goods like biscuits or *idli* mix. According to Roopa (2015), brown top millet can be used by combining it in a proportionate amount with other common cereal flours like wheat and rice. A key component in the creation of value-added ready-to-eat (RTE) and ready-to-cook (RTC) products is brown top millet (BTM). These include laddu, cookies, and dosa mix. BTM has been combined with other ingredients, such as drumstick, coconut, and chocolate, for making several products. The products along with their salient features and nutritional qualities are given in Table 9.1.

At least nine regional specialties, such as anna, dosai, kheer, roti, nippattu, chakkulli, haralu, nuchina mudda, kodubale, and kadabuappe, are made with the grains of southern India. Brown top millet can be polished and boiled to make gruel (kheer), and it is frequently used to make flour for flatbreads like roti or dosa. While the grains are kept for human use, the leftover straw and chaff from grain harvest are frequently used as animal feed. Rice is frequently surpassed by brown top millet, which is appreciated for its flavor. Furthermore, a few of these meal preparations have ritualistic applications and religious importance, which supports the continual growth of brown top millet (Klmata et al. 2000).

9.2.6 Constraints

Brown top millet is currently plagued by serious diseases like leaf blight (brown spot), leaf blast, and rust. According to Misra and Prakash (1972), *Bipolaris setariae*, which causes leaf spots, was the source of the first disease incidence in brown top millet to be documented. Numerous genotypes of brown top millet showed signs of leaf blight caused by the fungus *B. setariae* in field trials. It's unclear how much grain yield in brown top millet has been lost specifically as a result of leaf blight. Year 2019 saw the first reports of leaf blasts in brown top millet. While certain millets are well known to humans and easily found in local markets, others—like brown top millet, or BTM—are less well known and more difficult to find for the average person. Despite having a high nutritional value, BTM has not been fully investigated and there is little information based on scientific research at this time. For aspiring and seasoned researchers alike, this dearth of information poses serious

Table 9.1 Important food products with additional value made from brown top millet (Dayakar et al. 2021)

Product name	Salient characteristics	Protein (gm/100gm)	Fiber (gm/100gm)	Product shell impulse
BTM laddu	RTE convenient snack. Good snacks for children and adolescents	6.76	–	3
BTM breakfast Cookies	RTE. Gluten-free, rich in low trans-fat, and prepared from jaggery	8.72	10	5
BTM choco chip cookies	RTE. Gluten-free, rich in low trans-fat, and prepared from jaggery	7.71	9.5	5
BTM coconut cookies	RTE. Gluten-free, rich in low trans-fat, and prepared from jaggery	10.63	8.51	5
BTM butter cookies	RTE. Gluten-free, rich in low trans-fat, and prepared from jaggery	11	10.56	5
BTM dosa mix	Rtc. Good source of protein, fiber, and zinc	13.52	2.66	4
Anna	Eaten as boiled	–	–	–
Kheer	Prepared by mixing with sugar or jaggery. Good for nursing mothers	–	–	–
Nuchina Mudda	Semisolid porridge. Prepared by broken grains of BTM and finger millet flour	–	–	–
Roti	Unleavened bread, prepared by baking	–	–	–
Dosai	Unleavened thin pancake, prepared by frying fried Bengal gram and groundnut	–	–	–
Nippattu	RTE snack that is deep-fried. Its dough was made by combining groundnuts and fried Bengal gram.	–	–	–
Chakkulli	RTE deep-fried snack, made by combining flour of BTM with black gram flour or maida	–	–	–
Haralu	RTE deep-fried snack	–	–	–

obstacles to fully realizing the field's potential. Efforts to fully comprehend the benefits of thc grain, optimize its cultivation, and promote its usage on a larger scale are hampered by the lack of comprehensive studies and the grain's limited representation in conventional agriculture and food markets.

9.3 Status of Germplasm Collection

Germplasm resources are the foundation of any crop development initiative (Priyanka et al. 2021). But not much has been accomplished to gather and conserve brown top millet germplasm. Regarding the creation of genetic and genomic resources as well as breeding for increased production, brown top millet deserves more scientific attention. Crop was given a low research priority and scientists moved their focus to cash crops, which resulted in a major loss of diversified genetic resources (Vetriventhan et al. 2020).

The Brazilian Agricultural Research Corporation (EMBRAPA) maintains the largest collection of brown top millet germplasm at the International Center for Tropical Agriculture (CIAT) in Cali, Colombia (Keller-Grein et al. 1996). Brown top millet conservation and collecting in India started in 2018, when the plant was added to the All-India Coordinated Research Project (AICRP) system (Rahul et al. 2024). As of now, the AICRP project has collected around 170 brown top millet accessions, which are kept in several genebanks throughout India. To name a few, the National Bureau of Plant Genetic Resources (NBPGR), New Delhi, has 30 accessions; the Indian Institute of Millet Research (IIMR), Hyderabad, has 20; and the University of Agricultural Sciences (UAS), Bengaluru, has 180.

9.4 Germplasm Characterization and Evaluation

Germplasm characterization involves documenting distinctly identifiable heritable traits, distinguishing it from preliminary evaluation, which focuses on recording a select few agronomic traits vital for crop improvement. Sample picture of brown top millet spike and seeds are given in Fig. 9.2. The optimal exploitation of brown top millet is impeded by the scarcity of germplasm lines and inadequate knowledge regarding its genetic variability. Furthermore, genetic modification may provide a solution to the primary constraint of this crop's low yields. Few efforts have been undertaken to characterize and evaluate available brown top millet germplasm in sporadic manner. Shanthi Priya et al. (2022) evaluated 25 brown top millet germplasm lines using 8 quantitative parameters and have reported their average performance (Table 9.2).

The harvest index and fodder yield should be the main selection criteria for brown top millet to increase yield and yield components (Priya et al. 2022). Recent studies have demonstrated a notable growth in the amount of grain produced per plant along with significant levels of iron, zinc, and crude fiber contents. These distinct genotypes offer potential avenues for individual utilization in enhancing specific quality traits within brown top millet (Niharika et al. 2020).

Fig. 9.2 Panicle and seed of brown top millet crop

9.5 Genetic and Genomic Resources

Brown top millet is a type of cereal that is classified as a little millet because of its small seeds. Currently, genomic sequences for several related millets are available, such as fonio millet (Abrouk et al. 2020), tef (Paterson et al. 2009; Cannarozzi et al. 2014), Job's tear (Kang et al. 2020), finger millet (Hittalmani et al. 2017; Hatakeyama et al. 2018; Devos et al. 2023), and barnyard millet (Guo et al. 2017). Draft genome assemblies for the other millets are available in the NCBI database, while fully annotated genome sequences for foxtail millet, sorghum, and finger millet are available in the Phytozome database. However, no annotated or draft genome sequences for brown top millet are currently available. Thus, molecular markers from related millets' genome sequences—namely, sorghum, finger millet, and foxtail millet—can be used by brown top millet researchers for genetic diversity and gene mapping investigations. Through genetic diversity analysis in brown top millet, these molecular markers could be used to choose suitable parental lines for breeding operations. Furthermore, to improve the growth and productivity of brown top millet, it is essential to discover genes linked to biotic and abiotic stressors using the annotated genome sequences of these related millet. Researchers can better understand the functions of these stress-related genes in brown top millet by discovering and analyzing them, which could ultimately result in the creation of more resilient and productive cultivars.

In both biotic and abiotic stress situations, molecular markers are typically helpful tools for identifying certain plant traits. The absence of genomic resources and ESTs (Expressed Sequence Tags) has impeded the development of molecular markers for brown top millet. But for other millets, a multitude of QTL (quantitative trait

Table 9.2 Mean performance of 25 brown top millet germplasm lines for 8 quantitative characters

Accession	Days to 50% flowering	Days to maturity	Plant height (cm)	Number of productive tillers/ plant	Panicle length (cm)	Harvest index (%)	Fodder yield (t/ha)	Grain Yield (q/ha)
IC613554	61.0	91.5	144.9	10.4	16.4	9.1	3.9	12.3
IC613555	58.0	89.0	127.1	7.8	15.1	11.0	4.3	16.4
IC613556	59.0	90.0	139.8	9.6	16.4	9.7	3.9	13.5
IC613557	56.0	86.5	136.7	10.4	18.7	6.9	4.1	9.2
IC613558	54.0	84.5	138.6	11.3	17.4	11.9	5.0	20.4
IC613559	56.0	87.0	147.2	13.9	20.6	10.5	4.8	16.8
IC613560	54.5	85.5	147.7	7.9	18.6	11.4	4.6	18.1
IC613561	56.0	88.5	140.4	10.8	18.7	9.3	5.0	15.7
IC613562	53.0	85.5	148.2	10.0	19.6	5.5	5.5	9.4
IC617952	54.0	84.5	157.6	11.1	18.2	10.7	4.7	16.6
IC613546	52.5	84.0	149.4	13.9	18.2	10.1	4.8	15.8
IC617953	56.5	87.5	146.2	9.4	18.3	11.2	4.3	16.2
IC617954	59.5	90.0	141.5	11.1	16.6	8.9	4.1	12.3
IC617955	56.5	88.0	154.9	9.5	20.9	9.2	5.6	17.1
IC617956	56.0	88.0	130.2	7.6	16.6	7.9	6.2	16.0
IC617957	53.5	84.5	139.5	8.8	17.4	8.5	4.4	12.3
IC617958	55.0	87.5	143.5	10.5	18.5	8.5	6.1	17.3
IC617960	54.5	85.5	133.2	9.8	20.0	8.6	4.2	12.0
IC617961	53.5	83.5	142.5	13.8	17.2	8.8	4.4	13.1
IC617962	51.0	81.5	139.4	9.6	18.3	7.5	5.7	14.1
IC613548	53.5	85.5	140.4	11.9	13.1	10.9	3.8	14.3
IC613549	53.0	84.0	132.7	13.8	16.3	9.8	3.6	11.9
IC613550	56.5	89.0	152.5	9.1	20.0	9.2	5.0	15.2
IC613552	55.0	85.5	143.0	11.1	17.9	9.9	4.6	15.4
IC613553	54.0	85.0	147.2	11.4	17.8	8.3	5.1	14.3

loci) have been discovered using linkage and association mapping utilizing a variety of molecular markers, such as simple sequence repeats (SSR), single nucleotide polymorphisms (SNP), and diversity arrays technology (DArT). A unique QTL for rust disease was discovered in pearl millet using about 220 DArT and 70 SSR markers (Ambawat et al. 2016), and 5 QTL for downy mildew disease resistance were found using 88 SSR markers (Chelpuri et al. 2019). Finger millet research used 87 SSR markers to identify 8 QTL for variables linked to roots and shoots, and over 30 QTL for various physiological and biochemical traits under varying phosphate circumstances (Maharajan et al. 2023) (Ramakrishnan et al. 2017). Additionally, more than 70 SSR markers were employed to identify 11 QTLs for characteristics associated with drought stress (David et al. 2021). Moreover, DArT and SSR markers were employed to identify multiple QTLs for the iron and zinc content of pearl millet (Kumar et al. 2018b; Singhal et al. 2021). It has been successful in identifying

Table 9.3 Details of gene sequences available for brown top millet in the NCBI database

Genes	Protein ID	Role	Length of protein (bp)	Length of CDS (bp)
MatK	QBJ27163	It is proposed to aid the excision of seven different chloroplast group IIA introns that reside within precursor RNAs for essential elements of chloroplast function	249	747
	AZC11690		257	772
	AZC11689		269	809
	ATZ75369		401	1824
RBC	AZC11386	Involved in photosynthetic carbon fixation as it catalyzes the conversion of atmospheric CO2 into organic compounds	416	1248
			451	1353
NADH-PO	ATZ74998	Essential for plant growth and development during stress (especially oxidative stress) periods	402	1201

QTL linked to a variety of agronomic parameters across millets by using both linkage and association mapping (Maharajan et al. 2024).

Regarding brown top millet, the NCBI does not have species-specific genome assembly or EST sequences. However, an attempt has been to upload the protein and nucleotide sequences of a few millet genes to the NCBI database (Table 9.3) by various researchers as reported by Bhavani et al. (2024). Research on CDS and protein sequences is limited due to the paucity of accessible data compared to other millets.

9.6 Pre-breeding

Pre-breeding in brown top millet may involve utilizing genetic diversity from wild or unadapted germplasm to improve cultivated varieties for traits like drought tolerance, disease resistance, yield enhancement, and nutritional quality. Key steps include evaluating genetic resources, hybridization, and selecting promising genotypes for breeding programs. These efforts are essential to developing resilient and high-performing brown top millet varieties, especially for regions prone to challenging climatic conditions. However, pre-breeding approach is yet to be used in the brown top millet due to limited information about value rich or donor germplasm, difficult pollination mechanism. However, mutation breeding approach can be employed for creating variability.

9.7 Future Prospects

In a short amount of time, brown top millet can yield both good fodder and nutritious grains. It is notable for its adaptability, health advantages, and historical value. It thrives in a variety of agricultural systems, including rainfed and irrigated ones. The genetic variety of brown top millet restricts its successful use in regular breeding operations and its effective use in crop development programs, and there is a relatively limited amount of germplasm and information available. To find trait-specific resources, genes, and alleles that can be used in brown top millet breeding programs, it is crucial to prioritize the collection of germplasm. Selection of landrace offers resources to improve yield traits linked to domestication in brown top millet, characteristics like seed-type non-shattering and lodging resistance. Commercially grown brown top millet types have only been created thus far using conventional breeding techniques. The male sterility system, the utilization of heterotic pools, biotechnological tools, innovative mechanization techniques, and sound crop management methods can all help break the yield barrier in brown top millet. There are no genomic and transcriptomic resources for brown top millet to develop single nucleotide polymorphisms (SNP) and insertion/deletion (InDels) for breeding studies. It is important to understand better crossing techniques to produce true hybrids and to develop improved brown top millet genotypes. Several works need to be initiated urgently to conserve and enhance this crop. To speed up crop improvement, genomics-assisted breeding will make it easier to find new alleles and genes with improved agronomic performance and resilience to biotic and abiotic challenges. A full collection of millet germplasm may be molecularly characterized, thanks to next-generation sequencing techniques. In brown top millet, where genome decoding has not yet been completed, these methods can be applied. Despite the great nutritional potential of brown top millet, future studies are needed to fully comprehend the nutritional and non-nutritional aspects of the various genotypes of this grain.

References

Abrouk M, Ahmed HI, Cubry P, Šimoníková D, Cauet S, Pailles Y, Bettgenhaeuser J, Gapa L, Scarcelli N, Couderc M, Zekraoui L (2020) Fonio millet genome unlocks African orphan crop diversity for agriculture in a changing climate. Nat Commun 11(1):4488

Adekunle AA, Ellis-Jones J, Ajibefun I, Nyikal RA, Bangali S, Fatunbi AO, Angé A (2013) Agricultural innovation in sub-Saharan Africa: experiences from multiple stakeholder approaches. Forum for Agricultural Research in Africa (FARA), Accra, Ghana

Ambawat S, Senthilvel S, Hash CT, Nepolean T, Rajaram V, Eshwar K, Sharma R, Thakur RP, Rao VP, Yadav RC, Srivastava RK (2016) QTL mapping of pearl millet rust resistance using an integrated DArT-and SSR-based linkage map. Euphytica 209:461–476

Asharani VT, Jayadeep A, Malleshi NG (2010) Natural antioxidants in edible flours of selected small millets. Int J Food Prop 13(1):41–50

Ashoka P, Rajkumar GR (2020) Response of nutrient management on growth and yield attributes of Brown top millet [*Brachiaria ramose*] in red sandy soil. J Pharmacogn Phytochem 9(6):1298–1301

Bhat S, Prabhu CG, Nandini C, Prabhakar TV (2018) Browntop millet-a review. Agri. Res. & Tech.: Open Access J, 14(5), p.555937.
Bhavani P, Nandini C, Maharajan T, Ningaraju TM, Nandini B, Parveen SG et al (2024) Browntop millet: an overview of breeding, genetic, and genomic resources development for crop improvement. Planta 260(1):10
Cannarozzi G, Plaza-Wüthrich S, Esfeld K, Larti S, Wilson YS, Girma D, de Castro E, Chanyalew S, Blösch R, Farinelli L, Lyons E (2014) Genome and transcriptome sequencing identifies breeding targets in the orphan crop tef (Eragrostis tef). BMC Genomics 15:1–21
Chelpuri D, Sharma R, Durga KK, Katiyar P, Mahendrakar MD, Singh RB, Yadav RS, Gupta R, Srivastava RK (2019) Mapping quantitative trait loci (QTLs) associated with resistance to major pathotype-isolates of pearl millet downy mildew pathogen. Eur J Plant Pathol 154:983–994
Chivenge P, Mabhaudhi T, Modi AT, Mafongoya P (2015) The potential role of neglected and underutilised crop species as future crops under water scarce conditions in Sub-Saharan Africa. Int J Environ Res Public Health 12(6):5685–5711
Cooke M, Fuller DQ, Rajan K (2005) Early historic agriculture in southern Tamil Nadu: Archaeobotanical research at Mangudi, Kodumanal and Perur. Linden Soft
Clayton WD (2008) GrassBase-the online world grass Flora. http://www.kew.org/data/grasses-db.html
David RHA, Ramakrishnan M, Maharajan T, BarathiKannan K, Babu GA, Daniel MA, Agastian P, Caesar SA, Ignacimuthu S (2021) Mining QTL and genes for root traits and biochemical parameters under vegetative drought in South Indian genotypes of finger millet (*Eleusine coracana* (L.) Gaertn) by association mapping and in silico comparative genomics. Biocatal Agric Biotechnol 32:101935
Dayakar Rao B, Bhaskarachary K, Arlene Christina GD, Sudha Devi G, Vilas AT, Tonapi A (2017) Nutritional and health benefits of millets. ICAR_Indian Institute of Millets Research (IIMR) Rajendranagar, Hyderabad
Dayakar RB, Bhandari R, Tonapi VA (2021) White Papers on Millets – A Policy Note on Mainstreaming Millets for Nutrition Security. Rajendranagar, Hyderabad: ICAR-Indian Institute of Millets research (IIMR).
Devos KM, Qi P, Bahri BA, Gimode DM, Jenike K, Manthi SJ, Lule D, Lux T, Martinez-Bello L, Pendergast TH IV, Plott C (2023) Genome analyses reveal population structure and a purple stigma color gene candidate in finger millet. Nat Commun 14(1):3694
Dias-Martins AM, Pessanha KLF, Pacheco S, Rodrigues JAS, Carvalho CWP (2018) Potential use of pearl millet (*Pennisetum glaucum* (L.) R. Br.) in Brazil: food security, processing, health benefits and nutritional products. Food Res Int 109:175–186
Edge MS, Jones JM, Marquart L (2005) A new life for whole grains. J Am Diet Assoc 105(12):1856–1860
FAO (Food and Agriculture Organization) (2012) Economic and social department: the statistical division. Statistics Division. Available from FAO. Posted Sept 29, 2012
FAOSTAT (2022) Food and Agricultural Organization of United Nations
Fuller DQ (2006) Agricultural origins and frontiers in South Asia: a working synthesis. J World Prehist 20:1–86
Garí JA (2002) Review of the African millet diversity. In: International workshop on fonio, food security and livelihood among the rural poor in West Africa. IPGRI/IFAD, Bamako, Mali, pp 19–22
Guo L, Qiu J, Ye C, Jin G, Mao L, Zhang H, Yang X, Peng Q, Wang Y, Jia L, Lin Z (2017) Echinochloa crus-galli genome analysis provides insight into its adaptation and invasiveness as a weed. Nat Commun 8(1):1031
Harvey EL (2007) Early agricultural communities in Northern and Eastern India: an archaeobotanical investigation. University of London, University College London (United Kingdom)
Hatakeyama M, Aluri S, Balachadran MT, Sivarajan SR, Patrignani A, Grüter S, Poveda L, Shimizu-Inatsugi R, Baeten J, Francoijs KJ, Nataraja KN (2018) Multiple hybrid de novo genome assembly of finger millet, an orphan allotetraploid crop. DNA Res 25(1):39–47

Hittalmani S, Mahesh HB, Shirke MD, Biradar H, Uday G, Aruna YR, Lohithaswa HC, Mohanrao A (2017) Genome and transcriptome sequence of finger millet (Eleusine coracana (L.) Gaertn.) provides insights into drought tolerance and nutraceutical properties. BMC Genomics 18:1–16

Ibidapo O, Henshaw F, Shittu T, Afolabi W (2019) Bioactive components of malted millet (*Pennisetum glaucum*), Soy Residue "okara" and wheat flour and their antioxidant properties. Int J Food Prop 22(1):1886–1898

Kang SH, Kim B, Choi BS, Lee HO, Kim NH, Lee SJ, Kim HS, Shin MJ, Kim HW, Nam K, Kang KD (2020) Genome assembly and annotation of soft-shelled adlay (*Coix lacryma*-jobi Variety ma-yuen), a cereal and medicinal crop in the Poaceae family. Front Plant Sci 11:630

Keller-Grein G, Maass BL, Hanson J (1996) Natural variation in Brachiaria and existing germoplasma collections. p.17–42. In: Brachiaria: biology, agronomy, and improvement. Miles, J. W.; Maass, B. L.; and Valle, C. B., eds. CIAT/Embrapa, Colombia

Kingwell-Banham E, Fuller DQ (2014) Brown top millet: origins and development. In: Encyclopaedia of Global Archaeology. Springer, New York, pp 1021–1024

Klmata M, Ashok EG, Seetharam A (2000) Domestication, cultivation and utilization of two small millets, *Brachiaria ramosa* and *Setaria glauca* (Poaceae), in South India. Econ Bot 54:217–227

Kumar A, Tomer V, Kaur A, Kumar V, Gupta K (2018a) Millets: a solution to agrarian and nutritional challenges. Agric Food Secur 7(1):1–15

Kumar S, Hash CT, Nepolean T, Mahendrakar MD, Satyavathi CT, Singh G, Rathore A, Yadav RS, Gupta R, Srivastava RK (2018b) Mapping grain iron and zinc content quantitative trait loci in an iniadi-derived immortal population of pearl millet. Genes 9(5):248

Lei V, Jakobsen M (2004) Microbiological characterization and probiotic potential of koko and koko sour water, African spontaneously fermented millet porridge and drink. J Appl Microbiol 96(2):384–397

Maharajan T, Ajeesh Krishna TP, Rakkammal K, Ramakrishnan M, Ceasar SA, Ramesh M, Ignacimuthu S (2023) Identification of QTL associated with agro-morphological and phosphorus content traits in finger millet under differential phosphorus supply via linkage mapping. Agriculture 13(2):262

Maharajan T, Krishna TPA, Krishnakumar NM, Vetriventhan M, Kudapa H, Ceasar SA (2024) Role of genome sequences of major and minor millets in strengthening food and nutritional security for future generations. Agriculture 14(5):670

Maitra S (2020) Potential horizon of brown-top millet cultivation in drylands: a review. Crop Res 55:57–63. https://doi.org/10.31830/2454-1761.2020.012

Misra AP, Prakash O (1972) Helminthosporium species occurring on graminaceous hosts in India

Niharika V, Rao BG, Tushara M, Rao VS (2020) Studies on performance of brown top millet indigenous collections for grain yield and nutritional traits. J Pharmacogn Phytochem 9(5):2636–2638

Paterson AH, Bowers JE, Bruggmann R, Dubchak I, Grimwood J, Gundlach H, Haberer G, Hellsten U, Mitros T, Poliakov A, Schmutz J (2009) The Sorghum bicolor genome and the diversification of grasses. Nature 457(7229):551–556

Perumal A (2018) Production and consumption of minor millets in India- A structural break analysis. Ann Agriculture Res. 38:1–8.

Priya MS, Madhavilatha L, Kumar MH (2022) Genetic divergence, trait association and path analysis studies in brown top millet germplasm. Electron J Plant Breeding 13(3):1156–1161

Priyanka V, Kumar R, Dhaliwal I, Kaushik P (2021) Germplasm conservation: instrumental in agricultural biodiversity—a review. Sustain For 13(12):6743

Rahul GK, Bhavani P, Shyamalamma S, Nandini C (2024) Assessment of genetic diversity in germplasm collections of Brown top millet (*Brachiaria ramosa* (L.) Stapf.) using morphological traits. Electron J Plant Breeding 15(1):217–225

Ramakrishnan M, Ceasar SA, Vinod KK, Duraipandiyan V, Ajeesh Krishna TP, Upadhyaya HD, Al-Dhabi NA, Ignacimuthu S (2017) Identification of putative QTLs for seedling stage phosphorus starvation response in finger millet (*Eleusine coracana* L. Gaertn.) by association mapping and cross species synteny analysis. PLoS One 12(8):e0183261

Roecklein JC, Leung PS (eds) (1987) A profile of economic plants. Transaction Publishers

Roopa OM (2015) Nutritional analysis and development of value-added products from brown top millet. (Doctoral dissertation. University of Agricultural Sciences GKVK, Bengaluru

Sakamoto S (1987) A preliminary report of the studies on millet cultivation and its agro-pastoral culture complex in the Indian subcontinent. Kyoto University

Sheahan CM (2014) Plant guide for brown top millet (*Urochloa ramosa*). USDA-Natural Resources Conservation Service, Cape May Plant Materials Center, Cape May

Shivananjappa M (2018) Meats are malady and millets are magical. J Nutr Food Technol 1(1):14–15

Singh P, Raghuvanshi RS (2012) Finger millet for food and nutritional security. Afr J Food Sci 6(4):77–84

Singhal T, Satyavathi CT, Singh SP, Kumar A, Sankar SM, Bhardwaj C, Mallik M, Bhat J, Anuradha N, Singh N (2021) Multi-environment quantitative trait loci mapping for grain iron and zinc content using bi-parental recombinant inbred line mapping population in pearl millet. Front Plant Sci 12:659789

Sirisha KS, Tvhymavathi S, Rani RN (2022) Nutritional properties of brown top millet (*Brachiaria ramosa*). Pharm Innov 11:729–733

Vetriventhan M, Azevedo VC, Upadhyaya HD, Nirmalakumari A, Kane-Potaka J, Anitha S, Ceasar SA, Muthamilarasan M, Bhat BV, Hariprasanna K, Bellundagi A (2020) Genetic and genomic resources, and breeding for accelerating improvement of small millets: current status and future interventions. Nucleus 63:217–239

Yang X, Wan Z, Perry L, Lu H, Wang Q, Zhao C, Li J, Xie F, Yu J, Cui T, Wang T (2012) Early millet use in northern China. Proc Natl Acad Sci 109(10):3726–3730

Teff and Fonio

10

B. Amasiddha, Sangappa, A. V. Umakanth,
C. Tara Satyavathi, and G. P. Singh

Abstract

Teff and fonio, classified as "Neglected and Underutilized Species (NUS)," are also known as orphan crops. These crops have great potential to overcome food security and nutritional challenges due to their adaptability to extreme environmental conditions and rich nutrient profiles. They are probably the most ancient and culturally important cereal crops grown in Africa. The grains of teff are nutritionally rich with high dietary fiber and iron and gluten free. Fonio is rich in quality minerals, vitamins, fiber, and sulfur-containing amino acids like cysteine and methionine. Major constraints of teff and fonio production are use of less yielding varieties along with unimproved pre- and post-traditional practices, susceptibility to lodging, and socioeconomic constraints such as lack of awareness and the high cost and unavailability of seeds. This chapter discusses about status of germplasm diversity and collection of germplasm at both national and global level followed by conservation efforts in both the crops. The Institute of Biodiversity Conservation and Ethiopian Biodiversity Institute are majorly maintaining the vast germplasm collection of teff crop. Worldwide, 7235 teff accessions were conserved in various genebanks as ex situ conservation. The major collection of fonio germplasm is conserved at Institut de Recherche pour le Développement (IRD) in Montpellier and Genebank of Niaouli Agricultural Research Center of INRAB in Benin. In India, ICAR-NBPGR, New Delhi, has collected, characterized, and maintained 288 accessions of teff and 17 accessions

B. Amasiddha (✉) · Sangappa · A. V. Umakanth · C. Tara Satyavathi
ICAR-Indian Institute of Millets Research, Hyderabad, India
e-mail: amasiddha@millets.res.in

G. P. Singh
ICAR-National Bureau of Plant Genetic Resources, New Delhi, India

K. Tripathi et al. (eds.), *Plant Genebank Utilization for Trait Discovery in Millets*, https://doi.org/10.1007/978-981-96-4043-0_10

of fonio. The collection, augmentation, and conservation works are in progress considering their importance in nutritional security and climate resilience.

Keywords

Teff · Fonio · Plant genetic resources · Germplasm · Breeding · Utilization · Pre-breeding · Trait-specific germplasm

10.1 Introduction

Teff (*Eragrostis tef*) and fonio (*Digitaria exilis* and *Digitaria iburua*) are probably the most ancient and culturally important cereal crops grown in Africa. They have played a vital role for millions of people with their livelihoods and food security, especially in areas where agro-climate conditions make other cereals less feasible to grow. Even though these are minor cereals, they play a very important role in cultures that have maintained them under cultivation for millennia.

Teff (*Eragrostis tef* (Zucc.) Trotter) belongs to one of the biggest grass family *Poaceae* in the genera *Eragrostis* of the tribe *Eragrostideae* which is believed to be domesticated around 3000 years ago in Ethiopia (Vavilov 1951; Ketema 1997). It is also called by other common names such as Williams love grass, annual bunch grass, taf, and *Eragrostis tef*. The teff can be grown at the mean sea level of 3000 m and the best performance of the crop is observed at the MSL of 1100–2950 m height. Teff is a major food crop cultivated in Ethiopia, covering over a million hectares of land, and is adapted to waterlogged, heavy, and clayey soils. The area under teff cultivation is increasing in Ethiopia due to local and foreign demand for grains and various products. The grains of teff are nutritionally rich, with a high dietary fiber content, which plays a significant role in controlling diabetes. They are also rich in iron, which helps prevent anemia, and are gluten-free, making them beneficial for patients with gluten allergies or celiac disease. The grains are ground into flour and baked into pancake-like soft and sour bread, *injera*, which is the major dish and a national staple in Ethiopia (Demelash 2017). Teff is also consumed in the form of fermented or unfermented bread, known as *kita* or *anebabero*, native beer (*talla*), and more alcoholic cottage liquor (*katikalla*) (Assefa et al. 2015). Due to the attractive nutritional and functional properties of the grains, the crop is gaining importance worldwide. Its introduction and cultivation have begun in countries such as the United States, Canada, Australia, Switzerland, and the Netherlands (Gebru et al. 2020).

Teff offers several advantages over other major crops, including (1) broad adaptation to altitudes ranging from below sea level to 3000 meters above sea level and to varied agroecological and edaphic conditions; (2) reasonable tolerance to water stagnation and moisture stress; (3) suitability in various cropping patterns and crop rotation practices; (4) fewer pests and diseases compared to major crops; (5) relatively fewer storage pests and diseases; (6) *injera*, the fermented Ethiopian pancake, has good water-holding capacity, a long shelf life, a unique (slightly sour but pleasant) flavor, pliability, and a smooth, glossy texture; (7) the straw is used as dry

Table 10.1 Nutritional composition (%) of teff grains compared with other major cereals

Nutritional item	Teff	Wheat	Rice	Maize	Sorghum	Barley
Protein	11.0	11.0	9.7	9.4	8.6	8.5
Fat	2.6	1.9	1.6	4.4	3.8	1.5
Fiber	3.5	1.9	5.8	2.2	1.9	4.5
Carbohydrates	73.0	69.3	64.7	69.2	71.3	67.3
Mineral ash	3.0	1.7	5.0	1.3	2.4	2.6

Source: Assefa and Chanyalew (2018)

fodder and as a binder for plastering the walls of houses in villages; (8) the milling percentage of teff grains is higher (99%) compared to wheat (60–80%), enhancing consumer preference; and (9) teff grains are rich in minerals, vitamins, calcium, magnesium, iron, and zinc (Assefa and Chanyalew 2018) (Table 10.1).

Fonio (*Digitaria exilis* Stapf) belongs to the *Poaceae* family, within the *Panicoideae* subfamily, the *Panicaceae* tribe, and the *Digitaria* genus. This genus includes up to 325 annual and perennial species native to tropical and subtropical regions (Clayton and Renvoize 1986; Henrard 1950). Fonio is an ancient grain from West Africa that dates back to around 5000 BC and derives its name from the word "foño" (Cruz et al. 2016). It is also known by other names such as hungry rice, acha, and fundi. The two main species of fonio are white fonio (*Digitaria exilis*) and black fonio (*Digitaria iburua*) (Small 2015). Both species are well adapted to the semiarid conditions of West Africa. Fonio is drought tolerant and can thrive on infertile soils and marginal lands (Ayenan et al. 2018; Osundinakin et al. 2023). As one of the oldest domesticated crops, fonio originated in what is now Guinea and Mali, and it remains a staple food in several West African countries due to its exceptional nutritional properties. It is rich in quality minerals, vitamins, fiber, and sulfur-containing amino acids such as cysteine and methionine.

Fonio is a free-tillering annual plant, reaching heights of 80 cm (for *D. exilis*) or 150 cm (for *D. iburua*). It exhibits C_4 plant anatomy; produces slender, glabrous culms; and matures in 70–150 days depending on the species. Fonio is primarily cultivated on marginal soils at altitudes ranging from 400 to 1500 meters above sea level, with temperatures between 25 and 30 °C and an annual rainfall of 150–3000 mm (Abdul and Jideani 2019). As a gluten-free cereal, fonio is rich in carbohydrates, dietary fiber, and essential minerals such as iron and zinc (Zhu 2020). Due to its nutritional benefits, fonio is often referred to as a superfood and the "new quinoa" (Thomson Reuters Foundation 2019). In West Africa, fonio is commonly used for baking, making couscous, and preparing porridge. Its popularity is growing globally, particularly in the United States and the United Kingdom, due to its enormous health benefits (Jarial et al. 2024).

Both teff and fonio are classified as "Neglected and Underutilized Species" (NUS) and are often referred to as orphan crops. These crops hold great potential to address food security and nutritional challenges because of their adaptability to extreme environmental conditions and their rich nutrient profiles (Tadele 2009). This chapter will elaborate the history, origin, taxonomy, domestication, and production of these crops, as well as their uses and the challenges they face. It will also explore germplasm collection and diversity at both national and global levels, discuss available genetic and genomic resources, highlight the value of pre-breeding initiatives, and conclude with potential future developments for enhancing agricultural productivity and sustainability.

10.2 The Crop: History, Origin, Taxonomy, Domestication, Production, Use and Consumption, and Constraint

10.2.1 History and Origin

Ethiopia is widely recognized as one of the Vavilovian centers of crop origin and diversity; however, it is also the origin and diversity of teff and was domesticated in between 4000 and 1000 BC (Vavilov 1951). The genus *Eragrostis* is one of the largest within the grass family, comprising around 350 species (Watson and Dallwitz 1992). Of these, approximately 43% are thought to have originated in Africa, 18% in South America, 12% in Asia, 10% in Australia, 9% in Central America, 6% in North America, and 2% in Europe (Costanza 1974). In Ethiopia, 14 out of the 54 species of *Eragrostis* are native to the country, accounting for about 26% of the total species found there (Cufodontis 1974).

Fonio, an ancient grain from West Africa, is believed to have originated around 5000 BC (Cruz et al. 2016). There are four species of fonio grown worldwide, contributing to its overall biodiversity. In West Africa, the two main species cultivated are *Digitaria exilis* (white fonio) and *Digitaria iburua* (black fonio). Additionally, *Digitaria cruciata*, also called "raishan," is grown in India and Vietnam, while *Digitaria sanguinalis*, known as European millet or red manna, is cultivated in Eastern Europe (Froment and Renard 2001; Vodouhè and Achigan Dako 2006).

10.2.2 Taxonomy

According to taxonomy, teff is a member of the kingdom *Plantae*, order *Poales*, family *Poaceae* (formerly known as *Gramineae*), subfamily *Chloridoideae*, tribe *Eragrostideae*, subtribe *Eragrostoideae*, and genus *Eragrostis*. *Eragrostis tef* (Zucc.) Trotter is currently most commonly used as its scientific name or binomial nomenclature (Jifar et al. 2022). Generally, teff is an allotetraploid species ($2n = 4x = 40$) which was obtained by hybridization of two different species, followed by diploidization (Assefa and Chanyalew 2018). Teff is a highly self-pollinated species with autogamous nature of pollination; however, an outcrossing frequency of 0.2–1.0% was observed (Woldeyohannes et al. 2020). Here, florets size is very small and typically opens in the early morning hours (6:45–7:45 a.m.).

Fonio belongs to the kingdom *plantae*, order *poales*, family *poaceae*, and genus *Digitaria*. There are two primary domesticated species; *Digitaria exilis* (white fonio) and *Digitaria iburua* (black fonio) which are indigenous to West Africa (Kanlindogbe et al. 2020). Fonio species are probably self-fertilized crops (Sarker et al. 1993); however, the potential for an outcrossing system has been reported as well (Hilu et al. 1997). Fonio is a tetraploid species (2n = 4x = 36) with a highly inbreeding reproductive system (Adoukonou-Sagbadja et al. 2007a, b; Barnaud et al. 2017). Fonio is an annual herbaceous C_4 plant that yields small grains (~1 mm) covered by a husk (Cruz et al. 2016). The tetraploid annual weed *D. longiflora*, which is found throughout tropical Africa, is most likely the wild progenitor of cultivated fonio (Abrouk et al. 2020).

10.2.3 Domestication

Teff and fonio were domesticated by the selective breeding of cultivars with desired characteristics such as drought tolerance, rapid growth, and soil-type adaptability. To ensure effective harvesting in the case of teff, farmers chose plants with non-shattering seeds. The current teff cultivars that are now available were developed through selection process over generations (Assefa et al. 2011).

The fonio was also domesticated for its resilience since it was a staple in areas with poor soil and limited rainfall. Because of its short growing season (usually less than 3 months), it can thrive in regions with erratic rainfall patterns (Porteres 1976). Furthermore, despite the small size of seeds, fonio was made easier to harvest and prepare for consumption by the development of traditional processing methods.

10.2.4 Production

In Ethiopia, teff crop stands first in the area cultivated and second in the production (Central Statistical Agency 2018). The productive potential of the teff crop is substantially lower than the commonly cultivated major crops in Ethiopian countries, which achieves yields in the range of 1–2 t/ha (Cochrane and Bekele 2018). Teff is mainly cultivated in Ethiopia; it accounted for 17.26% of the overall grain production, while cereals accounted for 87.48% of the total production quantity (Tilaye et al. 2023). During the main Meher season, 3.02 million hectares of land produced roughly 5.28 million tons of teff (Central Statistical Agency 2019). Since it is adaptable to a wide range of soil types and climates, it is cultivated in agroecological zones that ranged from lowland to highland regions (Gelaw and Qureshi 2020). The low productivity of the crop has not limited its popularity because of its cultural relevance and high demand marketability (Woldeyohannes et al. 2020).

Fonio is cultivated on about 300,000 hectares of land in West Africa each year, and 3–4 million people rely on it for their livelihood (National Research Council 1996). It is mainly grown in Guinea, Mali, Burkina Faso, Nigeria, Benin, Togo, Senegal, and Guinea Bissau (Cruz et al. 2016) regions of West Africa. Due to its excellent adaptation to marginal soils and drought, fonio is an essential crop for ensuring food security in these areas. In 2017, worldwide production of fonio grain was 671,000 t. The average yield was about 0.7 t/ha which ranged from 0.4 t/ha in Nigeria to 1.4 t/ha in Ivory Coast (FAO 2019). In Nigeria, the crop was grown during the last decade on approximately 173,773.40 ha, with a mean production estimated at approximately 84,363.30 tons (FAO 2022).

10.2.5 Use and Consumption

Teff floor is used for making various food products like "injera" (traditional, fermented pancake). Similarly, it has been demonstrated that blending teff flour to sorghum flour enhances the injera's sensory qualities (Yetneberk et al. 2005). Teff can also be used to make porridges, unleavened breads (kitta), gruels (atmit), and traditional alcoholic beverages such as tella and arake; however, its use is far less common than that of "injera" (Gebremariam et al. 2012). Straw is used for construction of houses mixed with mud, mainly used as cattle feed, and its fodder is having high values of in vitro dry matter digestibility, crude protein content, low lignin, and high energy values (Yami 2013).

In West Africa for hundreds of years, fonio has been consumed in various forms. In the fourteenth century, Ibn Battuta (1982) reported that in these lands, couscous was prepared with fonio. Fonio can be cooked into porridge and cream and eaten directly. In Nigeria, the fonio grains are processed to produce a flour that can be used in making bread or porridges (the thick, unfermented "tuwo acha" and the thin, fermented "kunu acha"), either alone or in combination with other flours (Heuzé et al. 2019). Additionally, it can be converted into an oat-like breakfast cereal, eaten with vegetable soup, or prepared as a staple dish (Morales-Payán et al. 2002). Fonio grain can be fermented to prepare the beer "tchapalo" (Cruz et al. 2016). Although fonio grain is mostly used for food, its high energy and methionine content makes it a potential valuable feed for farm animals (NRC 1996). Its use has been noted in Togo for poultry feeding (Adoukonou-Sagbadja et al. 2006). Additionally, straw is used in stuffed mattresses or as a thatch to make kitchen and barn roofs (Ibrahim Bio Yerima et al. 2024).

10.2.6 Constraints

One of the major constraints of teff is relatively low production with a national average yield of approximately 1.5 t/ha in 2013 (CSA 2016). This may be due to use of less yielding varieties along with unimproved pre- and post-traditional practices (Ferede 2013). Susceptibility to lodging is also one of the major constraints, where lodging lowers the percentage and rate of seed germination by 41% and 44%, respectively, the grain yield per panicle by 51%, the 1000-kernel weight by 35%, and the grain yield by 11–22% (average = 17%) (Ketema 1983). Socioeconomic constraints include lack of awareness and the high cost and unavailability of seeds (Tefera and Ketema 2001). Efforts to improve teff's yield through selective breeding and modern agricultural practices are ongoing but have not yet reached the levels seen with crops such as wheat or rice.

As a "minor" millet, fonio is still highly underutilized and often ignored. Lack of improved seeds, infestation by insect pests, and difficult postharvest processes are the major constraints to its production (Animasaun et al. 2023). Some other constraints limiting the fonio production are inadequate capital, lack of improved technology, inaccessibility to credit, and inadequate land (Onuwa et al. 2022). In West Africa, factors affecting fonio production are low yields from lack of improved cultivars, poor management techniques, susceptibility to biotic and abiotic stresses, and laborious postharvest and processing technologies (Ibrahim Bio Yerima et al. 2024). Shattering at maturity is one of the abiotic stresses that causes up to 25% of yield loss and is a significant disadvantage when harvesting is delayed (Vodouhe et al. 2003). Furthermore, limited research and development investment have hampered the crop's integration into larger food systems, both domestically and internationally.

10.3 Status of Germplasm Collection and Conservation (National/Global Genebanks)

The germplasm conservation phenomenon plays an important role which further aids in maintaining the genetic basis required for breeding. Genebanks are one of the most significant places for collecting germplasm for short-, medium-, and long-term conservation. Through in situ and/or ex situ methods, millions of germplasm accessions have been gathered and preserved.

The Institute of Biodiversity Conservation is maintaining 4395 teff accessions, and these germplasm accessions are shared with research institutes in Ethiopia (Zeid et al. 2012). In teff, approximately 300 species have been identified, of which only *E. teff* is cultivated and used for human consumption. The genebank in Ethiopia holds over 5000 accessions of teff collected from various geographical regions, diversified in terms of climate and elevation (Assefa et al. 2015). Local farmers in Ethiopia and Eritrea have contributed immensely over millennia to the development of the thousands of landraces in the rich gene pool of teff available to the present generation (Central Statistical Agency 2016). The systematic collection, conservation, evaluation, and distribution of teff germplasm began in Ethiopia in the 1950s. The Ethiopian Biodiversity Institute (EBI) is maintaining a large number of both in situ and ex situ accessions of teff sampled from different agroecologies of the country. The total collection at EBI comprises 3850 accessions, of which complete passport information is available for 1754 accessions (Woldeyohannes et al. 2020). Jifar et al. (2020) reported that nearly 6000 accessions of teff have been collected and conserved at EBI. A study on the diversity of both qualitative and quantitative traits revealed that trait expression varies with collection altitudes, administrative zones, and regional states of Ethiopia. Globally, 7235 teff accessions are conserved in various genebanks as part of ex situ conservation efforts (Woldeyohannes et al. 2020). Details of the number of accessions conserved and the holding institutes are provided in Table 10.2.

Table 10.2 Summary of 7235 teff accessions currently conserved ex situ in germplasm banks worldwide, with an estimate of passport data completeness

Country	Holding institute/code	Institute home page	Number of accessions	Passport information %	
				Collection site	GPS
Ethiopia	EBI	https://www.ebi.gov.et/	6407	68	48
Israel	ISR002	NA	376	NA	NA
United States of America	USA022	https://www.ars-grin.gov/	373	91.4	3
Germany	DEU146	https://www.ipkgatersleben.de/	32	3	12.5
Australia	AUS165	NA	20	NA	NA
Australia	AUS167	https://www.pir.sa.gov.au/	12	8.3	NA
Ethiopia	ETH013	https://www.ilri.cgiar.org/	3	NA	NA
Hungary	HUN003	https://www.rcat.hu/	3	NA	NA
Kenya	KEN212	https://www.genetic.kalro.org/	3	66.6	33
Austria	AUT001	https://www.genbank.at/	2	50	NA
United Kingdom	GBR016	https://www.igergru.ibers.aber.ac.uk/	2	NA	NA
Bulgaria	BGR001	https://www.genebank.hit.bg/	1	NA	NA
Czech Republic	CZE122	https://www.vurv.cz/	1	NA	NA

Source: Woldeyohannes et al. (2022)
NA Not available, *EBI* Ethiopia Biodiversity Institute, Ethiopia, *ETH013* International Livestock Research Institute, Ethiopia, *ISR002* Israel GeneBank for Agricultural Crops, Agricultural Research Organization, Volcani Center, Israel, *USA022* Western Regional Plant introduction Station, USDA-ARS, Washington State University, USA, *DEU146* Leibniz Institute of Plant Genetics and Crop Plant research, Germany, *AUS165* Australian Grains Genebank, Department of Economic Development Jobs Transport and Resources, Australia, *AUS167* Australian Pastures Genebank, Australia, *HUN003* Institute for Agrobotany, Hungary, *KEN212* Genetic Resources Research Institute, Kenya, *AUT001* AGES Linz-Austrian Agency for Health and food Safety, Austria, *GBR016* Genetic Resources Unit, Institute of Biological, Environmental and Rural Sciences, Aberystwyth University, UK, *BGR001* Institute for Plant Genetic Resources K. Malkov, Bulgaria, *CZE122* GeneBank, Czech Republic

The ICAR-National Bureau of Plant Genetic Resources, New Delhi, India, has conserved 288 accessions of teff and 17 accessions of fonio. The collection and conservation processes for both crops are ongoing, with improvement programs focused on diversifying the genetic base of these crops through recombination breeding efforts.

The germplasm of teff and fonio is mainly collected from West African countries, specifically Ethiopia. Diversity studies conducted in various experiments using different sets of germplasm lines and farmer-grown landraces have largely indicated significant diversity for traits relevant to plant breeding. Breeding efforts aimed at overcoming yield stagnation and improving adaptability to both biotic and abiotic stresses hold great potential for success. The number of germplasm accessions collected at EBI is sufficient to meet the priority needs for improving grain yield in teff and fonio. The expansion of crop cultivation to countries like the United States, India, Austria, Australia, Bulgaria, and Germany, owing to the nutritional value, adaptability traits, and suitability to changing climates of these crops, is an added advantage.

A survey was conducted in 1995 to make germplasm exploration and collection expedition in Bauchi, Kaduna, Kebbi, Niger and Plateau states, FCT, and Abuja, and a collection base of 138 accessions were produced from survey (Kwon-Ndung et al. 1998). Collection of 641 *D. exilis* accessions were done from six countries (Benin, Niger, Togo, Mali, Burkina-Faso, and Guinea) (Clément and Leblanc 1984) and are preserved at the Institut de Recherche pour le Developpement (IRD) in Montpellier. A total of 95 accessions were collected from the two zones of Togo and were duplicated and conserved ex situ at the Faculty of Sciences (University of Lomé) and in the Genebank of Niaouli Agricultural Research Center of INRAB in Benin in order to preserve its genetic diversity ex situ for sustainable use by current and future generations (Adoukonou-Sagbadja et al. 2004). Adoukonou-Sagbadja et al. (2006) surveyed the Togo regions in Guinea to document the ethnobotanical and indigenous knowledge related to the production, diversity, use, and conservation of fonio crop. The 42 landraces maintained and managed by the farmers were found to originate in local domestication or introduction from neighboring countries.

The National Cereals Research Institute (NCRI) began research and development on acha (fonio) in 1995, initially planting 13 accessions gathered from various locations in Plateau State at Badeggi (Kwon-Ndung and Ochigbo 2007). The Genetics and Plant Breeding team of the Biosciences Laboratory at the University Joseph KI-ZERBO collected and preserved 60 fonio accessions (Balma et al. 2022). A total of 2372 fonio accessions, representing the entire crop range, were gathered or shared between 1977 and 2021 as part of various research programs involving collaborations between French and African research organizations. These seed collections were preserved by national genebanks such as GBioS/UAC in Benin and Montpellier's GAMeT Resource Center, ARCAD, and IRD in France (Kaczmarek et al. 2023).

To preserve genetic diversity and ensure future crop advancements for teff (*Eragrostis tef*) and fonio (*Digitaria exilis* and *Digitaria iburua*), germplasm conservation and utilization are essential. This section discussed about status of germplasm diversity and collection of germplasm at both national and global level followed by conservation efforts in both the crops. According to Conservation of Biological Diversity (CBD), Article 2, which was established at the Rio de Janeiro Earth summit, genetic diversity is a key element of biodiversity and diversity among species, between species and of ecosystems (Govindaraj et al. 2015).

10.4 Characterization and Evaluation

Assessment and characterization of germplasm is the main step in crop improvement program as it provides genetic diversity. Genetic diversity can be divided based on morpho-agronomic traits, biochemical and molecular markers, or their combinations, which can be further utilized for detection of potential germplasm lines (Carvalho 2004) and their utilization.

Ebba (1975) has extensively characterized the teff germplasm accessions of more than 1000 collected from field and markets in Ethiopia and observed array of variation. A study was conducted to evaluate 20 morphological, phenological, and agronomic characters to assess diversity of 320 germplasm lines. The analysis revealed that there is a highly significant variation among genotypes and 14 major complexes were formed by cluster analysis. Overall, trait diversity is observed in teff germplasm lines (Assefa et al. 1999). Woldeyohannes et al. (2021) characterized 366 Ethiopian farmer varieties for agro-morphological and molecular diversity. They reported six nonoverlapping clusters indicating the good amount of diversity, and the three PCs based on the SNP data yielded about 15% of the genetic variability in the teff panel. Here, teff clustering was associated with the local environmental adaptations and agroecological zones of Ethiopia. Further, a total of 120 germplasm lines were collected from eight regions of Ethiopia and were evaluated for genetic diversity with respect to pheno-morphic and agronomic traits, and the study revealed that there is significant variation among genotypes. Estimates of heritability were highest for panicle length. On the basis of principal component analysis, about 67% of the total phenotypic diversity in the germplasm was explained (Assefa et al. 2001). Similarly, based on the evaluation of 60 teff germplasm accessions collected from six western and southern regions of Ethiopia with respect to regions and altitude zones, Assefa et al. (2002) noted high Shannon-Weaver diversity indices for anther color and number of apical spikelet florets. Significant variations were observed for lemma color, number of culm internodes, and counts of middle and basal spikelet florets. Field view and panicle type of teff crop is represented in Fig. 10.1.

In another experiment, assessment of genetic diversity was carried out in four *Eragrostis tef* (teff) and 14 *Eragrostis pilosa* accessions by using radiolabeled and silver-stained Amplified Fragment Length Polymorphism (AFLP) markers and morphological traits. A total of 897 markers were obtained, out of which 395 were polymorphic, revealing significant diversity. Significant variation was observed for morphological traits, such as plant height and flowering. Cluster analysis revealed that some *E. pilosa* accessions were closely related to teff while others were distantly related (Ayele et al. 1999). The diversity of agro-morphological traits at two locations in Ethiopia was assessed using a total of 188 teff genotypes; highly significant differences across genotypes, locations, and genotype × environment interaction were noted, while for panicle length, peduncle length, number of fertile tillers, and spikelets per panicle, moderate values of GCV and PCV were observed. Based on 12 traits, the genotypes were separated into 6 different clusters using cluster analysis (Jifar et al. 2018).

Fig. 10.1 Field view showing plant features and panicle type in teff crop

Characterization of 3850 teff germplasm accessions, collected from different agroecologies across the country and conserved at Ethiopian Biodiversity Institute (EBI) revealed that the accessions were cultivated in a broad range of bioclimatic features of wide environmental conditions of Ethiopia and exhibited different trait values for phenotypic and pedoclimatic diversity in relation to agroecologies. Species distribution model analysis confirmed that the suitability of teff cultivation in the different agroecologies of the country diminished by 2070. Jifar et al. (2020) assessed genetic diversity in 144 teff accessions from Northern and Central Ethiopia using qualitative and quantitative traits and by estimating Shannon-Weaver diversity index and cluster analysis. The study revealed that panicle forms and seed color showed the highest mean Shannon diversity, and significant variation was observed in teff accessions collected from different altitudes, administrative zones, and regional states. SSR markers were used to identify genetic diversity for 49 teff genotypes (36 improved varieties, 10 farmers' varieties, 3 elite lines) and 2 wild relatives. The results showed mean polymorphic information content of 0.69 and 0.73 of gene diversity values. About 29% of genetic variation in the SSR molecular data was obtained by principal coordinate analysis. In conclusion, the farmer's varieties were distant from the improved varieties and represent more genetic diversity (Fikre et al. 2018).

A research study was conducted to determine the genetic relatedness among 118 landraces of fonio collected from Benin, Burkina Faso, Guinea, Mali, and Togo, the major fonio-growing regions in West Africa. In *D. exilis* landraces, the chromosome number was determined to be 2n = 36, and the results revealed that this landrace was closely related to *D. longiflora* and *D. ternata*, while *D. lecardii* and *D. ciliaris* were found to be distinct. The intraspecific variations observed within the species were slight and insignificant, suggesting genome size stability primarily within the cultivated gene pool (Adoukonou-Sagbadja et al. 2007a, b). The genetic diversity of 122 accessions collected from five countries (Benin, Burkina Faso, Guinea, Mali, and Togo) was assessed using AFLP markers. The study classified *D. exilis* and *D. iburua* into distinct groups. Moderate levels of genetic diversity were detected in

D. exilis, which were unequally distributed, with the majority observed in the Upper Niger River basin, while very low diversity was present in the Atacora mountain zone (Adoukonou-Sagbadja et al. 2007a, b). Dansi et al. (2010) surveyed 15 villages in the Benin region of Guinea to collect information on the fonio crop, preserve its genetic diversity, and utilize this diversity in developing improved cultivars to meet the needs of farmers and consumers. They collected 15 farmer-named landraces and grouped them into five morphotypes, of which four belonged to *D. exilis* and one to *D. iburua*. Farmers in the Benin region reported that *Digitaria horizontalis* Wild. and *Digitaria longiflora* (Retz.) Pers. were the putative parents of cultivated fonio.

10.5 Genetic and Genomic Resources

In the present advancement of agricultural research technologies, generation of genetic and genomic resources has become possible. Molecular markers are the efficient tools to fasten the breeding work and to enhance the accuracy of selection at genomics level. Biotechnological tools help the researchers to estimate the true genetic potential of the genotype, its breeding value, identification of gene/QTLs governing the traits, transfer of traits into the cultivated and popular lines to correct the specific defects, etc. In teff, several researchers have reported the use of molecular markers like RAPD (Bai et al. 2000), AFLP markers (Ayele et al. 1999), ISSR markers (Assefa et al. 2003), and SSR (Chanyalew 2007; Zeid et al. 2012; Abraha et al. 2016). Zeid et al. (2012) used SSR markers to assess the genetic diversity and relationship among the 326 accessions with 39 SSR markers. Their study revealed that plant breeders have been using a broad range of germplasm in their teff improvement programs as the detected alleles were present in the generated breeding material in teff crop. Assefa et al. (2003) also reported genetic similarity of 0.26 to 0.86 among 92 studies teff accessions. Abraha et al. (2016) studied the genetic relationship of 60 diverse teff genotypes with 10 SSR markers and observed the 63% of the genetic variation within species and 35% between species. An average value of 0.84% of PIC value of markers indicated power of SSR markers in discriminating the genotypes of different species in teff crop. Fikre et al. (2020) investigated the genetic diversity at molecular level with nine SSR markers and reported higher percentage of variation within the groups and among the groups, diversity was low. They have identified a total of 52 alleles amplified with a mean of 6 alleles per locus indicating vast genetic diversity in the studied set of genotypes.

The draft genome of the teff cultivar "Tsedey" (DZ-Cr-37) is available, but its application for genomic selection, marker-trait association, and the use of biotechnology tools to accelerate breeding activities remains limited. The chromosome-scale assembly of the subgenomes teff A and teff B, reported by VanBuren et al. (2019), provided insights into subgenome interactions and divergence patterns during evolution using advanced sequencing and chromatin capture techniques. However, the draft genome assembly of "Tsedey" is highly fragmented, with incomplete annotation, posing challenges for functional genomics, marker-assisted breeding, and genetic studies in teff. The subgenomes of teff exhibit near-perfect synteny, indicating a lack of homeologous exchanges, which suggests the presence of mechanisms preventing genome exchange. Research on teff genetics and improvement has progressed from generating a draft genome sequence to achieving a high-quality genome sequence, alongside developing advanced recombination and mutagenized breeding populations.

"Tsedey," an early maturing variety, was selected for genome sequencing, with two transcriptomes analyzed to explore genetic expression. The draft genome, based on flow cytometry, contains 672 Mbp, representing 87% of the total genome size. Comparative genomics revealed that among cultivated species, finger millet is the closest relative to teff which is not sequenced while foxtail millet and sorghum are its closest sequenced relatives. VanBuren et al. (2020) used the teff landrace "Dabbi" to develop a chromosome-scale assembly of teff genome using a combination of long-read single-molecule real-time sequencing and long-range high-throughput chromatin capture (Hi-C) techniques. At 85× coverage of the estimated 622 Mb genome, the PacBio reads were error corrected and assembled and contigs were polished to generate a total assemble size of 576 Mbp representing 92.6% of the estimated genome size in teff. CM05836, an accession of fonio from the Mopti region of Mali, was selected to create a high-quality *D. exilis* reference genome assembly. A genetic structure approach based on full-length long terminal repeat retro-transposons (fl-LTR-RT) was used. Length of genome assembly of fonio obtained was 716.4 Mbp. Two orthologous genes showed signatures of selection associated with seed shattering (Sh1 ortholog of sorghum) and seed size (GS5 ortholog of rice). Syntenic relationships and two distinct subgenomes with other cereal genomes were found by comparative research, and these findings provided valuable insights into the genetic basis of domestication and potential improvement of fonio crop (Abrouk et al. 2020).

Animasaun et al. (2018) conducted an experiment on 11 cultivated fonio genotypes to analyze the genetic diversity using ISSR markers. A total of 95 loci were generated by using 13 ISSR primers, and about 53% of polymorphism was observed implying significant genetic variation. Cluster analysis of these 11 genotypes made two distinct groups, *D. exilis* and *D. iburua*. In another study, 10 fonio accessions were selected to demonstrate the genetic diversity with the use of 15 RAPD markers. Fifteen primers showed polymorphism and a dendrogram was constructed through cluster analysis which divided the 10 accessions into two groups. The findings demonstrated genetic diversity without duplication and provided a foundation for future research by using advanced tools like RAPD (Kuta et al. 2005). According to the research conducted by Nyam et al. (2017), 11 genotypes including 3 species (*D. iburua, D. exilis*, and *D. barbinodis*) seem to be clearly separated based on RFLPs, which show their genetic variations at the molecular/DNA level. High percentage of polymorphism was observed for primer De 01. In terms of phenotype, *D. iburua* differs from *D. exilis* and *D. barbinodis*. Hilu et al. (1997) used 114 RAPD markers to examine the genetic diversity and evolution of 10 *D. exilis* (Kipp.) Stapf accessions and 4 accessions from other species. It was found that true fonio had a very high level of genetic diversity. According to the study, *D. ternata*

(A. Rich.) Stapf and *D. longiflora* (Halz) Pers. may have been the ancestors of black fonios and true fonios, respectively.

10.6 Future Prospects

Climate change is influencing and affecting the natural phenomenon, specifically agriculture in many ways. Variations in temperature and changes in distribution pattern of precipitation are increasing the pressure on agroecologies and natural resources degradation and posing serious threat to the agriculture production system. Teff and fonio are the unexplored and underexploited grain crops that need increased attention by researchers for many purposes like food and nutritional security, healthy food availability, small-scale farmer's income generation for better livelihood, and protecting and preserving the environmental resources in the current depleting plant resources and rapidly increasing human population. Average productivity of teff is very low, ranging from 1.0 to 1.5 t/ha, and the genetic potential for grain yield is up to 4.5 t/ha, which can be achieved by following improved cropping technologies, non-lodging conditions, appropriate sowing condition and seed curation, management of soil fertility and agronomic practices, pest and disease management, and effective weed management. Transdisciplinary approaches may efficiently propel untapped NUS farmer varieties into modern breeding to foster more resilient and sustainable cropping systems. Combining data-driven research approaches in a multidisciplinary characterization may accelerate teff and fonio breeding efforts to unlock its full potential. The generation of genetic and genomic resources and utilization of the advanced molecular breeding tools and techniques in these crops enhance the breeding efficiency and greatly support the cultivar development programs with increased grain yield. Studying the physiological parameters to exploit their inherent stress tolerance mechanisms and improving the agronomic practices to realize the potential yield are utmost necessary in both the crops. This not only helps in uplifting the farmer's interest but also expands the area under cultivation. The processing of grains and value addition is another area, which in untouched in these nutritionally rich crops.

References

Abdul SD, Jideani AI (2019) Fonio (*Digitaria* spp.) breeding. In: Advances in plant breeding strategies: cereals, vol 5. Springer, Cham, pp 47–81

Abraha MT, Shimelis H, Laing M, Assefa M, Amelework B (2016) Assessment of the genetic relationship of tef (*Eragrostis tef*) genotypes using SSR markers. S Afr J Bot 105:106–110

Abrouk M, Ahmed HI, Cubry P, Šimoníková D, Cauet S, Pailles Y et al (2020) Fonio millet genome unlocks African orphan crop diversity for agriculture in a changing climate. Nat Commun 11(1):4488

Adoukonou-Sagbadja H, Dansi A, Vodouhè R, Akpagana K (2004) Collecting fonio (*Digitaria exilis* Kipp. Stapf, D. iburua Stapf) landraces in Togo. Plant Genetic Resour Newslett 139:63–67

Adoukonou-Sagbadja H, Dansi A, Vodouhè R, Akpagana K (2006) Indigenous knowledge and traditional conservation of fonio millet (*Digitaria exilis, Digitaria iburua*) in Togo. Biodivers Conserv 15(8):2379–2395

Adoukonou-Sagbadja H, Schubert V, Dansi A, Jovtchev G, Meister A, Pistrick K et al (2007a) Flow cytometric analysis reveals different nuclear DNA contents in cultivated Fonio (*Digitaria* spp.) and some wild relatives from West Africa. Plant Syst Evol 267:163–176

Adoukonou-Sagbadja H, Wagner C, Dansi A, Ahlemeyer J, Daïnou O, Akpagana K et al (2007b) Genetic diversity and population differentiation of traditional fonio millet (*Digitaria* spp.) landraces from different agro-ecological zones of West Africa. Theor Appl Genet 115:917–931

Animasaun DA, Awujoola KF, Oyedeji S, Morakinyo JA, Krishnamurthy R (2018) Diversity level of genomic microsatellite among cultivated genotypes of *Digitaria* species in Nigeria. Afr Crop Sci J 26(2):305–313

Animasaun DA, Adedibu PA, Olawepo GK, Oyedeji S (2023) Fonio millets: an underutilized crop with potential as a future smart cereal. In: Neglected and underutilized crops. Academic Press, London, pp 201–219

Assefa K, Chanyalew S (2018) Agronomics of tef. In: The economics of Tef: exploring Ethiopia's biggest cash crop. International Food Policy Research Institute (IFPRI), Washington, DC, pp 39–70

Assefa K, Merker A, Tefera H (2003) Inter simple sequence repeat (ISSR) analysis of genetic diversity in tef (Eragrostis tef (Zucc.) Trotter). Hereditas 139: 174–183

Assefa K, Ketema S, Tefera H, Nguyen HT, Blum A, Ayele M et al (1999) Diversity among germplasm lines of the Ethiopian cereal tef [*Eragrostis tef* (Zucc.) Trotter]. Euphytica 106:87–97

Assefa K, Tefera H, Merker A, Kefyalew T, Hundera F (2001) Variability, heritability and genetic advance in phenotypic and agronomic traits of tef [*Eragrostis tef* (Zucc.) Trotter] germplasm from eight regions of Ethiopia. Hereditas 134(2):103–113

Assefa K, Merker A, Tefera H (2002) Qualitative trait variation in tef [*Eragrostis tef* (Zucc.) Trotter] germplasm from Western and Southern Ethiopia. Euphytica 127:399–410

Assefa K, Yu JK, Zeid M, Belay G, Tefera H, Sorrells ME (2011) Breeding tef [*Eragrostis tef* (Zucc.) Trotter]: conventional and molecular approaches. Plant Breed 130(1):1–9

Assefa K, Cannarozzi G, Girma D, Kamies R, Chanyalew S, Plaza-Wüthrich S et al (2015) Genetic diversity in tef (*Eragrostis tef* (Zucc.) Trotter). Front Plant Sci 6:177

Ayele M, Tefera H, Assefa K, Nguyen HT (1999) Genetic characterization of two Eragrostis species using AFLP and morphological traits. Hereditas 130:33–40

Ayenan MAT, Sodedji KAF, Nwankwo CIK, Olodo F, Alladassi MEB (2018) Harnessing genetic resources and progress in plant genomics for fonio (*Digitaria* spp.) improvement. Genet Resour Crop Evol 65:373–386

Bai GH, Ayele M, Tefera H, Nguyen HT (2000) Genetic diversity in tef [*Eragrostis tef* (Zucc.) Trotter] and its relatives as revealed by random amplified polymorphic DNAs. Euphytica 112:15–22

Balma P, Kabore B, Kabre VN, Kiebre M, Kiebre Z, Sanou J (2022) Genetic diversity and indigenous knowledge of fonio (*Digitaria exilis* Stapf) produced in Burkina Faso. Afr J Agric Res 18(10):840–849

Barnaud A, Vigouroux Y, Diallo MT, Saidou SI, Piquet M, Barry MB et al (2017) High selfing rate inferred for white fonio [*Digitaria exilis* (Kippist.) Stapf] reproductive system opens up opportunities for breeding programs. Genet Resour Crop Evol 64:1485–1490

Carvalho MA (2004) Germplasm characterization of *Arachis pintoi* Krap. and Greg. (Leguminosae). PhD thesis, University of Florida, USA

Central Statistical Agency (2016) Agricultural sample survey 2015/2016. Central Statistical Agency, Federal Democratic Republic of Ethiopia, Addis Ababa

Central Statistical Agency (CSA) (2018) Agricultural sample survey, area planned and production of major crops, Meher season. Volume I. CSA, Addis Abeba

Central Statistical Agency (CSA) (2019) Agricultural sample survey, area planned and production of major crops, Meher season. Volume I. CSA, Addis Abeba

Chanyalew S (2007) Phenotypic evaluation and molecular study of parents and recombinant inbred lines of tef (*Eragrostis tef* [Zucc.] Trotter). PhD thesis, Haramaya University, Haramya

Clayton WD, Renvoize SA (1986) Genera *graminum*, grasses of the world. Kew Bulletin Additional Series XIII

Clément J, Leblanc JM (1984) Collecte IBPGR-ORSTOM de 1977 au Togo. In: Prospection des *Digitaria exilis* (Fonio) en Afrique de l'Ouest. ORSTOM, Marseille, pp 3–7

Cochrane L, Bekele YW (2018) Contextualizing narratives of economic growth and navigating problematic data: economic trends in Ethiopia (1999–2017). Economies 6(4):64

Costanza SH (1974) Literature review and numerical taxonomy of tef (*Eragrostis tef*). MSc thesis, University of Illinois, Urbana, Illinois, USA

Cruz JF, Béavogui F, Dramé D, Diallo TA (2016) Fonio, an African cereal. CIRAD

Cufodontis G (1974) Enumeration plantarum aethiopiae spermatophyta. Jard Bot, Brussels 77:262–266.

Dansi A, Adoukonou-Sagbadja H, Vodouhe R (2010) Diversity, conservation and related wild species of Fonio millet (Digitaria spp.) in the northwest of Benin. Genetic Resources and Crop Evolution 57:827–839. https://doi.org/10.1007/s10722-009-9522-3.

Demelash A (2017) Screening of teff (Eragrostis tef) varieties for genotypic and phenotypic traits in Dejen Woreda, East Gojjam Zone. Int J Biodivers Conserv, 9(7): 239–245.

Ebba T (1975) Tef cultivars: morphology and classification, Part II, Experiment Station bulletin 66. Addis Ababa University, College of Agriculture, Dire Dawa

Engelmann F (1997) In-vitro germplasm conservation. In: International symposium on biotechnology of tropical and subtropical species part 2, vol 461. International Society for Horticultural Science, Belgium, pp 41–48

FAO (2019) FAOSTAT. Food and Agriculture Organization of the United Nations, Rome

FAO (2022) FAOSTAT statistical database. FAO, Rome. Retrieved from http://faostat.fao.org/

Ferede S (2013) Technological change and economic viability in tef production. EIAR, University of Bern, Bern

Fikre T, Tesfaye K, Assefa K (2018) Genetic diversity of Ethiopian tef (*Eragrostis tef* [Zucc.] Trotter) released and selected farmers' varieties along with two wild relatives as revealed by microsatellite markers. J Crop Sci Biotechnol 21(5):367–374

Fikre T, Assefa K, Tesfaye K (2020) Extent and pattern of genetic diversity for pheno-agro-morphological traits in Ethiopian improved and selected farmer's varieties of Tef (Eragrostis tef (Zucc.) Trotter. Afr J Agric Res, 16(6): 892–901.

Froment D, Renard C (2001) Fonio. In: Raemaekers RH (ed) Agriculture en Afrique tropicale. DGCI, Bruxelles, pp 37–43

Gebremariam MM, Zarnkow M, Becker T (2012) Teff (*Eragrostis tef*) as a raw material for malting, brewing, and manufacturing of gluten-free foods and beverages: a review. J Food Sci Technol 49(6):267–276. https://doi.org/10.1007/s13197-012-0745-5

Gebru YA, Sbhatu DB, Kim K-P (2020) Nutritional composition and health benefits of Teff (*Eragrostis tef* (Zucc.) Trotter). J Food Qual 2020:1. https://doi.org/10.1155/2020/9595086

Gelaw AM, Qureshi AS (2020) Tef (*Eragrostis tef*): a superfood grain from Ethiopia with great potential as an alternative crop for marginal environments. In: Emerging research in alternative crops. Springer, Cham, pp 265–278

Govindaraj M, Vetriventhan M, Srinivasan M (2015) Importance of genetic diversity assessment in crop plants and its recent advances: an overview of its analytical perspectives. Genetics Res Int 2015(1):431–487

Henrard JT (1950) Monograph of the genus Digitaria. Leiden University Press, Leiden

Heuzé V, Tran G, Hassoun P, Lebas F (2019) Fonio (*Digitaria exilis*) grain. Feedipedia, a programme by INRAE, CIRAD, AFZ and FAO. https://feedipedia.org/node/228

Hilu KW, M'Ribu K, Liang H, Mandelbaum C (1997) Fonio millets: ethnobotany, genetic diversity and evolution. S Afr J Bot 63:185–190. https://doi.org/10.1016/S0254-6299(15)30742-0

Ibn Battuta (1982) Voyages III. Inde, Extrême-Orient, Espagne et Soudan, collection FM/La Decouverte, Librairie Francois Maspero, Paris, 456 p

Ibrahim Bio Yerima AR, Oselebe H, Nnamani CV, Ifekwe C, Adje CO, Kwon-Ndung EH et al (2024) Stakeholders' perceptions of and preferences for utilizing fonio (Digitaria exilis) to enrich local diets for food and nutritional security in Nigeria. Genet Resour Crop Evol 71(3):999–1011

Jarial S, Kaur J, Singh J, Kaushik R, Kumar S, Abdullahi A (2024) Teff and Fonio. In: Cereals and nutraceuticals. Singapore, Springer Nature, pp 145–161

Jifar H, Dagne K, Tesfaye K, Assefa K, Tadele Z (2018) Agro-morphological traits diversity in tef [*Eragrostis tef* (Zucc.) Trotter] genotypes from various sources. Ethiop J Agric Sci 28(3):131–148

Jifar H, Assefa K, Tesfaye K, Dagne K, Tadele Z (2020) Diversity in qualitative and quantitative traits reveals huge potential for the improvement of an orphan crop Tef [*Eragrostis tef* (Zucc.) Trotter]. J Exp Agri Int 42(10):60–75

Jifar H, Chanyalew S, Tadele Z, Assefa K (2022) Tef taxonomy, origin, distribution and genetic resources. In: Principles and practice of Tef improvement. Ethiopian Institute of Agricultural Research and Agricultural Transformation Agency, Addis Ababa, pp 27–42

Kaczmarek T, Causse S, Abdul SD, Abraham S, Achigan-Dako EG, Adje C et al (2023) Towards conservation and sustainable use of an indigenous crop: A large partnership network enabled the genetic diversity assessment of 1539 fonio (Digitaria exilis) accessions. Plants People Planet 10. https://doi.org/10.23708/TX5DE2

Kanlindogbe C, Sekloka E, Kwon-Ndung EH (2020) Genetic resources and varietal environment of grown fonio millets in West Africa: challenges and perspectives. Plant Breed Biotechnol 8(2):77–88

Ketema S (1983) Studies of lodging, floral biology breeding techniques in tef [*Eragrostis tef* (Zucc.) Trotter]. PhD thesis, London University, London

Ketema S (1997) Tef, *Eragrostis tef* (zucc.) Trotter. Promoting the conservation and use of underutilized and neglected crops series no. 12. Institute of Plant Genetics and Crop Plant Research/International Plant Genetic Resources Institute, Gatersleben/Rome

Kuta DD, Kwon-Ngung E, Dachi S, Bakare O, Ogunkanmi LA (2005) Optimization of protocols for DNA extraction and RAPD analysis in West African fonio (*Digitaria exilis* and Digitaria iburua) germplasm characterization. Afr J Biotechnol 4(12)

Kwon-Ndung EH, Ochigbo AA (2007) Acha (fonio) genetic diversity, distribution and conservation in Nigeria. In: Vodouhe SR, Atta-Krah K, Achigan-Dako GE, Eyog-Matig O, Avohou H (eds) Plant Genetic Resources and Food Security in West and Central Africa. Regional conference, 26–30 April 2004. Bioversity International, Rome

Kwon-Ndung EH, Dachi SN, Misari SM (1998) Collecting germplasm of acha *Digitaria exilis* Kippis Stapf. accessions in Nigeria. IPGRI/FAO Plant Genetic Resour Newslett 116:30–31

Morales-Payán JP, Ortiz JR, Cicero J, Taveras F (2002) *Digitaria exilis* as a crop in The Dominican Republic. In: Supplement to: trends in new crops and new uses. ASHS Press, Alexandra, pp S1–S3

National Research Council (1996) Lost crops of Africa: volume I: grains. The National Academies Press, Washington, DC. https://doi.org/10.17226/2305

Nyam DD, Kwon-Ndung EH, Wuyep AP (2017) RFLP-based gel electrophoregram of DNA primers for Acha (Fonio) characterisation from Nigeria, vol 1, p 51

Onuwa GC, Binuyo G, Alamanjo CC, Dalla A, Holi H, Yitnoe G (2022) Benefit-cost and productivity analysis of Fonio (Digitaria Spp) cultivation. J Agri Res Pestic Biofertilizers 3(5)

Osundinakin M, Ogunkanmi A, Onuminya T, Rotimi T, Ogundipe O (2023) Physio-chemical properties responses of six Fonio genotypes subjected to drought stress conditions. Adv Res Life Sci 7(1):22–32

Porteres R (1976) African Cereals: EfeuJilJe, Fonio, Black Fonio, Teff, Brachiaria, Paspalum, Pennisetum and African Rice. In: Harlan JR, deWel JMJ, Siernler ASL (eds) Origins of African plant domestication. Moulton, The Hague, pp 409–452

Sarker S, Tyagi DVS, Islam MH (1993) Induced mutagenesis: an important tool for breeding small millets—a review. Agric Rev Karnal 14:61–65

Small E (2015) Teff & Fonio – Africa's sustainable cereals. Biodiversity 16(1):27–41. https://doi.org/10.1080/14888386.2014.997290

Tadele Z (2009) Role of orphan crops in enhancing and diversifying food production in Africa. Afr Technol Dev Forum J 6(3):9–15

Tefera H, Ketema S (2001) Production and importance of tef in Ethiopian agriculture. https://agris.fao.org/agris-search/search.do?recordID=ET2003000031. Accessed 25 Mar 2021

Thomson Reuters Foundation (2019) Look out superfood fans – fonio is the new quinoa. Retrieved from https://www.one.org/stories/look-out-superfood-fans-fonio-is-the-new-quinoa

Tilaye KA, Delele BT, Ogeto MA (2023) Adoption and impact of improved teff varieties adoption on food security: micro level evidence from North Eastern Amhara Regional state, Ethiopia. PLoS One 18(9):e0291434

VanBuren R, Wai CM, Pardo J, Yocca AE, Wang X, Wang H et al (2019) Exceptional subgenome stability and functional divergence in allotetraploid teff, the primary cereal crop in Ethiopia. BioRxiv 580720

VanBuren R, Man Wai C, Wang X, Pardo J, Yocca AE, Wang H et al (2020) Exceptional subgenome stability and functional divergence in the allotetraploid Ethiopian cereal teff. Nat Commun 11(1):884. https://doi.org/10.1038/s41467-020-14724-z

Vavilov NI (1951) The origin, variation, immunity and breeding of cultivated plants. Nutrition facts and functional potential of quinoa. Vega-Galvez-2010_JSFA, 72(6), 482

Vodouhè SR, Achigan Dako E (2006) In Brink M, Belay G (eds) Digitaria exilis (Kippist) Stapf. PROTA (Plant Resources of Tropical Africa/Ressources végétales de l'Afrique tropicale), Wageningen

Vodouhe S, Zannou A, Achigan Dako E (2003) Actes du premier atelier sur la diversité génétique du fonio (*Digitaria exilis* Stapf) en Afrique de l'Ouest, Conakry, Guinée, 4–6 août 1998

Watson L, Dallwitz MJ (1992) The grass genus of the world. CAB International, Wallingford

Woldeyohannes AB, Accotto C, Desta EA, Kidane YG, Fadda C, Pè ME, Dell'Acqua M (2020) Current and projected eco-geographic adaptation and phenotypic diversity of Ethiopian teff (*Eragrostis teff*) across its cultivation range. Agric Ecosyst Environ 300:107020

Woldeyohannes AB, Iohannes SD, Miculan M, Caproni L, Ahmed JS, Desta EA, Fadda C, Pè ME, Dell'Acqua M (2021) Data-driven, participatory characterization of traditional farmer varieties discloses teff (Eragrostis tef) adaptive and breeding potential under current and future climates. bioRxiv. https://doi.org/10.1101/2021.08.27.457623v1.

Woldeyohannes AB, Iohannes SD, Miculan M, Caproni L, Ahmed JS, de Sousa K, Desta EA, Fadda C, Pè, ME, Dell'Acqua M (2022) Data-driven, participatory characterization of farmer varieties discloses teff breeding potential under current and future climates. Elife, 2;11:e80009. https://doi.org/10.7554/eLife.80009.

Yami A (2013) Tef straw: a valuable feed resource to improve animal production and productivity. In: Assefa K, Chanyalew S, Zerihun T (eds) Achievements and prospects of Tef improvement: proceedings of the second international workshop, November 7–9, 2011, Debre Zeit, Ethiopia. Stämpfli AG, Bern, pp 233–254

Yetneberk S, Rooney LW, Taylor J (2005) Improving the quality of sorghum Injera by decortication and compositing with Teff. J Sci Food Agric 85(8):1252–1258

Zeid M, Assefa K, Haddis A, Chanyalew S, Sorrells ME (2012) Genetic diversity in tef [*Eragrostis tef* (Zucc.) Trotter] using SSR markers. Field Crop Res 127:64–70

Zhu F (2020) Fonio grains: physicochemical properties, nutritional potential, and food applications. Compr Rev Food Sci Food Saf 19(6):3365–3389

The manufacturer's authorised representative in the EU is Springer Nature Customer Service Centre GmbH, Europaplatz 3, 69115 Heidelberg, Germany. If you have any concerns regarding our products, please contact ProductSafety@springernature.com

Printed and bound by CPI Group (UK) Ltd, Croydon, CR0 4YY
30/09/2026
02225297-0013